Advances in Intel
Soft Computing

1ƆU

Editor-in-Chief

Prof. Janusz Kacprzyk
Systems Research Institute
Polish Academy of Sciences
ul. Newelska 6
01-447 Warsaw
Poland
E-mail: kacprzyk@ibspan.waw.pl

For further volumes:
http://www.springer.com/series/4240

Ljupco Kocarev (Ed.)

ICT Innovations 2011

Springer

Editor
Ljupco Kocarev
Macedonian Academy of Sciences and Arts
Skopje
Macedonia
Bio Circuits Institute
University of California San Diego
San Diego, California
USA

ISSN 1867-5662 e-ISSN 1867-5670
ISBN 978-3-642-28663-6 e-ISBN 978-3-642-28664-3
DOI 10.1007/978-3-642-28664-3
Springer Heidelberg New York Dordrecht London

Library of Congress Control Number: 2012932760

Printed on acid-free paper

Springer is part of Springer Science+Business Media (www.springer.com)

Preface

The ICT Innovations conference is the primary scientific action of the Macedonian Society in Information and Communication Technologies (ICT-ACT). The conference provides a platform for academics, professionals, and practitioners to interact and share their research findings related to basic and applied research in ICT.

The ICT Innovations 2011 conference gathered 146 authors from 12 countries reporting their valuable work and experience in developing solutions and presenting novel applications of technology in ICT. Only 33 papers were selected for this edition by 80 Program Committee Members, chosen for their reputation for decency and good decree in the field.

ICT Innovations 2011 was held in Skopje, at the Macedonian Academy of Science and Art, in the period September 14-16, 2011. The conference focused on variety of ICT fields: Complex networks and Cryptography, Wireless Communication, Mobile Application, Machine Learning, High Performance Computing, Software Implementation, and Hardware, Grid and Cloud Computing. The conference was open by Ivo Ivanovski, the Minister of Information Society and Administration of the Republic of Macedonia.

I would like to express sincere gratitude to the authors for submitting their works to this conference, the reviewers providing service for the editor but also for authors, readers, and the ICT community as a whole, and all colleagues from the Department of Computer Science and Engineering, especially Sonja Filiposka and Anastas Misev, for their support in organizing the conference and preparing the book proceedings.

Skopje
September 2011

Ljupco Kocarev

Organization

ICT Innovations 2011 was organized by the Macedonian Society of Information and Communication Technologies (ICT-ACT).

Conference and Program Chair

Ljupco Kocarev	Macedonian Academy of Sciences and Arts Bio Circuits Institute, University of California San Diego Faculty of Computer Science and Engineering, Skopje

Program Committee

Ana Madevska-Bogdanova	UKIM, Macedonia
Anastas Misev	UKIM, Macedonia
Andrea Kulakov	UKIM, Macedonia
Aneta Karaivanova	IICT BAS, Bulgaria
Antun Balaz	Institute of physics Belgrade, Serbia
Attila Szucs	University of California San Diego, San Diego, USA
Betim Cico	University of Tirana, Albania
Branko Marovic	University of Belgrade, Serbia
Dancho Davchev	UIT, Macedonia
Danilo Gligoroski	University of Trondheim, Norway
Dejan Gjorgjevikj	UKIM, Macedonia
Dejan Spasov	UKIM, Macedonia
Dilip Patel	LSBU, UK
Dimitar Trajanov	UKIM, Macedonia
Dragan Bosnacki	TUE, The Netherlands
Emanouil Atanassov	IPP BAS, Bulgaria
Francky Trichet	Nantes University, France
Francois Paris	University of Houston, USA

Gabor Vattay	Etovos Lorand University, Budapest, Hungary
Georgi Stojanov	American University of Paris, France
Gerg Duane	University of Colorado, Boulder, USA
Gianluca Setti	University of Bologna, Italy
Goce Armenski	UKIM, Macedonia
Goran Velinov	UKIM, Macedonia
Harold Sjursen	New York University, USA
Hrachya Astsatryan	IIAP NAS, Armenia
Igor Trajkovski	UKIM, Macedonia
Ivan Chorbev	UKIM, Macedonia
Jasen Markovski	Eindhoven University of Technology, The Netherlands
Jorge Marx-Gomez	Oldenburg University, Germany
Juergen Kurhs	Humboldt University, Berlin, Germany
Julien Bourgeois	University of Franche-Comté, France
Katerina Zdravkova	UKIM, Macedonia
Kosta Mitreski	UKIM, Macedonia
Lev Tsimring	University of California San Diego, San Diego, USA
Liane Haak	University Oldenburg, Germany
Ljiljana Trajkovic	Simon Fraser University, Vancouver, British Columbia, Canada
Ljupcho Antovski	UKIM, Macedonia
Lubomir T.-Chitkuchev	Boston University, USA
Luigi Fortuna	University of Catania, Italy
Margita Kon-Popovska	UKIM, Macedonia
Marija Mihova	UKIM, Macedonia
Mario Biey	Politecnico di Torino, Italy
Mario di Bernardo	University of Napoli, Italy
Marjan Gushev	UKIM, Macedonia
Mile Stojchev	University of Niš, Serbia
Milentijevic Ivan	University of Nis, Serbia
Mirjana Ivanovic	University of Novi Sad, Serbia
Natasha Jonoska	Univ. of South Florida, USA
Nevena Ackovska	UKIM, Macedonia
Nevenka Dimitrova	Philips Research, USA
Nikolai Rulkov	University of California San Diego, San Diego, USA
Noel Keenlyside	University of Bergen, Bergen, Norway
Norbert Grünwald	Hochschule Wismar, Germany
Pece Mitrevski	UKLO, Macedonia
Peter Parychek	Univ. Donau, Austria
Predrag Petkovic	University of Niš, Serbia
Sanja Lazarova-Molnar	UAE University, UAE
Sashko Celakovski	ITgma, Macedonia

Sasho Josimovski	UKIM, Macedonia
Saso Dzeroski	Institut Jozef Stefan, Ljubljana, Slovenia
Shushma Patel	LSBU, UK
Slobodan Bojanic	Universidad Politecnica de Madrid, Spain
Slobodan Kalajdziski	UKIM, Macedonia
Smile Markovski	UKIM, Macedonia
Sonja Filiposka	UKIM, Macedonia
Sonja Gievska-Krilu	GWU, USA
Stephan Olariu	ODU, USA
Suzana Loshkovska	UKIM, Macedonia
Tommaso Addabbo	University of Sienna, Italy
Vanco Litovski	University of Niš, Serbia
Verica Bakeva	UKIM, Macedonia
Vesna Dimitrova	UKIM, Macedonia
Visarath In	SPAWAR Systems Center Pacific, San Diego, USA
Vladimir Radevski	SEEU, Macedonia
Vladimir Trajkovic	UKIM, Macedonia
Wallace Tang	City University Hong Kong, Hong Kong, China
Yiannis Kompatsiaris	Centre for Research and Technology Hellas, Greece
Zamir Dika	SEEU, Macedonia
Zoran Hadzi-Velkov	UKIM, Macedonia

Contents

A Jumping Gene Evolutionary Approach for Multiobjective Optimization

Wallace K.S. Tang, Chris S.H. Yeung, and K.F. Man

Abstract. The phenomenon of Jumping genes was initially discovered by Nobel Laureate, Barbara McClintock, in her work on maize chromosome in fifties. The Jumping genes transpose from one position to another in horizontal fashion within the same chromosome or even to other chromosomes. In this paper, it is to present how this genetic transposition, after transforming into a computational method, can enhance the evolutionary multiobjective optimization. The fundamental concept, design of operations, performance justification and applications of the Jumping Gene evolutionary approach will be outlined.

1 Introduction

Genetic Algorithm (GA) was firstly proposed by John Holland [1] in late 1960s. As a major family of evolutionary algorithms inspired by the natural phenomenon of biological evolution, its success relies on the evolution of the chromosomes, following the principle of survival of the fittest.

GA is particularly well-known for its ability in handling multiobjective problems, thanks for the efforts made to effectively transform multiple objectives into a single fitness value [2, 3]. However, despite the numerous successful stories of applying GA, keeping diversity in the population during the GA optimization process is still a major problem.

To resolve this problem, many proposals have been made for various multiple objective genetic algorithms, including the use of high disruptive operations (For example, uniform crossover and multi-point crossover); diversity enhancement (e.g. mating restrictions [4]); and diversity perservation (e.g. crowding operator in NSGA2 [5], tournament selection and fitnesss sharing [6, 7]). However, it is still a

Wallace K.S. Tang · Chris S.H. Yeung · K.F. Man
Department of Electronic Engineering, City University of Hong Kong
e-mail: {eekstang, saiyeung, eekman}@cityu.edu.hk

L. Kocarev (Ed.): ICT Innovations 2011, AISC 150, pp. 1–14.
springerlink.com

very difficult task to obtain a widespread non-dominated solution set and the convergence speed is usually slow.

Recently, inspired by the biological mobile genes (called Jumping genes) [8], a new class of operations, called gene transposition, has aroused many attentions and is also the focus of this paper. It describes the transfer of Jumping gene (JG) existing in chromosome, which results in a migration of genetic information between chromosomes within a generation. This is also referred as horizontal transmission in biology as compared with the vertical transmission of genetic information appears from one generation to its descendants.

This paper is to give an introduction of how the transposition of JG contributes in evolutionary computation. The fundamental concept, design of operations, performance justification and the applications of JG, are to be discussed.

2 Evolutionary Multiobjective Optimization

2.1 Multiobjective Problems

Multiobjective problems (MOPs) are commonly encountered in real life. Unlike the single objective problems, there are two or more objective functions to be optimized in MOP at the same time. With such a nature, it is usually demanded to identify a set of tradeoff solutions, instead of a single global optimal one.

Figure 1 shows an example, in which two functions f_1 and f_2 are to be minimized. There exists a set of optimal solutions, called Pareto optimal set, which are not dominated by any other solutions.

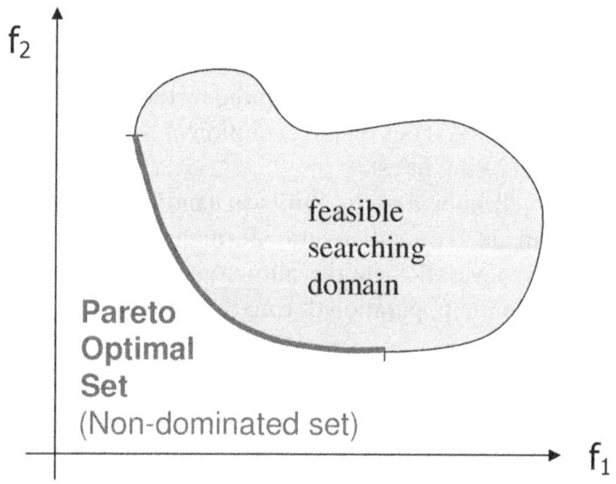

Fig. 1 Pareto optimal set

In mathematics, for a minimization problem (Remark: Similar definition can be given for maximization problem), we say a solution candidate u is dominated by another solution v (or v is more preferable than u) if

$$\begin{aligned} f_i(u) &\geq f_i(v) \quad \forall i = 1, 2, \cdots, n \quad \text{and} \\ f_i(u) &> f_i(v) \quad \exists j = 1, 2, \cdots, n. \end{aligned} \tag{1}$$

Due to optimization and practical reasons, it is more favourable to find the entire Pareto optimal set instead of a single solution. This is usually accomplished by population based methodologies, for which GA is one of them, so that only a single run is sufficient.

2.2 Overview of GA

GA is a population-based guided-search methodology, inspired by the rule of "survival of the fittest" and based on the biological DNA structure. As its simplest form, its flow can be described as below and a genetic operational cycle is graphically depicted in Fig. 2:

1. Population creation: Create an initial set of solutions for a problem by encoding each solution into a string of symbols, for example, a binary string, which is called as chromosome.
2. Fitness evaluation: Evaluate each of the chromosomes by the problem specific objective function(s) and a positive fitness value is to be assigned.
3. Selection: Select a subset of chromosomes based on their fitness values. The subset is called a mating pool.
4. Reproduction: New chromosomes (called offspring) are produced by mating chromosomes in the mating pool in pair (called parents). They are created by mixing the genetic information of the parents (the operation is known as crossover) and are then mutated by altering some genes in a random sense (the operation is known as mutation). The set of all offspring created is referred as the subpopulation.
5. Fitness evaluation: Evaluate the chromosomes in the sub-population as in Step 2.
6. Replacement: Some chromosomes in the original population are replaced by the offspring, usually based on their fitness values.
7. Termination: Steps 3 to 6 are repeated until some termination conditions are reached, for example the maximum number of generations is lasted.

Even with some variations, most of the GAs will have the similar above operations. They mimic the natural selection at the Selection and Replacement steps, and simulates the sexual reproduction in the Reproduction step. Further details of GA can be referred to [2, 9].

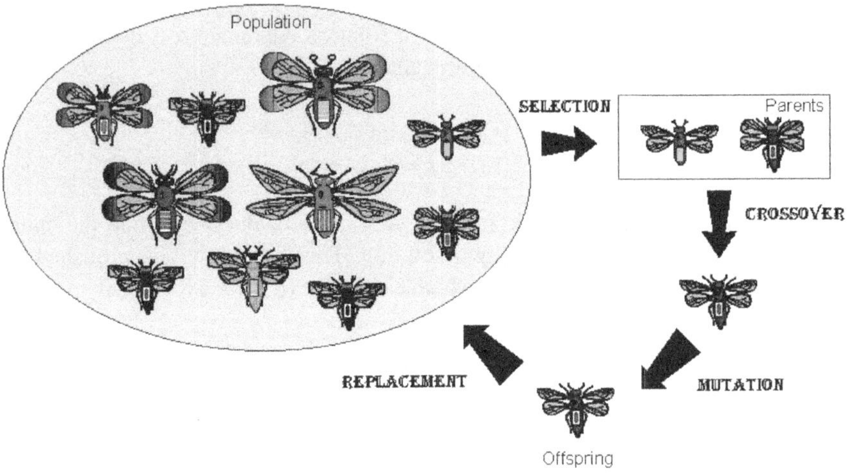

Fig. 2 Genetic algorithm cycle

2.3 *Multiobjective Ranking*

In order to achieve multiobjective optimization, the key is to assign the fitness of an individual according to its performance on the multiple objectives (i.e. Step 2 in Sec. 2.2). There are two commonly used Pareto ranking schemes which were proposed by Goldberg [2] and Fonseca & Fleming [3] respectively. They are briefly explained in the following subsections.

2.3.1 Goldberg Multiobjective Ranking

The rank of each individual is determined by the following procedures:

1. Set $p = 1$;
2. Determine all the non-dominated individuals in the population, and they are assigned as rank-p
3. Ignore those individuals with ranks having been assigned; increment p by 1 and go to Step 2 until all individuals have been processed.

An example of the ranking result is given in Fig. 3 (a), where the integer in the bracket shows the rank of the corresponding individual.

2.3.2 Fonseca and Fleming Multiobjective Ranking

The rank of each individual is $(1 + q)$ where q is the number of other individuals in the population that dominate it. An example is given in Fig. 3 (b).

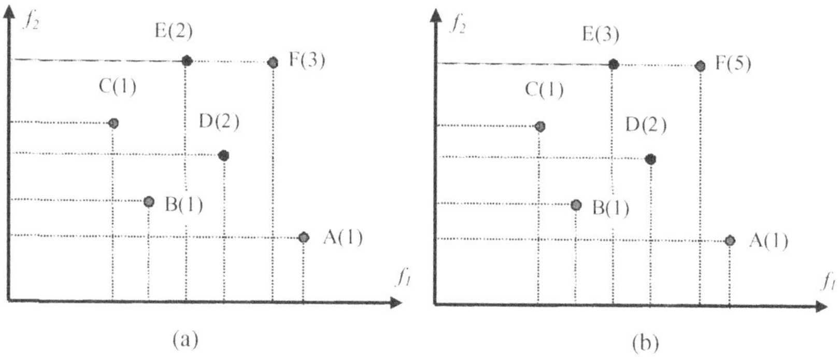

Fig. 3 Multiobjective ranking for minimizing f_1 and f_2: (a) Goldberg method (b) Fonseca and Fleming method.

By using the ranking as the fitness in the conventional GA described in Sect. 2.2, the basic framework of a mutliobjective GA can be formed. However, due to the stochastic nature and the processes, final solutions are likely to be crowded in a small region of the whole Pareto optimal set.

In order to explore the entire Pareto optimal set, diversity has to be maintained in the population. Different enhancements, such as high disruptive crossover, tournament selection, fitness sharing, crowding and so on, have been suggested. However, despite of many efforts being paid, it is still a difficult and challenging problem to find a widely spread solution set, especially the extreme ones, in MOP.

3 Computational Jumping Gene Operations

In 2004, inspired by the biological observation of jumping genes, an innovative optimization technique was proposed [10] to enhance the searching ability of a GA, in particular for MOPs. Two distinct computational operations are designed to mimic the biological gene transposition process.

3.1 Biological Background and Operational Design

The design of computational genetic transpositing operations is rooted onto the biological observation made by the Nobel Laureate in Physiology or Medicine, Barbara McClintock. In her pioneering work on maize chromosome, mobile genetic elements (called Jumping genes) were noticed and their transpositioning consequently leading to the changing patterns of coloration in maize kernels was reported [8].

Similar gene transposition processes have later been reported in different classes of organisms, including bacteria, plants and animals [11, 12]. The mechanism of transposition can be classified as conservative or replicative one. For conservative transposition, the mobile genetic element moves from its original locus to a new one without any replication, while it transposes by replication for the replicative case.

Based on these two mechanisms, two computational Jumping gene (JG) operations, called cut-and-paste and copy-and-paste, were proposed [10]. The schematic diagrams of these two operations are depicted in Figs. 4 (a) and (b), respectively.

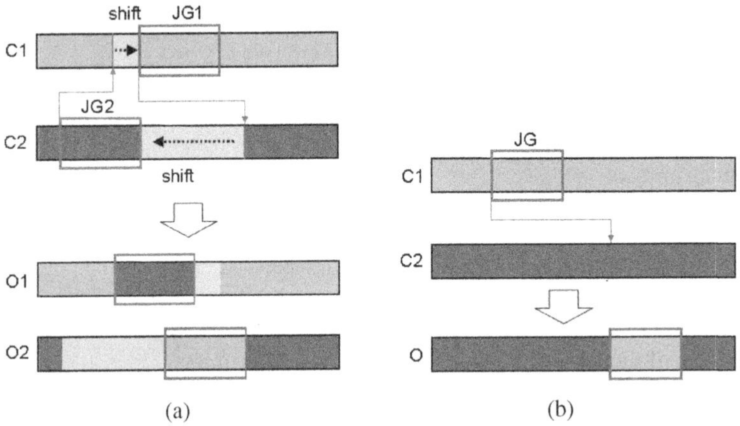

Fig. 4 (a) Cut-and-paste operation (b) Copy-and-paste operation

As shown in Fig. 4 (a), a section of genes (denoted as JG1 and JG2) is removed (or cut) from the original location of each of two chromosomes and then pasted into a new position of another chromosome when cut-and-paste is undergone. For copy-and-paste (see Fig. 4 (b)), the information of the JG is remained unchanged in the original chromosome, while a duplicate is pasted on another chromosome. In either case, the chromosomes and the JG locations are randomly chosen. Thus, these operations may occur within the same chromosome or two different ones.

3.2 Schemata Evolution Equations of JG Operations

In this section, it is to explain why these two JG operations can enhance the optimization process. A mathematical justificaiton is provided using the concept of schema.

Unlike the Holland's or other approximate approaches, the destruction and reconstruction of schemata under the two JG operations are derived, and hence exact schema evolution equations can be obtained [13, 14]. Similar approaches were adopted by Stephens and Waelbroeck, which gave an exact formulation (rather than a lower bound in the approximate approaches) for selection, single-point crossover and mutation [15, 16]. A major advantage of such an exact theorem is that the expected number of strings matching a schema can be predicted over multiple generations, which clearly overcomes the shortcoming of the approximate approaches. Moreover, this has also been used as a starting point for many other results on the behavior of a GA over multiple generations, based on the assumption of infinite population size.

Consider a schema ξ with L symbols, each symbol belongs to the set B={0, 1, *}, where 0 and 1 are called the actual bits while the wild character * is interpreted as a don't-care bit which can be either 0 or 1. Therefore, a schema represents a set of bit strings (i.e. these strings belong to the schema), for example, the bit strings 0100, 0110, 1100, and 1110 all belong to $\xi = *1*0$.

In the followings, some major definitions are given:

Definition 1. A map f_L is defined as $f_L : B^L \to V$, such that for $v \in V$, $v = f_L(\xi)$ returns a vector specifying the locations of all the actual bits in schema $\xi \in B^L$. V is a vector of non-negative integers, and it is assumed that the location begins with zero from left to right.

Definition 2. A map f_T is defined as $f_T : B^L \times V \to B^i$, such that for $\zeta \in B^i$, $\zeta = f_T(\xi, v)$ is formed by copying the symbols from schema $\xi \in B^L$ according to the locations specified in $v \in V$ where $size(v) = i$.

Definition 3. The primary schemata competition set of schema ξ is defined as

$$\Psi_\xi = \{\xi_i \in B^L : f_L(\xi_i) = f_L(\xi)\} \tag{2}$$

For example, given $\xi = *1*0$, $f_L(\xi) = [1,3]$ and $\Psi_\xi = \{*0*0, *0*1, *1*0, *1*1\}$.

Definition 4. For $x_i, y_i \in B$, the bit distance between two symbols x_i and y_i is defined as:

$$d(x_i, y_i) = \begin{cases} 1 & x_i = * \\ 1 & x_i \neq *, y_i \neq *, x_i = y_i \\ 0 & x_i \neq *, y_i \neq *, x_i \neq y_i \\ 0.5 & x_i \neq *, y_i = * \end{cases} \tag{3}$$

which describes how close y_i matches with x_i.

Definition 5. For $\zeta_1, \zeta_2 \in B^m$, the regional similarity of two strings ζ_1 and ζ_2 is specified by

$$\delta(\zeta_1, \zeta_2) = \prod_{j=0}^{m-1} d(\zeta_1(j), \zeta_2(j)) \tag{4}$$

where $\zeta_i(j)$ is the j-th bit of ζ_i.

Definition 6. For $v_1, v_2 \in V$ and $size(v_1) = size(v_2)$, $\Delta(\xi_1, v_1; \xi_2, v_2)$ is the regional similarity of ξ_1 in region v_1 and ξ_2 in region v_2, defined as

$$\Delta(\xi_1, v_1; \xi_2, v_2) \equiv \delta(f_T(\xi_1, v_1), f_T(\xi_2, v_2)) \tag{5}$$

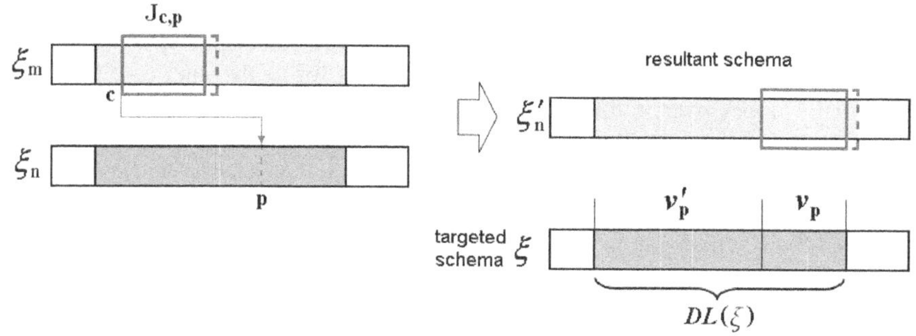

Fig. 5 Copy-and-paste operation with ξ_m and ξ_n, while ξ_n' is the resultant schema and ξ is the targeted schema.

By performing copy-and-paste operation, we can then derive the evolution equation of a particular schema ξ in each generation by studying its construction and destruction possibilities. The following schema evolution equation can be obtained [14]:

$$P(\xi,t+1) = (1-p_1)P(\xi,t) + \frac{p_1}{(L-L_g+1)^2} \sum_{\xi_m \in \Psi_\xi} \sum_{\xi_n \in \Psi_\xi} \sum_{c=0}^{L-L_g} \sum_{p=0}^{L-L_g} \left[\Delta(\xi, v_p; \xi_m, J_{c,p}) \Delta(\xi, v_p'; \xi_n, v_p')\right] P(\xi_m,t)P(\xi_n,t) \tag{6}$$

where p_1 is the copy-and-paste operational rate; $P(\xi,t)$ is the proportion of ξ at generation t; L_g is the bit length of the JG, Ψ_ξ is the primary schemata competition set of ξ; $J_{c,p}$ specifies the bit locations of selected JG in ξ_m which fall into the defining-length region of the schema ξ (denoted as $DL(\xi)$) after pasted; v_p specifies the bit locations in $DL(\xi)$ where the JG is pasted; and v_p' specifies the bit locations in $DL(\xi)$ but not belong to v_p. The definitions of different regions under copy-and-paste operation are given in Fig. 5.

Similarly, the evolution equation of schema ξ by cut-and-paste operation can be derived [14] as below:

$$P(\xi,t+1) = (1-p_2)P(\xi,t) + \frac{p_2}{(L-L_g+1)^3} \sum_{\xi_m \in \Psi_\xi} \sum_{\xi_n \in \Psi_\xi} \sum_{c_m=0}^{L-L_g} \sum_{c_n=0}^{L-L_g} \sum_{k_n \notin \kappa(c_n)} \Delta(\xi, I_{c_n,p_n}; \xi_m, J_{c_m,c_n,p_n}) \Delta(\xi, S_{c_n,p_n}; \xi_n, S'_{c_n,p_n}) \Delta(\xi, R_{c_n,p_n}; \xi_n, R_{c_n,p_n}) P(\xi_m,t)P(\xi_n,t) \tag{7}$$

where p_2 is the cut-and-paste operational rate, J_{c_m,c_n,p_n} indicates the bit locations of the selected JG cut from ξ_m which fall into $DL(\xi)$ after pasted into ξ_n; I_{c_n,p_n} indicates the bit locations within $DL(\xi)$ where the JG is pasted onto ξ_n, S_{c_n,p_n} indicates the bit locations within $DL(\xi)$ where some bits of ξ_n are shifted, S'_{c_n,p_n} indicates the

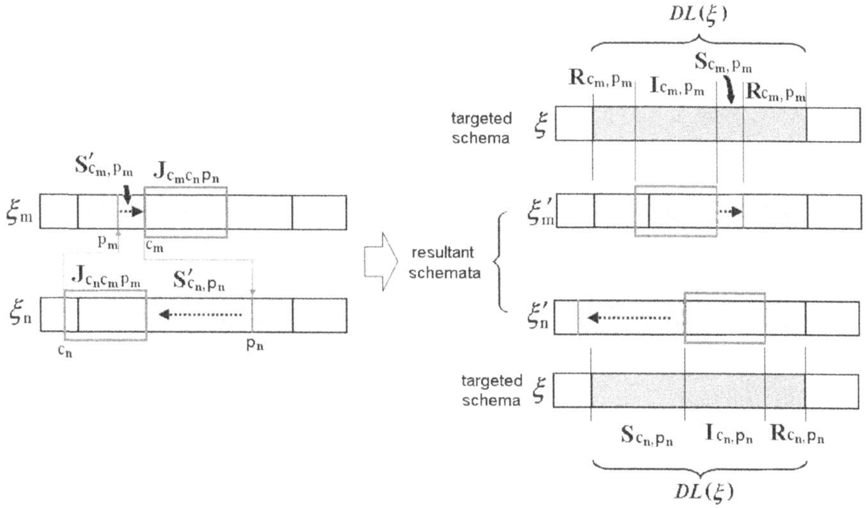

Fig. 6 Cut-and-paste operation with ξ_m and ξ_n, while ξ'_m and ξ'_n are the resultant schemata which are to be compared with the targeted schema ξ.

bit locations in ξ_n shift to S_{c_n,p_n} in ξ'_n, R_{c_n,p_n} indicates the bit locations within where $DL(\xi)$ remains unaffected by the operation, and $\kappa(c_n) = (c_n, c_n + L_g]$. Figure 6 depicts the definitions of different regions under cut-and-paste operation.

To further analyze the evolution equations (6) and (7), let's consider any primary competitive group $\Psi = \{\xi_1, \xi_2, \cdots, \xi_k\}$ with $k = 2^{o(\Psi)}$ (Remark: The order of Ψ, $o(\Psi)$, is defined as the number of actual bits in ξ for any $\xi \in \Psi$.) and assume that $p_1 = p_2 = 1$. Define $P(\Psi,t) = \left[P(\xi_1,t) \; P(\xi_2,t) \; \cdots \; P(\xi_k,t) \right]$, (6) or (7) can be rewritten in the form of

$$P(\xi_i, t+1) = P^T(\Psi,t) A_{\xi_i} P(\Psi,t) \tag{8}$$

where A_{ξ_i} is a constant matrix with size of $k \times k$ and $\xi_i \in \Psi$. This constant matrix can be obtained when the JG length L_g and the primary competitive group Ψ are fixed, and it is different for copy-and-paste and cut-and-paste.

Equation (8) defines a discrete-time system of $P(\Psi,t)$ in a quadratic form, and the following theorem can be mathematically proved.

Theorem 1. *For any primary schemata competition group Ψ with $0 < o(\Psi) < L$, the proportion of every schema in Ψ globally asymptotically converges to $\frac{1}{2^{o(\Psi)}}$ with the use of copy-and-paste or cut-and-paste operation. [17, 18]*

Details of the proof can be referred to [13].

3.3 Discussions

Theorem 1 implies that either the copy-and-paste or the cut-and-paste operation will force a uniform distribution of schemata in each primary schemata competition set, despite of its initial distribution, i.e. $P(\xi,\infty) = \frac{1}{2^{o(\Psi)}}\ \forall \xi \in \Psi$. This ensures the occurrence of all the solutions in the population if the population size is sufficiently large, or in the other word, global search is guaranteed.

For a finite population, this driving force also enhances the searching of solutions. An illustrative example is given below, in which the possible outcomes of a 16-bit string generated by different operations are recorded. The initial population only contains 100 chromosomes and typical genetic operations are tested. It can be noticed in Fig. 7 that the cut-and-paste is the best and obtains all the possible outcomes within the smallest generations. The performances of copy-and-paste and mutation with high operational rate are good and similar.

In addition, the inclusion of copy-and-paste and cut-and-paste operation helps to maintain the diversity in a finite population. A simple case is given in Fig. 8. The simulation is based on a simple genetic algorithm with a population size of 100. The chromosome is in 16-bit string while the fitness is assigned only according to the values of 3 specified bit locations in a chromosome. The result clearly demonstrates that the dominance of the best schema is slow-down by the JG operations, while the cut-and-paste presents a more significant effect as compared with copy-and-paste (see Fig. 8 (b) and (c)). This matches with what we obtain from the previous example, where cut-and-paste is more powerful in terms of locating the solutions and promoting the diversity. More simulation results can be found in [14, 18].

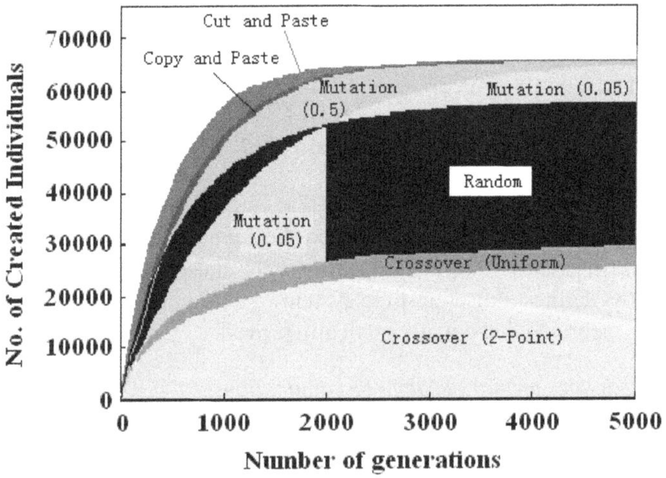

Fig. 7 Example showing the searching abilities of different operations with initial population of size 100 randomly generated.

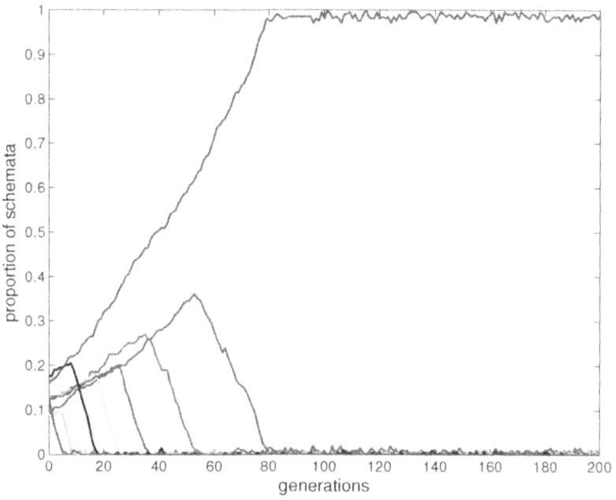

(a)

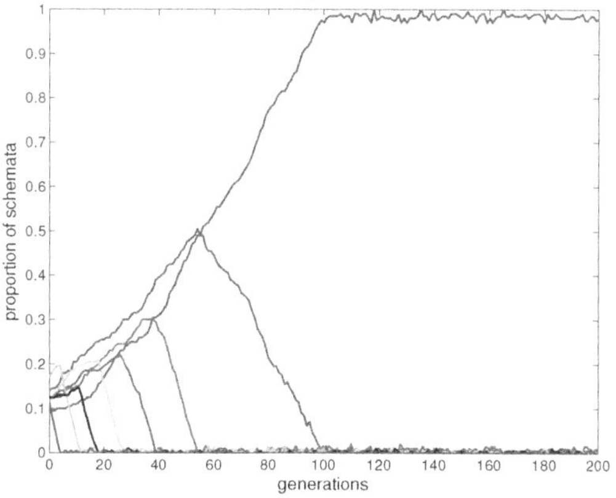

(b)

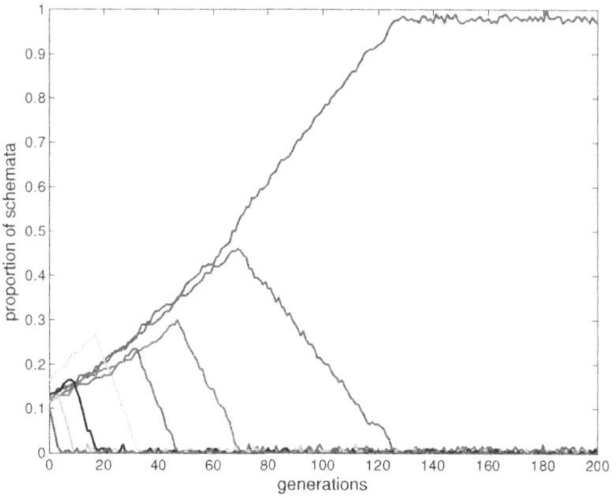

(c)

Fig. 8 Example showing the slow-down of the dominance of the best schema when cut-and-paste or copy-and-paste is applied in a standard GA. Proportion of schemata with (a) GA using crossover and mutation (b) GA using crossover, mutation and copy-and-paste (c) GA using crossover, mutation and cut-and-paste

4 Applications

The performance of JG-incorporated GA (known as JGGA) has been numercially studied in [19]. It is noticed that a rich set of non-dominated tradeoff solutions with better convergence and diversity can be obtained as compared with other MO evolutionary algorithms. Its outperformance is also reflected by its long list of applications for many real-world MOPs. Some of them are briefly summarized in the followings while the details can be referred to the cited references.

JGGA has been used to solve the radio-to-fiber repeater placement problem in wireless local loop systems [20]. Optimal locations for the repeaters can be duly determined so that the total repeater cost and total link cost are minimized simultaneously. A similar problem but targetting for placing base-stations for a wireless local area network in an IC factory is given in [21]. By finding a suitable number of base stations and identifying their corresponding locations, a specified quality of service is guaranteed. In [22], the resource management problem in wideband CDMA systems has been considered. It is concluded that JGGA can effectively optimize the total transmission power and the total transmission rate, and its performance is better than those common multiobjective algorithms.

Besides the resource management problems, JGGA has also been successfully applied in power voltage control [23], industrial product designs [24], and the designs of radio-frequency components and devices, such as the trapeizform U-slot folded patch feed antenna [25], the industrial, scientific and medical (ISM) band folded-patch antenna [26], the quarter-wave patch antenna [27], the planar monopole ultrawideband antenna [28] and microwave branch line hybrid coupler [14].

JGGA is capable of handling combinatorial optimization problems as well. In [29], JGGA has been applied to solve the job-shop scheduling problem whose goal is to find the best schedule of activities such that the length of schedule is minimized and the shared resources are efficiently used. In [30], a hybrid JGGA has been proposed to solve a scheduling problem in multiple destination routing. This is a large complex problem as it incorporates the scheduling and routing process of a set of requests which has a single source and multiple destinations in a network.

5 Conclusions

This paper is to summarize the fundamental concept, design of operations, performance justification and applications of the Jumping Gene approach. An overview has been presented and readers are encouraged to refer to the corresponding references for further details.

Acknowledgments. The first and third authors greatly appreciate the contributions from their former research students T.M. Chan, R.J. Yin and H.M. Ma.

References

1. Holland, J.H.: Adaptation in Natural and Artificial Systems. University of Michigan Press, Ann Arbor (1975)
2. Goldberg, D.E.: Genetic algorithms in search, optimization, and machine learning. Addison-Wesley (1989)
3. Fonseca, C.M., Fleming, P.J.: Multiobjective optimization and multiple constraint handling with evolutionary algorithms - Part I: A unified formulation. IEEE Trans. Syst. Man Cybern. Part A - Syst. Hum. 28, 26–37 (1998)
4. Fonseca, C.M., Fleming, P.J.: Multiobjective genetic algorithms made easy: Selection, sharing and mating restriction. In: Proc. Genetic Algorithms Engineering Systems: Innovations and Applications, pp. 45–52 (1995)
5. Srinivas, N., Deb, K.: Multiobjective function optimization using nondominated sorting genetic algorithms. Evol. Comput. 2(3), 221–248 (1994)
6. Goldberg, D.E., Richardson, J.: Genetic algorithms with sharing for multimodal function optimization. In: Proc. 2nd Int. Conf. on Genetic Algorithms (1987)
7. Oei, C.K., Goldberg, D.E., Chang, S.J.: Tournament selection, niching and the preservation of diversity. Technical Report (94002), Illinois GA Lab, University of Illinois (1994)
8. Fedoroff, N., Botstein, D. (eds.): The dynamic genome: Barbara McClintock's ideas in the century of genetics, Cold Spring Harbor, New York (1992)
9. Man, K.F., Tang, K.S., Kwong, S.: Genetic algorithms: concepts and applications. IEEE Trans. Ind. Electron. 43(5), 519–534 (1996)
10. Man, K.F., Chan, T.M., Tang, K.S., Kwong, S.: Jumping-genes in evolutionary computing. In: Proc. Annu. Conf. IEEE Ind. Electron. Society, vol. 2, pp. 1268–1272 (2004)
11. Finnegan, D.J.: Transposable elements in eukaryotes. Int. Rev. Cytology 93, 281–326 (1985)
12. Kleckner, N.: Transposable elements in prokaryotes. Annu. Rev. Genet. 15, 341–404 (1981)
13. Yin, R.J.: Schema Theorem for Computational Gene Transposition and Performance Analysis. PhD Thesis, City University of Hong Kong (2010)
14. Tang, K.S., Yin, R.J., Kwong, S., Ng, K.T., Man, K.F.: A theoretical development and analysis of jumping gene genetic algorithm. IEEE Trans. Ind. Inform. 7(3), 408–418 (2011)
15. Stephens, C.R., Waelbroeck, H.: Effective degrees of freedom in genetic algorithms and the block hypothesis. In: Proc. 7th Int. Conf. Genet. Algorithms, pp. 34–40 (1997)
16. Stephens, C.R., Waelbroeck, H.: Analysis of the effective degrees of freedom in genetic algorithms. Phys. Rew. D 57, 3251–3264 (1998)
17. Yin, J.J., Yeung, S.H., Tang, W.K.S., Man, K.S., Kwong, S.: Enhancement of multiobjective search: A Jumping-genes approach. In: Proc. IEEE Int. Symp. Ind. Electron., Vigo, pp. 1855–1858 (2007)
18. Tang, W.K.S., Kwong, S., Man, K.F.: A jumping genes paradigm: theory, verification, and applications. IEEE Circ. Syst. Mag. 8(4), 18–36 (2008)
19. Chan, T.M., Man, K.F., Kwong, S., Tang, K.S.: A jumping gene paradigm for evolutionary multiobjective optimization. IEEE Trans. Evolut. Comput. 12(2), 143–159 (2008)
20. Chan, T.M., Man, K.F., Tang, K.S., Kwong, S.: Multiobjective optimization of radio-to-fiber repeater placement a jumping gene algorithm. In: Proc. IEEE Int. Conf. Ind. Technol., pp. 291–296 (2005)
21. Chan, T.M., Man, K.F., Tang, K.S., Kwong, S.: A jumping-genes paradigm for optimizing factory WLAN network. IEEE Trans. Ind. Inform. 3(1), 33–43 (2007)

22. Chan, T.M., Man, K.F., Tang, K.S., Kwong, S.: A jumping gene algorithm for multiobjective resource management in wideband CDMA systems. Comput. J. 48(6), 749–768 (2005)
23. Ma, H.M., Ng, K.T., Man, K.F.: Multiobjective coordinated power voltage control using jumping genes paradigm. IEEE Trans. Ind. Electron. 55(11), 4075–4084 (2008)
24. Zheng, S.Y., Yeung, S.H., Chan, W.S., Man, K.F., Tang, K.S.: Design of broadband hybrid coupler with tight coupling using jumping gene evolutionary algorithm. IEEE Trans. Ind. Electron. 56(8), 2987–2991 (2009)
25. Yeung, S.H., Man, K.F., Luk, K.M., Chan, C.H.: A trapeizform u-slot folded patch feed antenna design optimized with jumping genes evolutionary algorithm. IEEE Trans. Antennas Propag. 56(2), 571–577 (2008)
26. Yeung, S.H., Man, K.F., Chan, W.S.: Optimised design of an ISM band antenna using a jumping genes methodology. IET Microw. Antennas Progag. 2(3), 259–267 (2008)
27. Sun, S.H., Man, K.F., Wang, B.Z., Wong, T.P.: An optimized wideband quarter-wave patch antenna design. IEEE Antennas Wirel. Propag. Lett. 4(1), 486–488 (2005)
28. Yang, X.S., Ng, K.T., Yeung, S.H., Man, K.F.: Jumping genes multiobjective optimization scheme for planar monopole ultrawideband antenna. IEEE Trans. Antennas Propag. 56(12), 3659–3666 (2008)
29. Ripon, K.S.N., Tsang, C.H., Kwong, S.: Multi-objective evolutionary job-shop scheduling using jumping genes genetic algorithm. In: Proc. Int. Joint Conf. Neural Netw., pp. 3100–3107 (2006)
30. Rahman, M., Mondol, S., Hossain, G.S., Dey, A.K.: A hybrid jumping genes genetic algorithm based request scheduling approach in multiple destination routing. In: Proc. Int. Conf. Inf. Commun. Technol., pp. 331–335 (2007)

Distributed Intelligent MEMS: Progresses and Perspectives

Julien Bourgeois and Seth Copen Goldstein

Abstract. MEMS research has until recently focused mainly on the engineering process, resulting in interesting products and a growing market. To fully realize the promise of MEMS, the next step is to add embedded intelligence. With embedded intelligence, the scalability of manufacturing will enable distributed MEMS systems consisting of thousands or millions of units which can work together to achieve a common goal. However, before such systems can become a reallity, we must come to grips with the challenge of scalability which will require paradigm-shifts both in hardware and software. Furthermore, the need for coordinated actuation, programming, communication and mobility management raises new challenges in both control and programming. The objective of this article is to report the progresses made by taking the example of two research projects and by giving the remaining challenges and the perspectives of distributed intelligent MEMS.

1 Introduction

Microelectromechanical systems (MEMS) have reached a state of design maturity which has led to some interesting prototypes and profitable products. While most MEMS devices have been used as independent elements of a larger system, this article deals with distributed MEMS systems composed of many MEMS devices which work together to achieve a global goal. The distinguishing feature of MEMS devices is that they are small and that they can be efficiently mass-produced. This naturally engender thinking of how they can be used together as a distributed system. Due to their small size, their low-cost and the fact that they can be mass-produced,

---

Julien Bourgeois
University of Franche-Comté, 1 cours Leprince-Ringuet, 25200 Montbeliard, France
e-mail: `Julien.Bourgeois@univ-fcomte.fr`

Seth Copen Goldstein
Carnegie Mellon University, Pittsburgh, PA 15213, USA
e-mail: `seth@cs.cmu.edu`

L. Kocarev (Ed.): ICT Innovations 2011, AISC 150, pp. 15–25.
springerlink.com

millions of units can be used in very small space. For example, a volume of less than 1 m^3 of 1mm-diameter silicon balls contains as many nodes as in the Internet. This characteristic requires paradigm-shifts both in hardware and software parts.

Past research focused on challenges of the engineering process, future challenges will consist in adding embedded intelligence to MEMS systems, so that they will be able to collaborate efficiently. This will require embedding MEMS sensors/actuators, electronics, communication capabilities, control of actuators and programs in the same unit. We suggest the use of the phrase "distributed intelligent MEMS (diMEMS)" when referring to such systems. DiMEMS systems will certainly contain heterogeneous units. However, to simplify the programming challenge, we consider in this article only systems composed of homogeneous units.

Designing and managing diMEMS inherently requires multiple disciplines (e.g. hardware and software research). The challenges are therefore present in every field of research as well as in the integration of all the parts. In the 90's, DARPA Information Science and Technology funded a study on the state of the art and the perspectives of distributed MEMS. The conclusions of this report [7], published in 1997, were that the challenges involved in realizing diMEMS were mainly in controlling large numbers of MEMS sensors and actuators, the emergence of distributed intelligence, the use of MEMS devices as computational elements and the multiple-energy-domain simulation, analysis, and design. This article examines these challenges and the new ones that have been identified since 1997 using the results of two projects that have been conducted in the field.

2 Challenges

Many of the challenges raised by diMEMS have been studied in isolation in different research fields. However, in diMEMS they must be examined together and they become even more extereme. The scale of diMEMS needs new software paradigms as well as requiring new hardware capabilities. In the main, scaling up is the main concern of software challenges, while scaling down is the main concern of hardware challenges.

2.1 Software Challenges, Scaling Up

Scalability

Scalability is the main concern of diMEMS as the number of units will likely number in the millions. Scalability therefore impacts the way units will communicate. Systems using synchronous communications can't scale as well as those using asynchronous communications because the constraints of synchronization lower the system efficiency [6]. Synchronous communication is therefore difficult to achieve and asynchronous communications have already shown better results.

To ensure scalability, the programming model and the language must hide the complexity from the programmer and the compiler should enable programming the system as a single ensemble. Within the Claytronics project, new languages like LDP [11] and Meld [4] been have developed to cope with this challenge.

As scalability has to be tested up to millions of units, simulation tools also have to scale up. Dprsim [25] which has been developed within the Claytronics project has successfully simulated millions of units.

Uncertainty Tolerance

Faulty behavior is inherent to any diMEMS system. This is due to several factors. The batch process used in MEMS fabrication creates different levels of reliability. While some of the devices will have no defects, most of them have a high percentage of failed units. On the software side, this characteristic has to be handled and fault-tolerance has to be implemented.

In the case of mobile distributed MEMS a logical topology has to be maintained in order to communicate between the units. Maintaining a logical topology over a physical one is the concern of many research topics like P2P [1], swarm intelligence [18, 23], ad hoc networks [26] or wireless sensor networks [29]. Mobile distributed intelligent MEMS is even more complex than these examples. Mobility, scalability, fault-tolerance and limited processing capability are the main challenges to solve in order to create and maintain a logical topology.

Communications

The tradeoff between computation/communication/sensing is a challenge that has already been studied in wireless sensor networks but it needs some adaptation to take into account the scalability factor which is inherent to diMEMS.

Each diMEMS project has its own communication model directly linked to the application. The question here would be to study the cost and the interest of having some abstraction layer.

Control

When each unit of a system is mobile, the changes in the physical topology modify the logical topology by changing network connectivity. This is one of the main concerns of MANET. The inverse is also true, the logical topology can drive the mobility. This is usually done by covering an area which needs to be sensed [27] but it can also be used for modifying the logical topology, for example to keep connectivity in a sensor network [14].

DiMEMS are composed of actuators which needs control and a degree of synchronization. Three synchronization schemes can be used: no synchronization between MEMS units, means that the control loop doesn't have to synchronize with other units, local synchronization means that a MEMS unit has to be synchronized with its neighbors while global synchronization means that all the MEMS units have to act synchronously.

Having actuators to control requires real-time deadlines. Some applications need a very high frequency from the controler. If the control is fully decentralized like it is the case in distributed intelligent MEMS and that different modules requires a local or a global synchronization, the time to communicate has to be very short.

Reliability through Properties Verification

Reliability is difficult to achieve in any information technology project. The approach taken to achieve it often uses modularity which allows one to define interfaces and to segment the causes of failures. In diMEMS, this modularity is limited. Methods to model the whole system with VHDL-AMS and UML/SysML would allow one to verify some properties of the system and to increase its reliability.

CPS and IoT, Relations with Macro World

DiMEMS are systems that can interact with other intelligent systems. This interaction is the focus of cyber-physical systems (CPS) [22] and Internet of Things (IoT) [5]. The new challenge with distributed inteligent MEMS systems is to manage the different density of communication between the macro-objects (low density) and the micro-objects (high density).

2.2 Hardware Challenges, Scaling Down

Seamless Integration of MEMS and Logic

Integrating MEMS with CMOS is still a challenge in the fabrication process [30, 15]. Most of the MEMS-CMOS integration follow a hybrid integration through wire-bonding but this approach is not well-suited for diMEMS which requires too many connections. DiMEMS requires a monolithic integration for two reasons. First of all, only a monolithic approach can guarantee scalable and affordable fabrication process whereas hybrid approaches often requires manual intervention. Secondly, the weight of a hybrid system is more important than a monolithic one. MEMS actuation requires higher voltages than logic which can create problems. Some solutions have been proposed [20] to tackle this problem but a real voltage difference management between actuation and logic is still a challenge.

Designing Robust MEMS

Due to their size, MEMS are very sensitive to external factors (e.g., dust and air quality) which can change the behavior of certain types of MEMS actuators. Modeling and simulation have proven to be efficient to solve design issues [10] but new solutions have to be found to increase MEMS robustness.

Building Micro-communication Devices

DiMEMS needs communication capabilities, but integrating communication and MEMS is still a significant challenge. Some of them are linked to the previous challenges described here like the voltage difference between the actuation and the logic and the integration of MEMS and logic but scalability is also an issue. One-way communication to the MEMS has been implemented, for example, in DMD where each micro-mirror has to be oriented is the right direction [19]. While this approach works for a fixed topology and even for a regular network of actuators, it doesn't scale for more complex devices or mobile units. Building a real micro-communication device is still a challenge that haven't been solved yet.

3 Examples

Many projects have been built around MEMS but no killer application has yet been realized. Two active projects which are studying diMEMS are the Smart projects (Smart Surface and Smart Blocks) [9] and Claytronics [17].

3.1 The Claytronics Project

3.1.1 Introduction

The past several decades have demonstrated the remarkable power of *programmability*: general-purpose computers have enabled new application domains that were not anticipated when these computers were designed. As impressive as cyberspace is these days, it is useful to remember that we live in a world that is physical and not merely virtual. To enable new classes of exciting applications that may stretch our imaginations (e.g., shape-shifting medical instruments, hands-on interactive 3D design, etc.), the claytronics project aims to bring the power of general-purpose programmability to everyday physical artifacts in a fundamental way through a new form of *programmable matter*. In the long term, our goal is to construct the programmable matter such that its shape, motion, appearance, and response to human touch can be arbitrarily controlled by software.

Our vision for realizing programmable matter is to harness the collective power of a vast number (perhaps millions) of tiny (e.g., millimeter sized) spherical robots that can stick together and move around each other to form an overall material with somewhat fluid properties that we call "claytronics". The name "claytronics" was inspired by the word "claymation," since the material might resemble a form of modeling clay that can shape itself. We refer to the individual robots that make up this ensemble as "catoms", which is short for "claytronics atoms".

3.1.2 Hardware Results

At first glance the ability to create a coherent ensemble of millions of units appears fantastical. But, if we step back and examine it, the question is not "if" we can manufacture it, but "when." It is clearly possible to do so in principle, e.g., biology builds ensembles of units which coordinate together to form dynamic 3D shapes which can interact in the real world. And, we already have MEMS processes which creates 3D devices. Reid has constructed spherical shapes by first printing a

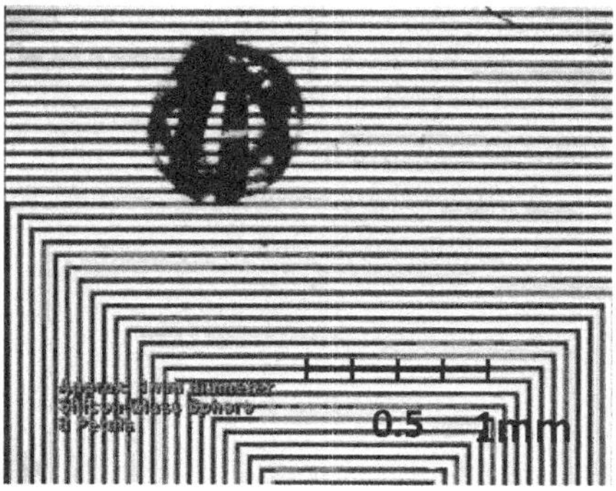

Fig. 1 An example of a spherical shell (with diameter of ~0.9mm) made using standard photolithography and stress induced curling. The shell (in the upper left) is sitting on a circuit board which can move the shell using electrostatic forces. Included with the permission of Rob Reid.

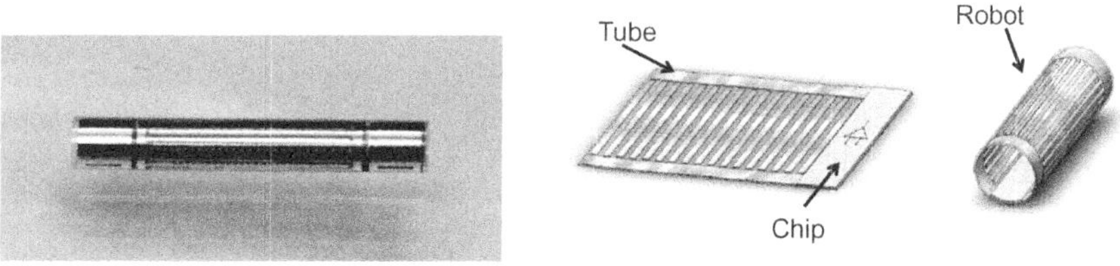

Fig. 2 An example of a complete cylindrical catom. On the left if a cad drawing of the silicon shell and compelted robot. On the right, the realized result: you can see the logic die inside the exterior silicon shell. Included with the permission of M. Karagozler.

projection of the sphere and then, by harnessing the inherent stresses in thin film silicon dioxide, causes the projection to self-assemble into a sphere (See Figure 1) [24, 28]. This same process can be applied to a pre-fabricated CMOS wafer to create 3D units with integrated processors and actuators.

3.1.3 Software Results

In pursuit of our goal we are exploring different programming approaches and have developed two new programming languages: LDP [12, 2] and Meld [3]. Both of these languages are declarative in nature and result in programs which are about 20x shorter than equivalent imperative programs. They each take an ensemble perspective, allowing a programmer to create simple, concise, programs that are automatically compiled down to programs that run on each unit.

One of the advantages of concise programs is the ability of the programmer to focus on the program logic, facilitating correctness and affording greater opportunity for algorithm enhancements (like optimizations). In the case of larger programs we often see better performance in the Meld implementation than an equivalent C++

one. The main reason for this is that Meld and LDP make it easier to write a parallel program than a sequential one, utilizing much of the latent parallelism inherent in the algorithm, while the C++ implementation is limited to the parallelism that the programmer can manage and explicitly encodes.

3.2 The Smart Surface Project

3.2.1 Introduction

The objective of the Smart Surface project is to design a distributed and integrated micro-manipulator based on an array of micro-modules in order to realize an *automated positioning and conveying surface*. Each micro-module will be composed of a micro-actuator, a micro-sensor and a processing unit. The cooperation of these micro-modules thanks to an integrated network will allow to recognize the parts and to control micro-actuators in order to move and position accurately the parts on the Smart Surface. The parts are small, they cover a few numbers of micro-modules (e.g. 4×4).

Figure 3 shows one prototype of Smart Surface. The rectangular holes seen on the front-side are the air nozzles. Air-flow comes through a micro-valve in the back-side of the device and then passes through the nozzle. The advantage of this solution is that the micro-actuators, the most fragile part of the surface, are protected.

Fig. 3 Smart Surface prototype with grouped control of actuators

3.2.2 Hardware Results

Three prototypes have been built within the Smart Surface project, these prototypes have different kinds of actuators, but their main difference can be found in the way the actuators can be controled.

Remotely Placed Actuators Surface

Each cell has oriented holes in 4 directions which means that each can create an airflow in the 4 directions. There is only one actuator per column so that all the cells of a column have the same behavior. The cells are therefore passives as they don't have an actuator.

Grouped Control of Actuators

This type of prototype is composed of actuators that can create airflow in 2 directions (see figure 3)and as the previous one, the actuators are controlled by column.

Individual Control of Actuators

The more advanced prototype has the same actuators as the previous one, i.e. 2 directions but each actuator can be controlled directly which allows a finer control. This last prototype is the more complex to build as each actuator as to be connected to the controller.

3.2.3 Software Results

Even in a technological project, software plays an major role. Four aspects have been studied within this project, sensor feedback, communications, control and modeling.

Sensor Feedback

Each cell that compose the surface comprises a processing unit, communication capabilities, an actuator and sensing capabilites. Sensing gives an indication on the presence of an object on top of it. More precisely, each sensor sends a binary information to its processing unit regarding the presence or the absence of the object. The object is therefore highly discretized. The first work has been to work on the differentiation possibility of highly discretized objects and on the choice of the best criteria to do so. The Exhaustive Comparison Framework (ECO)[8] has been designed in order to test exhaustively the efficiency of different differentiation criteria, in term of differentiation efficiency, memory and processing power needed. The second work has been to set the optimal number of sensors that have to be embedded inside the surface. The Sensor Network Calibrator (SNC) [13] allows to test different number and organization of sensors.

Communications

The physical topology of the Smart Surface is a rectangle, and where each cell is connected to its four neighbours, the network topology is then a mesh 2D. As the network topology is known and fixed, the challenges are in the algorithmic part. A mathematical model of discrete state acquisition and several distributed iterative algorithms have been proposed and tested [6]. Synchronous and asynchronous state acquisition methods and the asynchronous ones have shown better results. Simple initial points and convergence results for distributed algorithms have been proposed. In both the synchronous and asynchronous cases, stopping criteria have been designed. The Smart Surface Simulator has been designed to evaluate and to validate experimentally the proposed distributed algorithms.

Control

Controlling the Smart Surface is very challenging as pneumatic actuation has many problems that need to be solved. The reinforcement learning method which has been implemented is decentralized and addresses the global-local trade-off [21]. An integration of sensing, communications and control has been proposed [9] and the experiments have shown good properties of the system.

Modeling

The VHDL-AMS model [10] that have been developed inside the Smart Surface project can simulate the behavior of the surface while the SysML model [16] gives a more higher-level description of the architecture. The SysML model is derived from the VHDL-AMS one and the objective is to link the SysML description of the hardware to the UML description of the software. This would allow properties verification for the whole system.

4 Conclusion

The concept of diMEMS has been proposed 15 years ago but only few of the challenges have been fully addressed. One reason is that diMEMS is a multidisciplinary field of research involving the topics of MEMS sensors/actuators design and fabrication, electronics, networking, programming and control. Having all these skills in one project is diffficult. Another reason is that the topic of diMEMS relies on MEMS technology and it couldn't exist until MEMS research reaches its complete maturity. This time of maturity has come and we think that the time for pushing further the topic of diMEMS has come as well.

Acknowledgements. We would like to thank all the Smart Surface and Claytronics groups and in particular Eugen Dedu, Kahina Boutoustous and Didier El Baz, Michael Ashely-rollman, and Mustafa Emre Karagozler. This work was supported in part by ANR (ANR-06-ROBO-0009), DARPA (FA87501010215), NSF (CNS-0428738) and Intel Corporation.

References

1. Andersen, D.G., Balakrishnan, H., Kaashoek, M.F., Morris, R.: Resilient Overlay Networks. In: Proc. 18th ACM Symposium on Operating Systems Principles (SOSP), Banff, Canada, pp. 131–145 (2001)
2. Ashley-Rollman, M.P., De Rosa, M., Srinivasa, S.S., Pillai, P., Goldstein, S.C., Campbell, J.D.: Declarative programming for modular robots. In: Workshop on Self-Reconfigurable Robots/Systems and Applications, IROS 2007 (2007), `http://www.cs.cmu.edu/claytronics/papers/ashley-rollman-derosa-iros07wksp.pdf`

3. Ashley-Rollman, M.P., Goldstein, S.C., Lee, P., Mowry, T.C., Pillai, P.: Meld: A declarative approach to programming ensembles. In: Proceedings of the IEEE International Conference on Intelligent Robots and Systems, IROS 2007 (2007), http://www.cs.cmu.edu/~claytronics/papers/ashley-rollman-iros07.pdf
4. Ashley-Rollman, M.P., Lee, P., Goldstein, S.C., Pillai, P., Campbell, J.D.: A Language for Large Ensembles of Independently Executing Nodes. In: Hill, P.M., Warren, D.S. (eds.) ICLP 2009. LNCS, vol. 5649, pp. 265–280. Springer, Heidelberg (2009)
5. Ashton, K.: That 'internet of things' thing. RFID Journal (2009)
6. Baz, D.E., Boyer, V., Bourgeois, J., Dedu, E., Boutoustous, K.: Distributed part differentiation in a smart surface. Mechatronics (2011)
7. Berlin, A., Gabriel, K.: Distributed mems: New challenges for computation. IEEE Computational Science and Engineering Journal 4(1), 12–16 (1997)
8. Boutoustous, K., Dedu, E., Bourgeois, J.: An exhaustive comparison framework for distributed shape differentiation in a MEMS sensor actuator array. In: International Symposium on Parallel and Distributed Computing (ISPDC), pp. 429–433. IEEE computer society press, Krakow (2008)
9. Boutoustous, K., Laurent, G.J., Dedu, E., Matignon, L., Bourgeois, J., Fort-Piat, N.L.: Distributed control architecture for smart surfaces. In: IEEE/RSJ International Conference on Intelligent Robots and Systems, pp. 2018–2024. IEEE, Taipei (2010)
10. Chapuis, Y.-A., Zhou, L., Casner, D., Ai, H., Hervé, Y.: FPGA-in-the-loop for control emulation of distributed MEMS simulation using VHDL-AMS. In: Proc. of the 1st Workshop on Hardware and Software Implementation and Control of Distributed MEMS (dMEMS2010), pp. 92–99. IEEE CPS (2010)
11. De Rosa, M., Goldstein, S.C., Lee, P., Campbell, J.D., Pillai, P.: Programming modular robots with locally distributed predicates. In: Proceedings of the IEEE International Conference on Robotics and Automation, ICRA 2008 (2008)
12. De Rosa, M., Goldstein, S.C., Lee, P., Campbell, J.D., Pillai, P.: Programming modular robots with locally distributed predicates. In: Proceedings of the IEEE International Conference on Robotics and Automation, ICRA 2008 (2008), http://www.cs.cmu.edu/~claytronics/papers/derosa-icra08.pdf
13. Dedu, E., Bourgeois, J., Boutoustous, K.: Simulation to help calibration of a mems sensor network. International Journal of Pervasive Computing and Communications 6(4) (2010)
14. Derbakova, A., Correll, N., Rus, D.: Decentralized self-repair to maintain connectivity and coverage in networked multi-robot systems. In: Proc. of IEEE International Conference on Robotics and Automation, ICRA (2011)
15. Ghosh, S., Bayoumi, M.: On integrated cmos-mems system-on-chip. In: The 3rd International IEEE-NEWCAS Conference, pp. 31–34 (2005)
16. Giorgetti, A., Hammad, A., Tatibouët, B.: Using SysML for smart surface modeling. In: dMEMS 2010, 1st Workshop on Design, Control and Software Implementation for Distributed MEMS, pp. 100–107. IEEE Computer Society Press, Besançon (2010)
17. Goldstein, S.C., Mowry, T.C., Campbell, J.D., Ashley-Rollman, M.P., De Rosa, M., Funiak, S., Hoburg, J.F., Karagozler, M.E., Kirby, B., Lee, P., Pillai, P., Reid, J.R., Stancil, D.D., Weller, M.P.: Beyond audio and video: Using claytronics to enable pario. AI Magazine 30(2) (2009)
18. Gro, R., Bonani, M., Mondada, F., Dorigo, M.: Autonomous self-assembly in swarm-bots. IEEE Transactions on Robotics 22(6), 1115–1130 (2006)

19. Hornbeck, L.J.: Digital Light Processing for high-brightness high-resolution applications. In: Wu, M.H. (ed.) Society of Photo-Optical Instrumentation Engineers (SPIE) Conference Series. Society of Photo-Optical Instrumentation Engineers (SPIE) Conference Series, vol. 3013, pp. 27–40 (1997)
20. Karagozler, M.E., Thaker, A., Goldstein, S.C., Ricketts, D.S.: Electrostatic actuation and control of micro robots using a post-processed high-voltage soi cmos chip. In: IEEE International Symposium on Circuits and Systems, ISCAS (2011)
21. Matignon, L., Laurent, G.J., Le Fort-Piat, N., Chapuis, Y.A.: Designing decentralized controllers for distributed-air-jet mems-based micromanipulators by reinforcement learning. Journal of Intelligent and Robotic Systems 145(2), 59–80 (2010)
22. Lee, E.A.: Cyber physical systems: Design challenges. In: IEEE International Symposium on Object-Oriented Real-Time Distributed Computing, vol. 0, pp. 363–369 (2008)
23. Lewis, A., Bekey, G.: The behavioral self-organization of nanorobots using local rules. In: Proceedings of the IEEE/RSJ International Conference on Intelligent Robots and Systems (1992)
24. Reid, J.R., Vasilyev, V.S., Webster, R.T.: Building micro-robots: A path to sub-mm^3 autonomous systems. In: Proceedings of Nanotech. (2008)
25. Rister, B.D., Campbell, J., Pillai, P., Mowry, T.C.: Integrated debugging of large modular robot ensembles. In: ICRA, pp. 2227–2234 (2007)
26. Royer, E., Toh, C.K.: A review of current routing protocols for ad hoc mobile wireless networks. IEEE Personal Communications 6(2), 46–55 (1999)
27. Schwager, M., Slotine, J.J., Rus, D.: Decentralized, adaptive control for coverage with networked robots. In: 2007 IEEE International Conference on Robotics and Automation, pp. 3289–3294 (2007)
28. Vasilyev, V.S., Reid, J.R., Webster, R.T.: Microfabrication of si/sio2-spherical shells as a path to sub-mm^3 autonomous robotic systems. In: MRS Fall Meeting (2008)
29. Wang, Y.: Topology control for wireless sensor networks. In: Li, Y., Thai, M.T., Wu, W. (eds.) Wireless Sensor Networks and Applications. Signals and Communication Technology, pp. 113–147. Springer, US (2008)
30. Witvrouw, A.: Cmos-mems integration: why, how and what? In: Proceedings of the 2006 IEEE/ACM International Conference on Computer-Aided Design, ICCAD 2006, pp. 826–827. ACM (2006)

Extracting Emergent Semantics from Large-Scale User-Generated Content

Ioannis Kompatsiaris, Sotiris Diplaris, and Symeon Papadopoulos

Abstract. This paper presents a survey of novel technologies for uncovering implicit knowledge through the analysis of user-contributed content in Web2.0 applications. The special features of emergent semantics are herein described, along with the various dimensions that the techniques should be able to handle. Consequently a series of application domains is given where the extracted information can be consumed. The relevant techniques are reviewed and categorised according to their capability for scaling, multi-modal analysis, social networks analysis, semantic representation, real-time and spatio-temporal processing. A showcase of such an emergent semantics extraction application, namely ClustTour, is also presented, and open issues and future challenges in this new field are discussed.

Keywords: emergent semantics, social media analysis.

1 Introduction

Social media sharing properties, such as Flickr, Facebook and Picasa host billions of images and videos that have been annotated and shared among friends, or published in groups covering a specific topic of interest. The fact that users annotate and comment on the content in the form of tags, ratings, preferences, etc. and that these are applied on a daily basis, gives this data source an extremely dynamic nature that reflects events and the evolution of community focus. Although current Web2.0 applications encourage annotations and feedback by the users, these are not sufficient for extracting such "hidden" knowledge, because they lack clear semantics and it is the combination of visual, textual and social context, which provides the ingredients for a more thorough understanding of social content. Therefore, there is a need for scalable

---

Ioannis Kompatsiaris · Sotiris Diplaris · Symeon Papadopoulos
Informatics & Telematics Institute, 6th Km Charilaou-Thermi road,
57001 Thessaloniki, Greece
e-mail: {ikom,diplaris,papadop}@iti.gr

L. Kocarev (Ed.): ICT Innovations 2011, AISC 150, pp. 27–37.
springerlink.com

and distributed approaches capable of handling the massive amounts of available data and capturing the emergent semantics, a procedure also called Collective Intelligence, that would enable the exploitation of the knowledge hidden in the user contributed content (UGC).

Recent advances of Web technologies have effectively turned ordinary people into active members of the Web: casual users act as co-developers and their actions and collaborations with one another have added a new social dimension on Web data. For example, Wikipedia motivates thousands of volunteers around the world to create the world's largest encyclopedia. An image in Flickr is annotated with descriptive tags, associated with the users that seem to like it and mark it as favorite, described by the visual features encoding its visual content, and often spatial and temporal information denoting its context. Even though all these facets of information are not combined naturally with each other, still they carry knowledge about the resource, with each facet providing a representation of the particular resource in a different feature space.

There already exist a number of approaches that are based on user-contributed content in order to provide useful information for various applications. For example, mobile location information and uploaded content is used to monitor online traffic and generate traffic patterns in [1], connect citizens in Boston [2], share nature experience [3], discover travel patterns and provide travel advice [4, 5], communicate problems in a city [6] and deal with climatic changes as in the Climate Collaboratorium [7] project of the MIT Center for Collective Intelligence[1]. The Collective Prediction effort, tries to make accurate predictions about future events such as product sales, political events, and outcomes of medical treatments [8]. However, the main characteristic of such applications is that they are mostly based on collecting well-structured contributions through specific applications, on shallow statistical processing of the contributions and their visualisation. Very few focus on analysis of implicit relations in UGC and feedback and on dealing with unstructured large-scale data, where an important source of knowledge is hidden.

This paper focuses on approaches that capture and analyse the emergent semantics lying under the large-scale multimedia content, which are extracted by using media resource and user relations, actions and interactions in social networking and sharing applications. A state-of-the-art survey is given by first defining the different features of existing applications dealing with emergent semantics extraction in terms of analysis dimensions and application domains (Section 2). Then the applications are reviewed and categorised according to the previously defined criteria (Section 3). We also present an example application, namely ClustTour, which encompasses many of the different aspects/dimensions discussed herein, and showcasing the capability of emergent semantics extraction to exploit different input data sources with combined analysis techniques for the delivery of enhanced services to the end users (Section 4). Finally, the paper concludes with a discussion on the identified issues and future challenges in developing emergent semantics extraction applications in the large scale (Section 5).

[1] http://cci.mit.edu

2 Social Media Analysis Dimensions and Applications

Understanding large-scale social media content involves the consideration of a multitude of features in the development of suitable analysis techniques.

A key issue to be addressed is the presence of *noisy and ambiguous data*. User contributed content is very noisy containing many non-relevant contributions either intentionally (spamming) or unintentionally. The lack of constraints with respect to tagging is the source of numerous annotation quality problems, such as spam, misspellings, and ambiguity of semantics. Moreover, the lack of (hierarchical) structure in the contributed information results in tag ambiguity, tag synonymy and granularity variation.

UGC can be viewed as a *rich multi-modal source of information* including attributes such as time, favorites and social connections. Social media analysis systems can exploit different modalities of input, ranging from single visual, textual, audio or user information to fused sets of such media sources, i.e. annotated images, audio or audiovisual content, or even the combination of such content with the graph structure of social networks. In such approaches, also called feature-level approaches, the need to obtain a joint, unique representation for the multimedia object demands techniques that manage to handle the very different characteristics exhibited by the different types of data. This is true both in terms of the nature of raw features (e.g., sparse, high-dimensional word co-occurrence vectors extracted from text descriptions, compared to usually dense and low-dimensional descriptors extracted from visual content), as well as in terms of their semantic capacity (e.g., while abstract concepts like "freedom" are more easily described with text, concrete concepts like "sun" are more easily grounded using visual information).

Scalability is an important issue since the discovery of implicit information is based on massive amounts of data. Such huge volumes of user generated data raise scalability issues that significantly compromise the performance (in terms of accuracy) of algorithms operating on such data. The situation gets worse in cases where either the employed algorithms aim at extracting knowledge patterns that only become stable after a specific usage period or processing needs to address *(near) real-time data*. These cases pose very demanding requirements in terms of algorithmic design, computational power, memory and storage. This includes parallelisation and distributed techniques, speed, storage, memory and relevant considerations, which are indispensable in order to advance to real-time systems offering emergent semantic discovery capabilities.

Another important aspect characterising how analysis techniques can unveil the hidden semantics is the presence or absence of *spatial processing* in the data sources. Following current web socialising and multimedia handling user trends, most users nowadays upload, geo-tag and localise their personal photos. Moreover, most Web2.0 systems make heavy use of geotagged resources. Therefore, many recommendation, presentation or prediction techniques are used for enhancing location-based systems and services.

A different dimension, orthogonal to the location aspect is time. The possibility of processing *temporal* features endows such systems with event processing capabilities.

When time aspects of the UGC are considered, it is possible to derive information about situations or events. Based on massive user contributions, the emergent semantic extraction results could range from the single representation of events, and expand to past/current events detection or even future events prediction.

More elaborate applications are able to capture knowledge in the combined *spatio-temporal* dimension. Considering both space and time information, it is possible to depart from mere presentations of geotagged collections and discover knowledge about routes or map areas with particular interest, as well as to detect events in the combined spatio-temporal dimension.

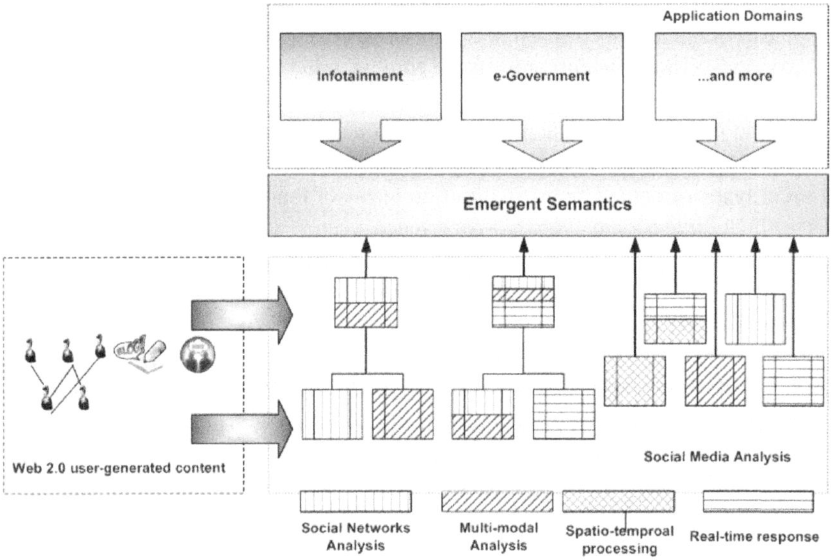

Fig. 1 Extracting emergent semantics from Web2.0 applications.

To summarise, analysis techniques that deal with Web2.0-based and social network content and structure can be categorised with respect to the different dimensions they embrace (multi-modality), their ability to process the underlying social network structure information, the different features they consider (visual, audio, textual or their fusion), the capability of scaling, as well as their potential for running in real-time.

Exploiting the information carried by social media can then help tackle a variety of issues in different disciplines, such as information retrieval (e.g., poor recall and precision), machine learning, data mining, and multimedia understanding, where for example, social media sites can be used as a rich source of weakly labelled data for solving large-scale artificial intelligence problems. They can also be applied to domains such as tourism, culture, social sciences, politics, economics, and marketing. In tourism and culture, uploaded media can reveal "off-the-beaten-path" points of interest and events, otherwise difficult to discover through usual Web sources. In economics, marketing and brand monitoring, the number of related media uploaded online can reflect the number and locations of products sold in the market. Fig. 1 depicts the procedure by which the emergent semantics are analyzed and consumed.

3 Approaches for Emergent Semantics Extraction

In this section we will review some of the social media analysis applications that take into account large-scale semi-structured UGC and apply computational methods in discovering implicit relations and therefore "hidden" knowledge. We also attempt to classify them according to the categorisation and the features these techniques consider, as discussed in Section 2.

3.1 Multi-modal Analysis Approaches

In [9] geo-location and tag information are used in order to generate representative city maps. In [10, 11] tags and visual information together with geo-location are used for object (e.g. monuments) and event extraction. The description of city cores can be derived automatically, by exploiting tag and location information [12]. The approach is able of distinguishing between administrative and vernacular uses of place names, thus avoiding the potential for confusion in the dispatch of emergency services. Tags from Flickr images and timestamp information are used in [13] to form a chronologically ordered set of geographically referenced photos and distinguish locals from tourist travelling. Another work in the line of multi-modality–aided localisation is VIRaL [14], a web-based tool which is used to identify and localise similar multimedia content under different viewpoints, applicable to any case that involves still image search and retrieval.

3.2 Community, Trend and Event Detection

Although still at the level of research, there are several applications that exploit the knowledge extracted from massive user contributions. For instance, it is common to derive community-based views of networks, i.e. networks of which the nodes correspond to the identified communities of the original networks and the edges to the relations between the communities. Such views are more succinct and informative than the original networks. It is for this reason that community detection has found applications in the field of recommendation systems [15-18], as well as for representing user profiles [19, 20]. Other applications that make use of the knowledge extracted from tag communities include sense disambiguation [21] and ontology evolution/population [16].

The patterns emerging often show deep interconnections with various world events [22, 23] in a way that the evolving world reality is captured at each instant.

A leap in the exploitation of UGC research is the work in [24] which explores global trends and sentiments that can be drawn by analyzing the sharing patterns of uploaded and downloaded social multimedia. Taking into account both spatial and temporal aspects of content item views and uploads in social media sites and aggregating them, the authors are trying to forecast future events impacting politics, economics and marketing.

In [25] the authors are using human sensors to detect real world events, and generate situation awareness. They describe how spatio-temporal-thematic data in various social media can be aggregated into 'social pixels'. They derive image-like representations which enable sophisticated data processing, offering to users a query algebra tool for posing sophisticated situational queries. The technique is applied in business analysis, seasonal characteristics analysis and political event analytics.

Further, other work shows that the actions of individual Web users, when properly pooled, can indicate macro trends. There are studies using Search Engine queries for influenza surveillance over the Web [26], such as Google Trends [27], search advertisement click through [28], Yahoo search queries [29] and health website access logs [30]. Specifically in [27], Google search engine queries and data from the Centers for Disease Control (CDC) are used to monitor influenza rates 1-2 weeks ahead of the CDC reports.

3.3 Real-time Applications

Finally, a separate class of applications involves the real-time aspect in the analysis. An early, but not quite scalable, tool [31] deals with the analysis of user profiles and query logs for the extraction of personalised touristic information (places and events suggestions) using a hierarchical semantic geospatial model as well as an event notification system. SCENT [32] is a framework for monitoring the evolution of multi-faceted social network data resulting from users' continuous online interactions. It enables very large scale data management and analysis by modeling social data in the form of tensor (multi-dimensional array) streams, tracking changes in the spectral properties of the tensor over time. It has mainly been used in recommendation and monitoring use cases.

Another tool [33] is developed to improve search quality (by reranking) and recommend supplementary information in query time (i.e., search-related tags and canonical images) analyzing visual content, high-level concept scores, time and location metadata. In [34] a real-time approach combining the contextual information from time duration, number of views, and thumbnail images with content analysis derived from color and local points is used in order to achieve near-duplicate elimination in video search in Web 2.0 applications. Finally, the authors in [35] exploit georeferenced toponyms found in community websites to model vernacular places in cities. The system is capable of answering queries in real time.

In the following section another example of emergent semantics extraction, namely ClustTour, is described in more detail. Table 1 presents a categorisation of the above techniques and applications with respect to the kind of analysis they employ (i.e. recommendation, clustering, localisation, monitoring, event detection, etc.) and the parameters discussed in Section 2.

It can be observed that most applications do not employ the spatial or spatio-temporal dimension; instead they mostly make use of textual annotations in order to provide recommendations, or represent events and situations.

Table 1. Categorisation of emergent semantic extraction applications.

Analysis Methods/ Dimensions	Scalability	Spatio-temporal	SNA	Multi-modal	Semantic representation	Real-time
Recommendation	[32-34]	[31, 32]	[15-20, 31, 32, 34]	[32-34]	[19, 20]	[31-34]
Clustering	[36]	[13, 36]	[13, 21, 36]	[10, 11, 36]	[13]	
Localisation	[12, 14, 35]	[14]	[14, 35]	[9, 14]	[9, 12]	[35]
Trend/event detection	[26-30, 36]	[22-25, 36]	[16, 22-24, 30, 36]	[9, 10, 22, 23, 36]	[9, 22, 23, 25]	

However, more advanced techniques and applications [9,10,13,22,23,24,25] have also been presented, capable of processing more kinds of input modalities enabling the spatio-temporal and situational dimension. Scalability is addressed by a wide range of applications; however the amount of works enabling the real-time aspect is still very limited. It is notable that very few real-time applications are also able to handle and analyze multi-modal information.

4 ClustTour: An Example Application

ClustTour brings to surface emergent semantics by using photo clusters corresponding to landmarks and events to assist the online exploration of a city. It employs an efficient graph-based scheme that clusters massive amounts of photos into clusters by use of a community detection method that is applied on a hybrid similarity graph encoding both visual and textual similarities between photos. Subsequently, it classifies these clusters into landmarks and events [11]. ClustTour leverages the obtained clusters through a map-based interface that enables users to navigate through a city. The application offers two modes of exploration: a city view depicting a high-level view of the most important areas in the city (Fig. 2) and an area view centered on the selected area and showing landmarks and events in its vicinity. ClustTour also provides a temporal content organisation layer on top of the detected areas and photo clusters. In that way, it endows users with enhanced content exploration and browsing capabilities, and at the same time, it improves upon the quality of the presented clusters. In the end, users obtain informative views over the interesting spots and areas in a city depending on the temporal context that is of interest to them.

With its ability to analyse massive visual and textual information from UGC (Flickr) and to present both landmarks and events in a spatio-temporal dimension, ClustTour is a typical case of implicit knowledge extraction application enabling the analysis of multi-modal input in the large scale.

5 Conclusions and Open Issues

The herein reviewed emergent semantics extraction techniques are trying to harness one or more forms of online user contributions in order to benefit end-users and organisations by employing large-scale recommendation, prediction, detection, representation or summarisation analysis techniques. However, such advanced efforts are still very limited, compared to the widespread usage of social media. Several aspects are currently covered by these methodologies, addressing issues such as scalability, efficiency, fusion and integration of multi-modal sources, as well as real-time analysis. Additional issues and dimensions relevant to social media analysis and data mining include:

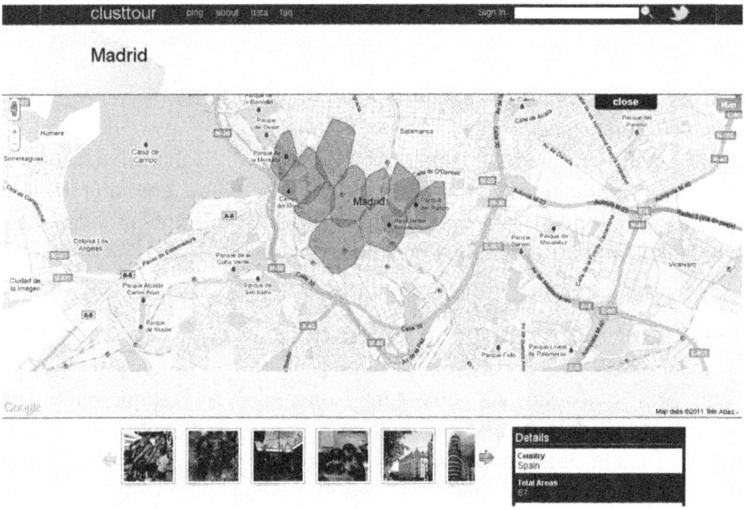

Fig. 2 Areas detected by spatial clustering of Flickr photos in Madrid.

- **Aggregation of multiple input sources.** Analysis of multiple Web2.0 sources by semantics extraction techniques is a step beyond the existing approaches that mostly consider single sources (e.g. Flickr photos).
- **Linking with Open Data.** Exposing, sharing, and connecting pieces of information extracted from social media, using URIs and RDF, with existing Open Data is a crucial step towards a more semantic description and representation of the extracted information.

- **User Interfaces and Visualisation** must be developed in order to support user interaction and understanding of the results of social media analysis.
- **Trust, security, privacy** is always a concern for users when they contribute content and especially when this is further analyzed by the application. The users should be guaranteed that their contribution remains anonymous and/or is used for an objective they consent to.
- **Social data sampling, modeling and representation.** In order to deal with the enormity of real-world social media datasets, appropriate sampling and data reduction techniques are needed, which manage to reduce the size of the data but at the same time they preserve their structure, topology and social context. These techniques are closely related to the representation approach used (for example, in graph-based representations they lead to sub-graph sampling) and there are still many open questions when dealing with social media-related sampling strategies.

Acknowledgment. This work is supported by the SocialSensor FP7 project, partially funded by the EC under contract number 287975.

References

1. Work, D., Blandin, S., Piccoli, B., Bayen, A.: A traffic model for velocity data assimilation. Applied Mathematics Research eXpress (2010)
2. Torres, L.H.: Citizen sourcing in the public interest. KM4D Journal 3(1), 134–145 (2007)
3. iSpot, your place to share nature, `http://ispot.org.uk/`
4. Mobnotes, `http://www.mobnotes.com/`
5. Dopplr, `http://www.dopplr.com/`
6. FixMyStreet, `http://www.fixmystreet.com/`
7. Malone, T.W., Klein, M.: Harnessing Collective Intelligence to Address Global Climate Change. Innovations 2(3), 15–26 (2007)
8. Kemp, C., Shafto, P., Berke, A., Tenenbaum, J.B.: Combining causal and similarity-based reasoning. In: Advances in Neural Information Processing Systems, vol. 19 (in Press)
9. Kennedy, L.S., Naaman, M., Ahern, S., Nair, R., Rattenbury, T.: How Flickr helps us make sense of the world: context and content in community-contributed media collections. ACM Multimedia, 631–640 (2007)
10. Quack, T., Leibe, B., Van Gool, L.: World-scale mining of objects and events from community photo collections. In: Proc. Int. Conf. on Content-Based Image and Video Retrieval, pp. 47–56 (2008)
11. Papadopoulos, S., Zigkolis, C., Kompatsiaris, Y., Vakali, A.: Cluster-based Landmark and Event Detection on Tagged Photo Collections. IEEE Multimedia 18(1), 52–63 (2011)
12. Hollenstein, L., Purves, R.S.: Exploring place through user- generated content: using Flickr to describe city cores. Journal of Spatial Information Science (2009)
13. Girardin, F., Calabrese, F., Dal Fiore, F., Ratti, C., Blat, J.: Digital footprinting: Uncovering tourists with user-generated content. IEEE Pervasive Computing 7(4), 36–43 (2008)
14. Kalantidis, Y., Tolias, G., Avrithis, Y., Phinikettos, M., Spyrou, E., Mylonas, P., Kollias, S.: VIRaL: Visual Image Retrieval and Localization. Multimedia Tools Appl. 51(2), 555–592 (2011)

15. Nanopoulos, A., Gabriel, H.-H., Spiliopoulou, M.: Spectral Clustering in Social-Tagging Systems. In: Vossen, G., Long, D.D.E., Yu, J.X. (eds.) WISE 2009. LNCS, vol. 5802, pp. 87–100. Springer, Heidelberg (2009)
16. Specia, L., Motta, E.: Integrating Folksonomies with the Semantic Web. In: Franconi, E., Kifer, M., May, W. (eds.) ESWC 2007. LNCS, vol. 4519, pp. 624–639. Springer, Heidelberg (2007)
17. Diederich, J., Iofciu, T.: Finding communities of practice from user profiles based on folksonomies. In: Proceedings of the 1st International Workshop on Building Technology Enhanced Learning Solutions for Communities of Practice, TEL-CoPs 2006 (2006)
18. Schifanella, R., Barrat, A., Cattuto, C., Markines, B., Menczer, F.: Folks in folksonomies: social link prediction from shared metadata. In: WSDM 2010: Proc. 3rd ACM Int. Conference on Web Search and Data Mining, pp. 271–280. ACM Press, New York (2010)
19. Au Yeung, C.M., Gibbins, N., Shadbolt, N.: A study of user profile generation from folksonomies. In: SWKM (2008)
20. Gemmell, J., Shepitsen, A., Mobasher, B., Burke, R.: Personalizing Navigation in Folksonomies Using Hierarchical Tag Clustering. In: Song, I.-Y., Eder, J., Nguyen, T.M. (eds.) DaWaK 2008. LNCS, vol. 5182, pp. 196–205. Springer, Heidelberg (2008)
21. Au Yeung, C.M., Gibbins, N., Shadbolt, N.: Contextualising tags in collaborative tagging systems. In: HT 2009: Proceedings of the 20th ACM Conference on Hypertext and Hypermedia, pp. 251–260. ACM Press, New York (2009)
22. Sakaki, T., Okazaki, M., Matsuo, Y.: Earthquake Shakes Twitter Users: Real-time Event Detection by Social Sensors. In: World Wide Web Conference (2010)
23. Signorini, A.: Swine Flu monitoring using twitter, `http://compepi.cs.uiowa.edu/alessio/twitter-monitor-swine-flu/`
24. Jin, X., Gallagher, A., Cao, L., Luo, J., Han, J.: The Wisdom of Social Multimedia: Using Flickr For Prediction and Forecast. In: MM 2010 Proceedings of the International Conference on Multimedia, Firenze, Italy (2010)
25. Singh, V.K., Gao, M., Jain, R.: Social Pixels: Genesis and Evaluation. In: MM 2010 Proceedings of the international conference on Multimedia, Firenze, Italy (2010)
26. Corley, C.D., Mikler, A.R.: A computational framework to study public health epidemiology. In: International Joint Conferences on System Biology, Bioinformatics and Intelligent Computing (IJCBS 2009), Shanghai, China (2009)
27. Ginsberg, J., Mohebbi, M.H., Patel, R.S., Brammer, L., Smolinski, M.S., Brilliant, L.: Detecting influenza epidemics using search engine query data. Nature 457, 1012–1014 (2009)
28. Eysenbach, G.: Infodemiology: tracking flu-related searches on the web for syndromic surveillance. In: AMIA 2006 Symposium Proceedings, pp. 244–248 (2006)
29. Polgreen, P.M., Chen, Y., Pennock, D.M., Nelson, F.D.: Using internet searches for influenza surveillance. Clinical Infectious Diseases (Supplement), 1443–1448 (2008)
30. Johnson, H.A., Wagner, M.M., Hogan, W.R., Chapman, W., Olszewski, R.T., Dowling, J., Barnas, G.: Analysis of web access logs for surveillance of influenza. Stud. Health Technol. Inform. 107(Pt 2), 1202'1C6 (2004)
31. Hinze, A., Voisard, A.: Location and Time-Based Information Delivery in Tourism. In: Hadzilacos, T., Manolopoulos, Y., Roddick, J., Theodoridis, Y. (eds.) SSTD 2003. LNCS, vol. 2750, pp. 489–507. Springer, Heidelberg (2003)

32. Lin, Y.R., Candan, K.S., Sundaram, H., Xie, L.: Scent: Scalable compressed monitoring of evolving multi-relational social networks. In: Review at ACM Trans. on Multimedia Computing (2010)
33. Yang, Y.H., Wu, P.T., Lee, C.W., Lin, K.H., Hsu, W.H., Chen, H.H.: ContextSeer: context search and recommendation at query time for shared consumer photos. In: Proc. 16th ACM Int. Conf. on Multimedia (MM 2008), pp. 199–208. ACM, New York (2008)
34. Wu, X., Ngo, C.W., Hauptmann, A.G., Tan, H.K.: Real-Time Near-Duplicate Elimination for Web Video Search With Content and Context. IEEE Transactions on Multimedia 11(2), 196–207 (2009)
35. Henrich, A., Lüdecke, V.: Determining geographic representations for arbitrary concepts at query time. In: Proceedings of the First International Workshop on Location and the Web (LOCWEB 2008), pp. 17–24. ACM, New York (2008)
36. Papadopoulos, S., Zigkolis, C., Kapiris, S., Kompatsiaris, Y., Vakali, A.: City exploration by use of spatio-temporal analysis and clustering of user contributed photos. In: Demo Paper Accepted in ACM International Conference on Multimedia Retrieval, ICMR (2011)

Protein Function Prediction by Clustering of Protein-Protein Interaction Network

Ivana Cingovska, Aleksandra Bogojeska, Kire Trivodaliev, and Slobodan Kalajdziski

Abstract. The recent advent of high throughput methods has generated large amounts of protein-protein interaction network (PPIN) data. When studying the workings of a biological cell, it is useful to be able to detect known and predict still undiscovered protein complexes within the cell's PPINs. Such predictions may be used as an inexpensive tool to direct biological experiments. Because of its importance in the studies of protein interaction network, there are different models and algorithms in identifying functional modules in PPINs. In this paper, we present two representative methods, focusing on the comparison of their clustering properties in PPIN and their contribution towards function prediction. The work is done with PPIN data from the bakers' yeast (Saccaromyces cerevisiae) and since the network is noisy and still incomplete, we use pre-processing and purifying. As a conclusion new progress and future research directions are discussed.

Keywords: Protein interaction networks, Graph clustering, Protein function prediction.

1 Introduction

The rapid development of genomics and proteomics has generated an unprecedented amount of data for multiple model organisms. As has been commonly realized, the acquisition of data is a preliminary step, and a true challenge lies in developing effective means to analyze such data and endow them with physical or functional meaning [1].

Ivana Cingovska · Aleksandra Bogojeska · Kire Trivodaliev · Slobodan Kalajdziski
Ss. Cyril and Methodius University, Faculty of Computer Science and Engineering, Ruger Boskovic 16, 1000 Skopje, Macedonia
e-mail:{ivana.chingovska,aleksandra.bogojeska,kiret,skalaj}@finki.ukim.mk

L. Kocarev (Ed.): ICT Innovations 2011, AISC 150, pp. 39–49.
springerlink.com

Significant amount of data used for computational function prediction is produced by high-throughput techniques. Methods like Microarray co-expression analysis and Yeast2Hybrid experiments have allowed the construction of large interaction networks. A protein-protein interaction network (PPIN) consists of nodes representing proteins, and edges representing interactions between proteins. Such networks are stochastic as edges are weighted with the probability of interaction. There is more information in a PPIN compared to sequence or structure alone. A network provides a global view of the context of each gene/protein. Hence, the next stage of computational function prediction is characterized by the use of a protein's interaction context within the network to predict its functions. A node in a PPIN is annotated with one or more functional terms. Multiple and sometimes unrelated annotations can occur due to multiple active binding sites or possibly multiple stable tertiary conformations of a protein. The annotation terms are commonly based on ontology, like Gene Ontology (GO) project [2].

One of the main characteristics of the protein interaction networks is that they contain regions or subnetworks densely connected within, but very sparsely interconnected between themselves. This is the main reason for development of methods that perform clustering over the graph representing the protein interaction network. The modular structure of the biological networks in general is proven in [3], where the protein interaction network is clustered using three different approaches. The first one finds the completely connected subgraphs of the network and considers them as clusters. The second one exploits super paramagnetic clustering for data which are not in some metric space. The third approach observes the clustering as an optimization problem, thus maximizing the density of connectedness.

One very important system for clustering protein interaction networks is the MCODE system described in [4]. It performs the clustering in three stages: (1) weightening the nodes in the graph, using the clustering coefficient of the node's neighbourhood, (2) the weighted graph is traversed recursively and molecular complex is formed out of the nodes which have weight above a certain threshold, (3) post processing the results. NetworkBlast [5] is a tool in which every subgraph of the graph of protein interactions is considered as a candidate for functional module. Its modularity is then evaluated by calculating the ratio between the likelihood that it can be set to previously created model of protein complex and the probability that the edges in it are random. An algorithm which detects densely connected subgraphs with n nodes and need at least $n/2$ edges to be deleted in order to break its connectivity is presented in [6]. Markov clustering is used in the algorithm proposed in [7].

The algorithm proposed in [8] is a typical member of the family of algorithms which represent the proteins in some metric space. It calculates the adjacency between two proteins as their probability to have m common neighbours. Afterwards, hierarchical clustering is applied to the obtained distance matrix. Other representative example is the system PRODISTIN [9], which assumes that the Czekanovski-Dice distance between two proteins, which is based on the number of their common neighbours, mirrors their functional distance as well.

The quality of the obtained clusters can be evaluated in couple of ways. One of the criteria rates the clustering as good if the proteins in a cluster are densely connected between themselves, but sparsely connected with the proteins in the rest of the network [10]. Some systems provide tools for generation of graphs with known clusters, which is modelled with the parameters of the explored network [11]. Then the clusters obtained with the clustering algorithm are compared to the known ones. The clustering method can also be evaluated by its ability to reconstruct the experimentally and biologically confirmed protein complexes or functional modules [3][10][12]. For a system for protein function prediction, the most useful property of a clustering method is the functional homogeneity of the clusters.

In this paper we set up a framework for predicting protein function by using clustering in PPIN. We use two clustering methods, one that take into account the graph theory adjusted for PPIN, and other that transforms the PPIN into metric space. The PPIN data we use are from the bakers' yeast (Saccaromyces cerevisiae).

2 Research Methods

The methods for protein function prediction by clustering of PPIN generally consist of three phases, as represented on figure 1.

The first phase is the dividing the network in clusters, using its topology or some other information for the nodes or the edges, if such an information is available. The compactness and the characteristics of the obtained clusters are then evaluated in the second phase. From physical aspect the clusters are assessed by the ratio of the number of edges within and between the clusters, and from biological aspect they are assessed by the functional and biological similarities of the proteins in the clusters. This second phase is not mandatory, but it is useful because it can point out what to expect from the function prediction itself. The prediction of the protein annotations for the proteins in the clusters is the task of the third phase.

2.1 Protein-Protein Interaction Data

High-throughput techniques are prone to detecting many false positive interactions, leading to a lot of noise and non-existing interactions in the databases. Furthermore, some of the databases are supplemented with interactions computationally derived with a method for protein interaction prediction, adding additional noise to the databases. Therefore, none of the available databases are perfectly reliable and the choice of a suitable database should be made very carefully.

For the needs of this paper the PPIN data are compiled, pre-processed and purified from a number of established datasets, like: DIP [13], MIPS [14], MINT [15], BIND [16] and BioGRID [17]. The functional annotations of the proteins were taken from the SGD database [18]. It is important to note that the annotations are unified with Gene Ontology (GO) terminology [2]. The GO consists of three

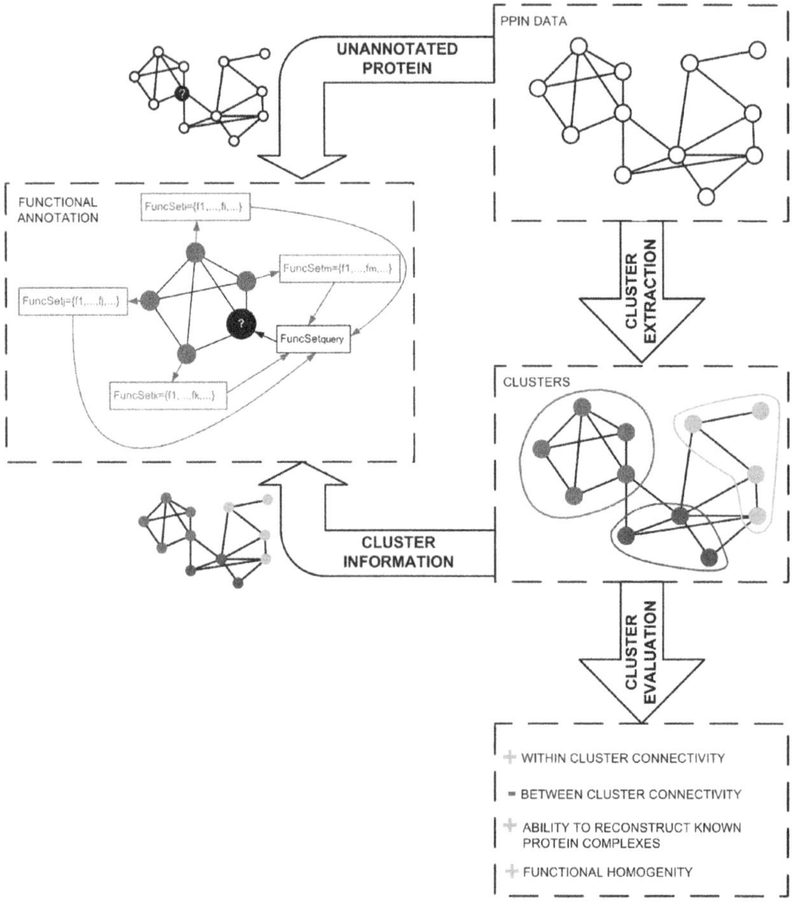

Fig. 1. General framework for protein function prediction by clustering in PPIN

structured dictionaries (ontologies): cellular component, biological process and molecular function. Due to the hierarchical structure of GO, the terms are linked between themselves with the relations: 'is_a', 'part_of', and 'regulates'.

The data for protein annotations are not used raw, but are preprocessed as proposed in [20]. First, the trivial functional annotations, like 'unknown cellular compartment', 'unknown molecular function' and 'unknown biological process' are erased. Then, additional annotations are calculated for each protein by the policy of transitive closure derived from the GO. The extremely frequent functional labels (appearing as annotations to more than 300 proteins) are also excluded, because they are very general and do not carry significant information.

After all the preprocessing steps, the used dataset is believed to be highly reliable and consists of 2502 proteins from the interaction of the baker's yeast, has 12708 interactions between them and are annotated with a total of 888 functional labels. For the purposes of evaluating the proposed methods, the largest connected component of this dataset is used, which consists of 2146 proteins.

2.2 Cluster Extraction

We use two different methods for cluster extraction from the PPIN data. The first one, edge-betweenness clustering, found its first use for clustering biological networks in [20], but the scope of this paper was to explore the dependence between the number of obtained clusters and the number of deleted edges for different datasets. The second method relies on spectral analysis of the PPIN. Similar algorithm is used in [21], but only for the purpose of predicting protein interactions rather than annotations.

2.2.1 Edge-Betweenness Clustering

The idea for clustering of networks using the concept of edge-betweenness was first proposed in [22] and is an extension of the concept of node-betweenness, which is an estimate of the centrality of a node in a network. Analogous to the definitions for node-betweenness, the betweenness of an edge is calculated as the number of shortest paths between any two nodes in the graph which pass through that edge. The edges which are between clusters, i.e. which connect two nodes of different clusters have higher betweenness then the edges which connect nodes that belong to the same cluster. By deleting the edges with the highest betweenness, after certain number of iterations the graph will be separated into several components which can be treated as clusters.

The main changing parameter of this algorithm is the number of edges that need to be deleted. In our research, this number in obtained empirically. The betweenness of each edge is recalculated after each iteration, which, regarding to [22] is better strategy then just calculating the edge betweenness of every edge only once at the beginning and then deleting the edges with highest betweenness.

If the number of nodes in the graphs is $|V|$ and the total number of edges between them is $|E|$, then the complexity of the algorithm is $O(|E|^2|V|)$.

2.2.2 Clustering Based on Spectral Analysis of the Protein-Protein Interaction Graph

One of the basic types of graph clustering, according to [23], is the spectral clustering, which performs spectral analysis of the graph's adjacency matrix or some of its derivatives, by finding its eigenvalues and eigenvectors. The first step in the spectral clustering is transforming the initial dataset into a set of points in an n-dimensional space, whose coordinates are elements of n selected eigenvectors. This change in the representation of the data enhances the characteristics of the clusters making them more distinctive. Then a classical clustering algorithm, like k-means for example, can be used, to cluster the data.

Although the initial idea for spectral analysis was intended directly to the adjacency matrix of the graph, the newer algorithms use the Laplacian matrix L, which is derived from the adjacency matrix A as in equation (1).

$$L = D - A . \tag{1}$$

In this equation, D is a diagonal matrix whose diagonal element D_{ii} equals the degree of the node i of the graph. Before spectral analysis, the Laplacian matrix needs to be normalized.

The main characteristic of the graphs' Laplacian matrix is the fact that the number k of zero eigenvalues equals to the number of connected components of the graph. The non-zero values of the corresponding eigenvectors are on the indices of the nodes that belong to the corresponding connected component. If those eigenvectors are put as columns of one $|V|$xk matrix, each row represents one node which has only one non-zero value: on the position of the eigenvector of the connected component it belongs to.

If the graph consists of only one connected component, that the Laplacian will have only one non-zero eigenvalue. Let the number of clusters that the graphs should be separated into be k. Taking the k eigenvectors that correspond to the k eigenvalues closest to 0, and transforming the nodes of the graph into the k-dimensional space that they form, all the nodes that belong to one cluster will be situated close in that space. This way the nodes will be brought into a form suitable for using any clustering algorithms, like k-means. The number of clusters is determined empirically.

2.3 Cluster Evaluation

One of the evaluation models which give a general overview of the qualitative differences between the clustering algorithms is proposed in [10] and it provides information whether a cluster has the character of a module or densely connected subgraph. This is highly important, because, according to [3], the term functional module is closely related to subgraphs rich in edges within it. The first necessary criteria for a cluster to be considered as a module is given with (2), where n is the number of nodes, k_{in} is the number of edges within the cluster and k_{out} is the number of edges from the cluster to nodes which don't belong to it.

$$\sum_{i=1}^{n} k_{in}^{i} > \sum_{i=1}^{n} k_{out}^{i} \tag{2}$$

The second criterion requires that collectively, the number of neighbors of each node within the cluster is higher than the number of neighbors from the module to the outside. This criterion is given by (3).

$$\{k_{in}^{1}, k_{in}^{2}, \ldots, k_{in}^{n}\} >> \{k_{out}^{1}, k_{out}^{2}, \ldots, k_{out}^{n}\} \tag{3}$$

Whether the module meets these criteria or not is determined by using the Wilcoxon non-parametric statistical test for comparing the distribution of two random variables.

2.4 Functional Annotation Using Clusters

After clustering the PPIN we set up a strategy for annotating the query protein with the adequate functions according to the functions of the other proteins in the cluster where it belongs. The simplest and most intuitive approach would be that each function is ranked by its frequency of appearance as an annotation for the proteins in the cluster. This rank is calculated by the formula (4) and is then normalized in the range from 0 to 1.

$$f(j)_{j\in F} = \sum_{i\in K} z_{ij} \tag{4}$$

where F is the set of functions present in the cluster K, and

$$z_{ij} = \begin{cases} 1, \text{if i - th protein from K is annotated with the j - th function from F} \\ 0, \text{otherwise} \end{cases} \tag{5}$$

3 Results and Discussion

Each protein in the PPIN is streamed through the prediction process one at a time as a query protein. The query protein is considered unannotated, that is we employ the leave-pone out method. Each of the algorithms works in a fashion that ranks the "proximity" of the possible functions to the query protein. The ranks are scaled between 0 and 1 as explained in 2.4. The query protein is annotated with all functions that have rank above a previously determined threshold ω. For example, for ω = 0, the query protein is assigned with all the function present in its cluster. We change the threshold with step 0.1 and compute numbers of true-positives (TP), true- negatives (TN), false-positives (FP) and false-negatives (FN). For a single we consider the TP to be the number of correctly predicted functions, and for the whole PPIN and a given value of ω the TP number is the total sum of all single protein TPs .

To compare performance between different algorithms we use standard measures as sensitivity and specificity (6).

$$sensitivity = \frac{TP}{TP+FN} \qquad specificity = \frac{TN}{TN+FP} \tag{6}$$

We plot the values we compute for the sensitivity and specificity using a ROC curve (Receiver Operating Curve). The *x*-axe corresponds to the false positive rate, which is the number of false predictions that a wrong function is assigned to a single protein, scaled by the total number of functions that do not belong to that particular protein. This rate is calculated with (7).

$$fpr = \frac{FP}{FP+TN} = 1 - specificity \tag{7}$$

The *y*-axe corresponds to the rate of true predictions that is the sensitivity. At last we use the AUC (Area Under the ROC Curve) measure as a numeric evaluator of the ROC curve. The AUC is a number that is equal to the area under the curve and its value should be above 0.5, which is the value that we get if the prediction process was random. The closer the value of AUC to 1, the better is the prediction method.

Before we evaluate the prediction performance of the proposed methods, first we assess their clustering properties on the PPIN. For each of the methods we use a changing parameter as explained in 2.2. For the edge-betweenness method we performed experiments using deletion of 1000 and 1400 edges. For the spectral clustering we experimented with different numbers of eigenvalues starting from 50 up to 300, with a changing step of 50. The results are presented in Table 1.

As can be concluded from Table 1, the number of clusters which have the nature of a module reduces as the cluster size decreases i.e. as the total number of clusters increases. However, for certain parameters for the both algorithms (1000 removed edges with the edge-betweenness method and 50, 150 and 200 eigenvalues for the spectral clustering method), the percentage of modules among the obtained clusters is sufficiently high. Thus, it is reasonable to presume that the clustering process has produced functional modules.

Table 1 Evaluation results of the clustering methods using method described in 2.3

	changing parameter	**number of clusters**	**clusters meeting module criteria (%)**
edge-betweenness	1000	103	85.44
	1400	217	54.84
spectral clustering	50	50	100.00
	100	100	94.00
	150	150	84.67
	200	200	65.50
	250	250	47.60
	300	300	35.67

After evaluating the clustering properties we move towards the evaluation of the function prediction when using the two clustering methods.

By comparing the AUC values of the results in Table 2 and Table 3, it can be concluded that the edge-betweenness method performs better when 1000 edges are removed, while the spectral clustering renders best results when only 50 eigenvalues are considered. The spectral clustering method is slightly superior over the edge-betweenness method according to the AUC values, but the edge-betweenness method achieves better sensitivity and false positive rate for $\omega = 0.1$. It is important to notice that for $\omega = 0$ the algorithms achieve very high sensitivity of over 77.78% for the edge-betweenness method and 82.35% for the spectral clustering method in best case. However, the profitability of this result for the spectral clustering method is questionable, because the false positive rate in that case is nearly 20%. Therefore, it would be useful to inquire what the permissible trade-off limit between correctly and incorrectly detected protein functions is. Graphical visualization of the function prediction results is given in Fig. 2.

Table 2 Function prediction evaluation when using edge-betweenness method

No. of deleted edges	ω =	0,1	0,3	0,5	0,7	0,9	AUC
1000	sens.	0,6693	0,4753	0,3266	0,2459	0,1445	0,8610
	fpr	0,0456	0,0136	0,0051	0,0027	0,0011	
1400	sens.	0,6651	0,5131	0,3741	0,2872	0,1623	0,8430
	fpr	0,0355	0,0118	0,0046	0,0025	0,0012	

Table 3 Function prediction evaluation when using spectral clustering method

No. of eigenvalues	ω =	0,1	0,3	0,5	0,7	0,9	AUC
50	sens.	0,6484	0,4436	0,3246	0,2082	0,1147	0,8644
	fpr	0,0565	0,0159	0,0077	0,0036	0,0014	
100	sens.	0,6702	0,4713	0,3376	0,2430	0,1404	0,8590
	fpr	0,0479	0,0142	0,0056	0,0028	0,0010	
150	sens.	0,6709	0,5053	0,3620	0,2688	0,1598	0,8531
	fpr	0,0400	0,0128	0,0050	0,0026	0,0012	
200	sens.	0,6783	0,5295	0,3859	0,3053	0,1870	0,8458
	fpr	0,0383	0,0116	0,0049	0,0027	0,0013	
250	sens.	0,6690	0,5434	0,3987	0,3175	0,2039	0,8381
	fpr	0,0329	0,0121	0,0047	0,0028	0,0015	
300	sens.	0,6538	0,5375	0,3952	0,3196	0,2143	0,8283
	fpr	0,0307	0,0121	0,0046	0,0028	0,0017	

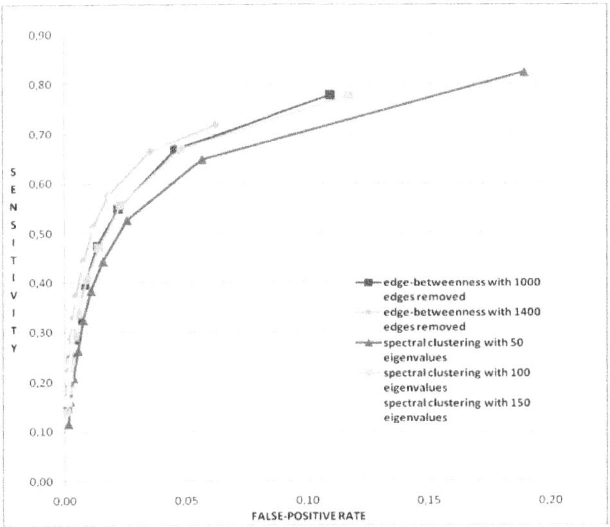

Fig. 2 ROC curves for the function prediction evaluation for the edge-betweenness and spectral clustering method

4 Conclusion and Future Directions

This paper exploits the ability of two graph clustering methods for detecting functional modules and predicting protein functions from PPIN. The methods were tested over one of the richest interactomes: the interactome of the baker's yeast. The first approach uses the edge-betweenness algorithm for graph clustering, while the second one performs spectral clustering over the Laplacian of the adjacency matrix of the PPIN. Due to the fact that the PPIN data contain a lot of false positive interactions, the dataset needs to be preprocessed and purified prior to the functional annotation. This paper also illustrates a general framework for the vast set of algorithms for protein function prediction which are based on clustering of the PPIN. The proposed approaches prove that utilizing clustering of the PPIN has high potential in the task of protein function prediction. The results show that both algorithms achieve high sensitivity and small false positive rate and they both have high AUC values, with some advantage of the edge-betweenness method which has smaller false positive rates. However there is one limitation of our current approach, that is, all of our analyses were performed on unweighted graphs, because our reference PPIN does not contain any information that would enable us to assign reliability values (weights) to the edges. It should be mentioned that if a method can deal with weighted graphs it would be likely to give better performances if the weights reflect the reliability of the links between proteins. Since spectral clustering can deal with weighted graphs, while the edge-betweenness clustering does not take in account any edge weight, future directions for using clustering for the aim of function prediction should follow the spectral clustering approach.

References

1. Yu, G.X., Glass, E.M., Karonis, N.T., Maltsev, N.: Knowledge-based voting algorithm for automated protein functional annotation. PROTEINS: Structure, Function, and Bioinformatics 61, 907–917 (2005)
2. The gene ontology consortium, Gene ontology: Tool for the unification of biology. Nature Genetics 25(1), 25–29 (2000)
3. Spirin, V., Mirny, L.A.: Protein complexes and functional modules in molecular networks. PNAS 100, 21 (2003)
4. Bader, G.D., Hogue, C.W.V.: An automated method for finding molecular complexes in large protein interaction networks. BMC Bioinformatics (4), 2 (2003)
5. Sharan, R., Ideker, T., Kelley, B., Shamir, R., Karp, R.M.: Identification of protein complexes by comparative analysis of yeast and bacterial protein interaction data. Computational Biology 12, 835–846 (2005)
6. Przulj, N., Wigle, D.A., Jurisica, I.: Functional Topology in a Network of Protein Interactions. Bioinformatics 20(3), 340–348 (2004)
7. Krogan, N.J., Cagney, G., Yu, H., Zhong, G., Guo, X., Ignatchenko, A.: Global Landscape of Protein Complexes in the Yeast Saccaromyces cerevisiae. Nature 440, 637–643 (2006)
8. Samanta, M.P., Liang, S.: Predicting protein functions from redundancies in large-scale protein interaction networks. PNAS 100, 22 (2003)

9. Brun, C., Chevenet, F., Martin, D., Wojcik, J., Guénoche, A., Jacq, B.: Functional classification of proteins for the prediction of cellular function from a protein-protein interaction network. Genome. Biology 5, R6 (2003)
10. Chen, J., Yuan, B.: Detecting Functional Modules in the Yeast Protein-Protein Interaction Network. Bioinformatics 22(18), 2283–2290 (2006)
11. Lancichinetti, A., Fortunato, S., Radicchi, F.: Benchmark Graphs for testing Community Detection Algorithms. Physical Review E78, 046110 (2008)
12. Rives, A.W., Galitski, T.: Modular Organization of cellular Networks. PNAS 10(3), 1128–1133 (2003)
13. Salwinski, L., Miller, C.S., Smith, A.J., Pettit, F.K., Bowie, J.U., Eisenberg, D.: The Database of Interacting Protein. Nucleic Acids Res. 32, D449–D451 (2004)
14. Guldener, U., Munsterkotter, M., Oesterheld, M., Ragel, P., Ruepp, A., Mewes, H.W.: MPact: the MIPS protein interaction resource on yeast. Nucleic Acids Res. 34, D436–D441 (2006)
15. Chatr-aryamontri, A., Ceol, A., Montecchi Palazzi, L., Nardelli, G., Schneider, M.V., Castagnoli, L., Cesareni, G.: MINT: the Molecular INTeraction database. Nucleic Acids Res. 35, D572–D574 (2007)
16. Bader, G.D., Hogue, C.W.V.: BIND–a data spec. for storing and describing biomolecular interactions, molecular complexes and pathways. Bioinformatics 16(5), 465–477 (2000)
17. Breitkreutz, B.J., Stark, C., Tyers, M.: The GRID: The General Repository for Interaction Datasets. Genome Biology 4(3), R23 (2003)
18. Dwight, S., Harris, M., Dolinski, K., Ball, C., BUnkley, G., Christie, K., Fisk, D., Issel-Tarver, L., Schroeder, M., Sherlock, G., Sethuraman, A., Weng, S., Botstein, D., Cherry, J.M.: Saccharomyces Genome Database (SGD) provides secondary gene annotation using Gene Ontology (GO). Nucleic Acids Research 30(1), 69–72 (2002)
19. Letovsky, S., Kasif, S.: Predicting protein function from protein/protein interaction data: a probabilistic approach. Bioinformatics 19, i197–i204 (2003)
20. Dunn, R., Dudbridge, F., Sanderson, C.M.: The Use of Edge-Betweenness Clustering to Investigate Biological Function in PIN. BMC Bioinformatics 6(1), 39 (2005)
21. Sen, Z., Kloczkowski, A., Jernigan, R.L.: Functional Clustering of Yeast Proteins from the Protein-Protein Interaction Network. BMC Bioinformatics 7, 355 (2006)
22. Girvan, M., Newman, M.E.J.: Community Structure in Social and Biological Networks. PNAS 99(12), 7821–7826 (2002)
23. Fortunato, S.: Community Detection in Graphs. Physics Reports 486, 75–174 (2010)

A Novel Fuzzy Decision Tree Based Method for Detecting Protein Active Sites

Georgina Mirceva, Andreja Naumoski, and Danco Davcev

Abstract. The knowledge of the functions of protein structures is essential for development of new drugs, better crops and synthetic biochemical. There are numerous experimental methods for determining the protein functions, but because of their complexity the number of protein molecules with undetermined functions is rapidly growing. Thus, there is an evident need for development of computer methods for determining the functions of the protein structures. In this study, we introduce the fuzzy theory for protein active sites detection. We propose a novel fuzzy decision tree (FDT) based method for predicting protein active sites that later could be used for determining the functions of the protein molecules. First, we extract several characteristics of the amino acids. Then, we induce FDTs that would be used to predict the protein active sites. We provide experimental results of the evaluation of the prediction power of the proposed method. Also, our method is compared with other machine learning techniques that could be used for this purpose.

1 Introduction

Many recent projects have a goal to annotate protein functions by analysis of their sequence and structure [1]. The knowledge of the protein functions is essential for development of new drugs, better crops and synthetic biochemical. Experimental methods for determining protein function exist, but because of their complexity the number of protein structures with undetermined functions is rapidly growing. Another drawback is that these methods are very expensive. Therefore, there is an evident need for computer methods for determining the functions of the protein molecules.

Some methods for protein function annotation analyze the structural and sequence homology [2] of the protein molecules. However, with this analysis we can

Georgina Mirceva · Andreja Naumoski · Danco Davcev
Ss. Cyril and Methodius University in Skopje, Skopje, Macedonia
e-mail: {georginamirceva,andrejanaumoski,dancodavcev}@gmail.com

L. Kocarev (Ed.): ICT Innovations 2011, AISC 150, pp. 51–60.
springerlink.com

determine only a global similarity between the protein structures. Other methods perform analysis of the protein-protein interaction networks [3, 4]. These methods take into account the information about the pairs of protein structures that interact. Nevertheless, they do not consider the sequences and structures of the protein molecules. Many other methods consider the conservation of the protein sequence and/or structure [5], which can be determined by aligning multiple protein sequences and/or structures [6]. The most appropriate methods for determining the functions of the protein molecules predict the active sites of the protein structures [7]. As an example, the BIND database [8] provides protein functions annotations that are determined by experimental prediction and analysis of the active sites of the protein structures. In this research we focus on detecting the active sites of the protein molecules.

In the process of prediction of the active sites of the protein structures we could consider many characteristics of the amino acid residues, including Accessible Surface Area (ASA) [9], depth index (DPX) [10], protrusion index (CX) [11] and hydrophobicity [12]. Then, the most significant characteristics of the protein active sites could be taken into account in order to determine the protein functions. There are many methods for protein function annotation that are based on different criteria, i.e. change in ASA [13], Van der Walls distance [14], physicochemical and shape complementarity [15] etc.

Various methods for prediction of the active site of the protein structures exist [14, 16, 17, 18, 19, 20]. However, they are sensitive to small changes of the characteristics of the amino acid residues. To overcome this problem, we introduce the fuzzy decision trees (FDTs) [21] for predicting the active sites of the protein structures.

There is an extensive research in FDTs, mainly due to their robustness of changes in the data and resistance to over-fitting. In [22], Wang and Mendel have presented an algorithm for inducing fuzzy rules by learning from examples. Classic decision trees [23] inspired many studies concerning FDTs. Different fuzzy decision tree inductions are presented by Janikow [21], Olaru and Wehenkel [24]. In [25] and [26] different optimizations of FDTs are presented, while [27] additionally points out the advantages and disadvantages of FDTs. In [28], a method for soft discretization is introduced, and it is applied for machine fault diagnosis.

In this paper, we propose a novel fuzzy decision tree based method for predicting the active sites of the protein structures. First, we extract several characteristics of the amino acids, and then we induce FDTs [28] that would be used for prediction of the active sites of the protein molecules. We provide experimental results concerning the prediction power of the proposed method. The proposed method is compared with other machine learning techniques that can be used for this purpose.

The rest of the paper is organized as follows. The proposed method is explained in section 2. Section 3 presents some experimental results concerning the prediction power of our method, while section 4 concludes the paper and gives directions for future work.

2 Our Fuzzy Decision Tree Based Protein Active Site Detector

In this research, we propose a novel method for detecting protein active sites. We extract several characteristics of the amino acids of the protein molecules. Then, we build fuzzy decision trees that are used for detection of the active sites of the protein structures.

2.1 *Characteristics of the Amino Acids*

We take into account the following characteristics of the amino acid residues: accessible surface area (ASA), depth index (DPX), protrusion index (CX) and hydrophobicity. ASA is first described by Lee and Richards [29] and is calculated using the rolling ball algorithm [9]. The algorithm uses a sphere (probe) with a specific radius to analyze the surface of the protein molecule. Usually, the probe has the same radius (1.4 Å) as the water molecule. ASA is expressed in $Å^2$ (square Ångstroms) and is defined as the surface area of a given sphere (atom) that can get in contact with the rolling probe.

We make consecutive thin slices over the protein surface to estimate the surface of the atoms that is accessible by the rolling probe. ASA per slice is calculated by (1), where L_i denotes the length of the intersection drawn in slice i, Z_i denotes the distance from the centre of the sphere to the slice i, ΔZ is the spacing between the slices, while $\Delta' Z$ is min($\Delta Z/2$, R-Z_i). The total ASA is calculated as a sum of the intersections of the probe and the atoms of the protein structure.

$$ASA = R / \sqrt{R^2 - Z_i^2} \times D \times L_i$$
$$D = \Delta Z / 2 + \Delta' Z \tag{1}$$

Based on value of ASA, we find the amino acids that are at the protein surface. According to [30], if at least 5% of the area of a given amino acid is accessible by the probe we consider that amino acid to be at the protein surface. In this research, we use the total surface area values for the amino acids given in [30]. In the second stage of the method, the classification stage, we consider only the surface amino acids, since amino acids buried in the protein interior could not be involved in interactions.

The depth index of the i -th atom (DPXi) [10] is calculated as an Euclidean distance to the nearest solvent accessible atom (2)

$$\text{DPX}i = \min(d_1, d_2, d_3, ..., d_n), \tag{2}$$

where d_1, d_2, d_3, ..., d_n are the Euclidean distances between the i-th atom and the solvent accessible atoms. Thus, the depth index of a solvent accessible atom equals zero, and is greater than zero for the atoms buried in the protein interior [10].

The protrusion index (CX) is calculated using the following procedure. For each non-hydrogen atom, we calculate the number of the heavy atoms within a given sphere with radius R. As suggested in [11], we used R=10 Å. Then, the volume occupied within a sphere (V_{int}) is calculated by multiplying the number of atoms within the sphere and the mean volume of the atoms (20.1 ± 0.9 Å) [11]. The protrusion index is thus calculated as V_{ext}/V_{int} [11], where V_{ext} is the difference between the volume of the sphere and the occupied volume within that sphere. For each amino acid, DPX and CX are calculated as a mean depth/protrusion index of the atoms in that amino acid.

The hydrophobicity of an amino acid presents its hydrophobic properties. This characteristic is very significant since hydrophobic amino acids are more likely to be found in the protein interior, while hydrophilic amino acids are more commonly found near the protein surface. Larger hydrophobicity corresponds to more hydrophobic amino acid. In this research we use the Kyte and Doolittle hydrophobicity scale [12].

2.2 Fuzzy Decision Trees

In classical decision tree [23], a crisp discretization is performed where the decision space is divided into non-overlapping subspaces in which each example is assigned to a certain class. On the other hand, FDTs [21], [28] give the possibility that a given example belongs to a certain class and there is an overlapping of the fuzzy partitions.

Assume that we have a set S with N examples and we want to select an attribute. In the process of inducing classical decision trees, we discretize an attribute A into intervals A_1= [min(A), T] and A_2= (T, max(A)] by finding an appropriate threshold T. We form an ordered sequence a_1, a_2, ...a_N by sorting the examples according to the values of the attribute. From every pair of consecutive data points in the ordered sequence we have a candidate threshold $T=(a_i+a_{i+1})/2$, and we obtain the corresponding partitions of the dataset S_1 and S_2. The uncertainty in a set S_i is expressed as

$$E(S_i) = -\sum_{j=1}^{k} p(c_j, S_i)\log(p(c_j, S_i)) \tag{3}$$

where $p(c_j, S_i)$ is a possibility that an example of the set S_i belongs to class c_j, and k denotes the number of classes. The information gain is calculated by (4), where N_1, N_2 and N denote the number of examples in S_1, S_2 and S correspondingly.

$$Gain(A,T,S) = E(S) - \left(\frac{N_1}{N}E(S_1) + \frac{N_2}{N}E(S_2)\right) \tag{4}$$

We select the threshold for an attribute which maximizes the information gain. According to the threshold, we obtain the child nodes, and then the same procedure is repeated for each child node until a stopping criterion is satisfied.

In soft discretization [28], instead classical information measures defined in probability domain we use their extended versions in possibility domain. Using a sharp characterization function $A_c(a)$ we can express each crisp dataset A_c. On the other hand, each fuzzy dataset A is described by a membership function $A(a)$ that present the possibility of A to take a value a.

For each soft discretization, we have to define a threshold for an attribute and membership functions of a fuzzy set pair A_1 and A_2. The threshold is selected by maximizing the fuzzy information gain (5), where the fuzzy entropy in a given set S_i is calculated by (6).

$$Gain_F(A,T,S) = E_F(S) - \left(\frac{N_F^{S_1}}{N_F^S} E_F(S_1) + \frac{N_F^{S_2}}{N_F^S} E_F(S_2) \right) \quad (5)$$

$$N_F^S = \sum_{i=1}^{|S|} (A_1(a_i) + A_2(a_i))$$

$$E_F(S_i) = -\sum_{j=1}^{k} p(c_j, S_i) \log(p(c_j, S_i))$$

$$p(c_j, S_i) = \frac{\sum_{a_{ki} \in c_j} A_i(a_k)}{\sum_{k=1}^{|S|} A_i(a_k)}, \; i = 1,2 \quad (6)$$

The membership functions are selected based on the features of the attribute. For example, if we want to obtain membership functions for an attribute with high uncertainty or noise, we should consider wide overlapping membership functions.

The induction of FDTs [28] can be summarized in the next steps:

1: Sort the examples of a given node (dataset) S according to the values of an attribute and find the candidate threshold points;
2: For each candidate threshold, find and evaluate the corresponding discretization;
3: Select the discretization that maximizes the information gain;
4: Repeat the previous steps for the remaining attributes until all attributes are examined;
5: Select the attribute with maximal information gain and generate the two child nodes;
6: For each branch, calculate the truth level as $\eta = N_F^{S_i} / N_F^S$. If a given branch has a truth level that is not greater than α, then delete that branch. Otherwise, calculate the truth level of that branch belonging to the j-th class (7).

$$\mu_{i,j} = \frac{\sum_{a_c \in c_j} A_i(a_c)}{N_F^{S_i}} \quad (7)$$

If $\max_{j=1}^{k} \mu_{i,j} \geq \beta$, then terminate this branch with leaf node and label the leaf as class c_j that maximizes the truth level. Otherwise, divide the set S in two subsets $S_1=\{s \mid A_1(a_i) \geq \lambda, a_i \in S\}$ and $S_2=\{s \mid A_2(a_i) \geq \lambda, a_i \in S\}$, and repeat the same procedure for both subsets.

Too small α and too high β will generate more complex tree and can even result in over-fitting, especially when the data is uncertain. In [28], it is recommended to use $\alpha = 0.1\sim0.2$, $\beta = 0.8\sim0.9$ and $\lambda = 0.5$. In this research we use $\alpha = 0.1$, $\beta = 0.8$ and $\lambda = 0.5$.

3 Experimental Results

Next, we perform evaluation of the proposed method using a part of the BIND database [8]. First, we select a representative dataset of protein chains so that each pair of protein chains has less than 40% sequence similarity. For this reason we use the selection criterion given in [32]. From these chains, we form the test set so that each pair of test chains has less than 20% sequence similarity [32], while the other chains are taken as training data. In this way we ensure that the most representative protein chains are considered in the test set.

In this way we obtain 1062 training protein chains, and 1858 test protein chains. The training chains contain 365862 amino acids from which 284168 are surface residues, while the total number of amino acids in the test set is 608434 from which 484637 are on the surface. From the surface amino acids in the training set 26889 are classified as protein active sites in the BIND database, while from the surface amino acids in the test set 47501 are classified as active sites in the BIND database. It is evident that the non-active class is dominant, thus in order to avoid favoring the non-active class we balance the training data so that the active amino acids are taken into account several times until the dataset is balanced. It is worth to mention that this balancing is made only on the training data. In this way we obtain 514550 training amino acids and 484637 test amino acids.

For evaluation, we used several standard measures that are widely used in information retrieval. TP is the number of the correctly classified positive examples (amino acids classified as active sites), FP is the number of negative examples classified as positive, TN is the number of correctly classified negative examples and FN is the number of positive examples classified as positive. Then, the *precision*=TP/(TP+FP) and *recall*=TP/(TP+FN) could be determined, as well as their harmonic mean *F-measure=2*precision*recall/(precision+recall)*. However, the *F-measure* do not considers TN, thus measures such as the *Matthews correlation coefficient* (*MCC*) are preferable to evaluate the performance of a binary classifier. *MCC* is defined by (7) and is considered as a balanced measure that can be used even when the classes have very different sizes, like it is in our case. The *MCC* is a correlation coefficient between the observed and predicted classifications and obtains a value between −1 and +1. If *MCC* equals +1 we have a perfect prediction, 0 corresponds to a random prediction, and −1 signifies an inverse prediction. In this research we used TP, FP, TN, FN, and *MCC* to evaluate the prediction power of the proposed method.

$$MCC = \frac{TP \times TN - FP \times FN}{\sqrt{(TP + FP)(TP + FN)(TN + FP)(TN + FN)}} \tag{7}$$

We examined the influence of the number *N* and type of membership functions on the prediction power of the method (see Table 1). We used the triangular, trapezoidal and Gaussian membership functions (MFs) [31].

Table 1 The results obtained by the proposed method using different types and different number *N* of membership functions.

	TP	FP	TN	FN	*MCC*
N=3 Triangular	19110	126030	311106	28391	0,074003
N=3 Trapezoidal	25766	207108	230028	21735	0,040853
N=3 Gaussian	19110	126032	311104	28391	0,074000
N=4 Triangular	17630	107750	329386	29871	0,084638
N=4 Trapezoidal	19042	125368	311768	28459	0,074164
N=4 Gaussian	19190	117880	319256	28311	0,088681
N=5 Triangular	25675	176613	260523	21826	0,082298
N=5 Trapezoidal	17251	101295	335841	30250	0,090923
N=5 Gaussian	**27080**	187615	249521	20421	0,084339
N=10 Triangular	23419	147366	289770	24082	**0,097035**
N=10 Trapezoidal	26521	174769	262367	20980	**0,095648**
N=10 Gaussian	23551	148294	288842	23950	**0,097307**

According to Table 1, as *N* increases, also *MCC* increases. For *N*=3, the trapezoidal MF obtains lowest value for MCC, but it does not forces the non-active class as it is with the triangular and Gaussian MFs. As *N* increases, *MCC* for the trapezoidal MF gets closer to the *MCC* for the other MFs. For *N*=5 using Gaussian MF, TP is the highest and this model favors the non-active class the less. According to the *MCC* measure, we can conclude that for *N*=10 we obtain the most accurate models. We want to point out that the number of membership functions *N* should not be set too large since the induction of the model would be very slow, and also the model would be very complex and could be over-fitted.

Additionally, we made comparison of the proposed method and several other classification methods where the same characteristics of the amino acids were considered. In this analysis we used C4.5 Tree [33], REPTree (variant of C4.5 where reduced-error pruning is used), ADTree [34] and its multi class version LADTree [35], Naïve Bayes [36], Bayes Net [37] and NBTree [38]. The results of this analysis are given in Table 2.

Table 2 The results obtained by several existing methods.

Method	TP	FP	TN	FN	*MCC*
C4.5	13281	95471	341665	34220	0,043613
REPTree	13029	96306	340830	34472	0,038398
ADTree	25542	159713	277423	21959	**0,105458**
LADTree	25542	159713	277423	21959	**0,105458**
Naïve Bayes	22414	145597	251539	25087	0,067052
Bayes Net	**29090**	197657	239479	18411	**0,095489**
NBTree	**29090**	197649	239478	18411	**0,095496**

It can be seen that C4.5 and REPTree favor the non-active class and they obtain lower *MCC*. Naïve Bayes showed as better than C4.5 and REPTree, but since the attributes are not independent, it is not the best choice. On the other hand, Bayes Net considers the dependences between the attributes and therefore it is much better than Naïve Bayes. Also, the NBTree that is a hybrid of the Naïve Bayes and Decision Trees showed as a good induction method. ADTree and its multi class version LADTree obtain highest *MCC* (0.105458), and since we solve a binary problem, these two methods give the same results. However, Bayes Net and NBTree obtain highest TP rate. Our method showed as much better than C4.5, REPTree and Naïve Bayes, while its *MCC* is comparable with the *MCC* obtained by the other methods considered in this research. It is also worth to mention that the induction of the models and the testing by the proposed method is faster than by the other methods.

In this analysis we used the basic MFs. We expect that using other more sophisticated MFs (sigmoidal, log-normal, bell etc.) the prediction power of the method would be increased.

4 Conclusion

In this paper, we proposed a novel fuzzy decision tree based method for detecting the active site of protein molecules. First, we extracted several characteristics of the amino acid residues, and then we induced fuzzy decision trees in order to predict the protein active sites.

The results showed that our method outperforms the C4.5 Tree, REPTree and Naïve Bayes, while it obtains comparable prediction power with the other examined methods. Therefore, this method showed as a promising approach for predicting the active sites of the protein structures.

However, in this study we used the basic MFs. As a future work, we plan to introduce other more sophisticated MFs in order to improve the method.

References

1. Hegyi, H., Gerstein, M.: The relationship between protein structure and function: A comprehensive survey with application to the yeast genome. J. Mol. Biol. 288(1), 147–164 (1999)
2. Todd, A.E., Orengo, C.A., Thornton, J.M.: Evolution of function in protein superfamilies, from a structural perspective. J. Mol. Biol. 307(4), 1113–1143 (2001)
3. Kirac, M., Ozsoyoglul, G., Yang, J.: Annotating proteins by mining protein interaction networks. Bioinformatics 22(14), e260–e270 (2006)
4. Sharan, R., Ulitsky, I., Shamir, R.: Network-based prediction of protein function. Mol. Sys. Bio. 3, 88 (2007)
5. Panchenko, A.R., Kondrashov, F., Bryant, S.: Prediction of functional sites by analysis of sequence and structure conservation. Protein Science 13(4), 884–892 (2004)
6. Leibowitz, N., Fligelman, Z.Y., Nussinov, R., Wolfson, H.J.: Automated multiple structure alignment and detection of a common substructure motif. Proteins 43(3), 235–245 (2001)

7. Tuncbag, N., Kar, G., Keskin, O., Gursoy, A., Nussinov, R.: A survey of available tools and web servers for analysis of protein-protein interactions and interfaces. Briefings in Bioinformatics 10(3), 217–232 (2009)
8. Bader, G.D., Donaldson, I., Wolting, C., Ouellette, B.F., Pawson, T., Hogue, C.W.: BIND: the Biomolecular Interaction Network Database. Nucleic Acids Res. 29(1), 242–245 (2001)
9. Shrake, A., Rupley, J.A.: Environment and exposure to solvent of protein atoms. Lysozyme and insulin. J. Mol. Biol. 79(2), 351–371 (1973)
10. Pintar, A., Carugo, O., Pongor, S.: DPX: for the analysis of the protein core. Bioinformatics 19(2), 313–314 (2003)
11. Pintar, A., Carugo, O., Pongor, S.: CX, an algorithm that identifies protruding atoms in proteins. Bioinformatics 18(7), 980–984 (2002)
12. Kyte, J., Doolittle, R.F.: A simple method for displaying the hydropathic character of a protein. J. Mol. Biol. 157(1), 105–132 (1982)
13. Jones, S., Thornton, J.M.: Analysis of protein-protein interaction sites using surface patches. J. Mol. Biol. 272(1), 121–132 (1997)
14. Aytuna, A.S., Gursoy, A., Keskin, O.: Prediction of protein-protein interactions by combining structure and sequence conservation in protein interfaces. Bioinformatics 21(12), 2850–2855 (2005)
15. Lawrence, M.C., Colman, P.M.: Shape complementarity at protein/protein interfaces. J. Mol. Biol. 234(4), 946–950 (1993)
16. Neuvirth, H., Raz, R., Schreiber, G.: ProMate: a structure based prediction program to identify the location of protein-protein binding sites. J. Mol. Biol. 338(1), 181–199 (2004)
17. Bradford, J.R., Westhead, D.R.: Improved prediction of protein-protein binding sites using a support vector machines approach. Bioinformatics 21(8), 1487–1494 (2005)
18. Murakami, Y., Jones, S.: SHARP2: protein-protein interaction predictions using patch analysis. Bioinformatics 22(14), 1794–1795 (2006)
19. Ogmen, U., Keskin, O., Aytuna, A.S., Nussinov, R., Gursoy, A.: PRISM: protein interactions by structural matching. Nucleic Acids Res. 33(2), W331–W336 (2005)
20. Jones, S., Thornton, J.M.: Prediction of protein-protein interaction sites using patch analysis. J. Mol. Biol. 272(1), 133–143 (1997)
21. Janikow, C.Z.: Fuzzy decision trees: issues and methods. IEEE Transactions on Systems, Man, and Cybernetics 28(1), 1–14 (1998)
22. Wang, L.X., Mendel, J.M.: Generating fuzzy rules by learning from examples. IEEE Transactions on Systems, Man, and Cybernetics 22(6), 1414–1427 (1992)
23. Quinlan, R.J.: Decision trees and decision making. IEEE Transactions on Systems, Man, and Cybernetics 20(2), 339–346 (1990)
24. Olaru, C., Wehenkel, L.: A complete fuzzy decision tree technique. Fuzzy Sets and Systems 138(2), 221–254 (2003)
25. Suárez, A., Lutsko, J.F.: Globally optimal fuzzy decision trees for classification and regression. IEEE Transactions on Pattern Analysis and Machine Intelligence 21(12), 1297–1311 (1999)
26. Wang, X., Chen, B., Olan, G., Ye, F.: On the optimization of fuzzy decision trees. Fuzzy Sets and Systems 112(1), 117–125 (2000)
27. Chen, Y.-L., Wang, T., Wang, B.-S., Li, Z.-J.: A Survey of Fuzzy Decision Tree Classifier. Fuzzy Information and Engineering 1(2), 149–159 (2009)

28. Peng, Y., Flach, P.: Soft Discretization to Enhance the Continuous Decision Tree Induction. In: Giraud-Carrier, C., Lavrac, N., Moyle, S. (eds.) Integrating Aspects of Data Mining, Decision Support and Meta-Learning, pp. 109–118 (2005)
29. Lee, B., Richards, F.M.: The interpretation of protein structures: Estimation of static accessibility. J. Mol. Biol. 55(3), 379–400 (1971)
30. Chothia, C.: The Nature of the Accessible and Buried Surfaces in Proteins. J. Mol. Biol. 105(1), 1–12 (1976)
31. Klir, G.J., Yuan, B.: Fuzzy sets and fuzzy logic: theory and applications, 1st edn. Prentice-Hall (1995)
32. Chandonia, J.-M., Hon, G., Walker, N.S., Conte, L.L., Koehl, P., Levitt, M., Brenner, S.E.: The ASTRAL Compendium in 2004. Nucleic Acids Res. 32, D189–D192 (2004)
33. Quinlan, R.: C4.5: Programs for Machine Learning. Morgan Kaufmann Publishers, San Mateo (1993)
34. Freund, Y., Mason, L.: The alternating decision tree learning algorithm. In: Sixteenth International Conference on Machine Learning, pp. 124–133 (1999)
35. Holmes, G., Pfahringer, B., Kirkby, R., Frank, E., Hall, M.: Multiclass alternating decision trees. In: 13th European Conference on Machine Learning, pp. 161–172 (2001)
36. John, G.H., Langley, P.: Estimating Continuous Distributions in Bayesian Classifiers. In: Eleventh Conference on Uncertainty in Artificial Intelligence, pp. 338–345 (1995)
37. Neapolitan, R.E.: Learning Bayesian Networks. Prentice Hall, Upper Saddle River (2004)
38. Kohavi, R.: Scaling Up the Accuracy of Naive-Bayes Classifiers: A Decision-Tree Hybrid. In: Second International Conference on Knowledge Discovery and Data Mining, pp. 202–207 (1996)

Comparison of Protein Descriptors Used in Hierarchical Multi-label Classification Based on Gene Ontology

Natasha Pavlovikj, Ilinka Ivanoska, and Slobodan Kalajdziski

Abstract. Proteins are the main, building cell blocks, responsible for the existing cell biological processes. Therefore, precise knowledge of protein function is of great significance. There are a lot of methods which are used for protein comparison and for determining protein function. Some of them use structure alignment, others use sequence alignment, while some use protein descriptors. Here, we use two protein descriptors: Voxel and Ray-based descriptors to encode the structural and biological features of proteins. In biology there is a trend to hierarchically organize the things, like protein functions, cell components and the whole living world. There are a lot of classification systems which generate proteins in tree structure. However, due to the fact that it often happens that one protein has more than one parent, the Directed Acyclic Graph (DAG) hierarchy is used. Gene Ontology (GO) is a system for structural and hierarchical representation of proteins and gene products which support DAG hierarchy. CLUS, however, is a system which deals with hierarchical data. In this paper, we present a comparison between the two previously mentioned protein descriptors for predicting protein function. Firstly, protein descriptors are extracted from the structural coordinates found in the Protein Data Bank (PDB) and proteins backbone, appropriately. Afterwards, GO class hierarchy is added to each protein which has descriptor data. This created file is used as an input to the CLUS system. CLUS generates a decision tree model which is trained from the protein structure. The results from this system are the GO classes in which the protein belongs. The generated output shows that the predicting protein function with the Voxel protein descriptor gives better results instead of predicting protein function with the Ray protein descriptor.

Keywords: Gene Ontology, CLUS, Voxel protein descriptor, Ray-based protein descriptor, Predicting protein function.

Natasha Pavlovikj · Ilinka Ivanoska · Slobodan Kalajdziski
Ss. Cyril and Methodious University, Faculty of Computer Science and Engineering, str."Ruger Boskovic" 16, 1000 Skopje, Macedonia
e-mail: natasha.pavlovikj@yahoo.com, {linka.ivanoska,slobodan.kalajdziski}@finki.ukim.mk

L. Kocarev (Ed.): ICT Innovations 2011, AISC 150, pp. 61–71.
springerlink.com

1 Introduction

Proteins are the most important parts of all living organism. They can be found in all living systems, starting from bacteria and viruses, to the human. In human's body there are about 100.000 proteins, and their interaction defines the human as an individual. Proteins are polymers composed of amino acids. Each polymer is consisted of a linear sequence of 20 different L-α amino acids. The three-dimensional representation of a residue sequence and the way this sequence folds in the 3D space are very important to understand the logic in which a protein function is based on. There are many ways to predict proteins function. For example, the function can be found by protein's sequence, where the sequence similarity with other molecules is checked, or some other specific protein's features are used[1]. However, the protein's structure is less likely to be changed than its sequence during the evolution process. That is why the protein structure allows better recognition of protein's homology. Once the protein's structure is known, the potential function of each protein can be predicted by comparing their structures. Two similar structures are very likely to have similar functions. There are several databases in which different information about the proteins and their structures, functions and features are stored. Protein Data Bank (PDB) is global database of proteins. PDB is public database of proteins where everyone has access to discovered molecules, and can download specific protein's information. PDB gives systemized description of protein's structures, and therefore is considered as the most important database for protein's data extraction[2]. There are other databases, such as GenPept[3], RefSeq[4], SwissProt[5], PIR[6]. GenPept is a databank composed of replicated genetic sequences. RefSeq database is non-redundant collection of annotated DNA, RNA and proteins sequences. SwissProt is the most appropriate basis for referencing functional proteins. PIR is used to store protein's sequences. Searching for proteins structure is closely related with describing and presenting the protein structure, and developing algorithms that will know to recognize if two protein structures are similar or not. There are several methods used for comparison of protein's molecules. Needlman–Wunch[7], BLAST[8], PSI–BLAST[8] are methods that compare proteins by comparing their sequence alignment. These are fast methods that compare the amino acids sequence, and search similarities between sequences. SSAP[9] is an algorithm used in CATH for direct comparison of protein's structure. DALI[10] algorithm is based on distance matrix alignment, which matrixes are mapped from 3D proteins structure. CE[11] is another algorithm for structure alignment, which aligns fragments from protein's 3D structure. VAST[12] is an algorithm which first decomposes the protein's structure on more triplets, and afterwards uses these triplets for protein's comparison. The above methods for structural and sequence alignment are too complex and expensive, thus in this paper protein descriptors are used.

There are several classification systems which organize proteins in tree structure. Such known systems are SCOP (Structural Classification of Proteins)[13], CATH (Class, Architecture, topology, Homologous SuperFamily)[14], FSSP (Families of Structurally Similar Proteins)[15]. SCOP performs classification of proteins on five main hierarchical levels–Class, Fold, SuperFamily, Family and

Domain. SCOP's objective is a protein's relevant hierarchical classification, no matter of time resources. CATH has eight hierarchical levels–Class, Architecture, Topology and Homologous SuperFamily which define the proteins similarity by structure and function, while other four levels merge the proteins according to their sequence from the higher levels. FSSP provides completely automated classification system, where the similarity between two proteins is based on their secondary structures. It often happens for a protein to have more parents, which is not allowed in tree structure. Therefore, it is recommended to use systems that support Direct Acyclic Graph (DAG) structure. Such system for genes structural and hierarchical representation is Gene Ontology[16]. Biologists all over the world tend to have hierarchical organization in biological classification systems. Today, there are many biological classification systems that consist of catalogues of gene functions, cellular components, species, anatomical and molecular structures. CLUS[17] is a system which deals with hierarchical data. In this paper, we present a comparison between two protein descriptors for predicting protein function. Firstly, protein descriptors are generated from the structural coordinates found in the Protein Data Bank (PDB), and appropriate GO terms representing the protein functions for that protein are added to each descriptor vector. This created file is used as an input to the CLUS system. CLUS generates a decision tree model, and the results from this model are the GO terms associated to an unknown protein passed to the system.

In Section 2 we present the classification process architecture and the used protein descriptors and classification methods. Section 3 presents the experimental results obtained by comparing the both used protein descriptors, while the Section 4 concludes the paper.

2 Gene Ontology Protein Classification Architecture Based on Structural Features and Decision Trees

The architecture of our system is composed of few steps. First, for each protein in PDB database a protein descriptor is generated. In this paper, Voxel and Ray-based protein descriptors are used. Voxel protein descriptor transforms the protein's tertiary structure into N–dimensional feature vector, and additionally gives some other protein structural features. Ray-based protein descriptor transforms the protein backbone into N–dimensional feature vector. Second, in this architecture the Gene Ontology classification hierarchy is used. In this way an ".ARFF" file, which is an input in the CLUS system, is generated. The other input file in the CLUS system is the ".S" file which is the settings file where the parameters needed for CLUS system are defined.

2.1 Protein Descriptor

Protein descriptor can be described as an N-dimensional vector of real numbers which describes the proteins structure. In this paper, for experimental comparison, Voxel[18] and Ray-based[19] protein descriptors are used.

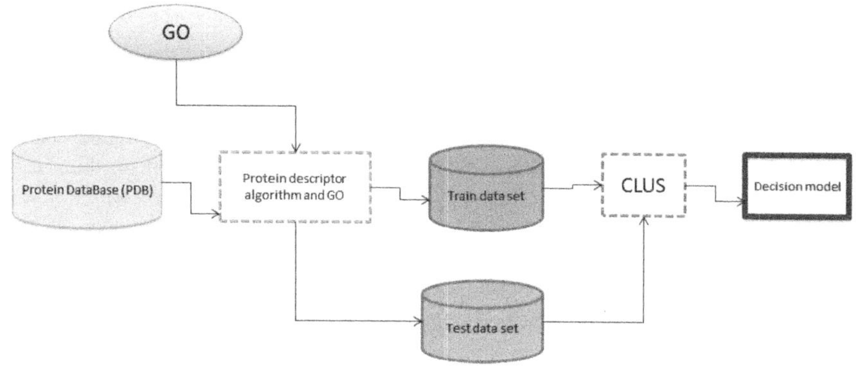

Fig. 1 Classification process architecture

2.1.1 Voxel Protein Descriptor

Descriptors consisting of 450 features (416 of them describe the protein's geometry, while 34 of them give information for the primary and secondary protein structure) are generated for each protein from the Protein Data Bank (PDB). Voxel descriptor relies on the geometric 3D structure of the proteins. The algorithm used for this descriptor consists of four phases: triangulation, normalization, voxelization of the 3D protein structure, and Discrete Fourier Transformation. The process of triangulation covers the triangular representation of protein's atom. The phase of normalization is used to ensure invariance to translation and scaling. Voxelization is the process of transforming the continual 3D space into discrete 3D space. Afterwards, 3D-DFT is used on the results to generate a global protein descriptor.

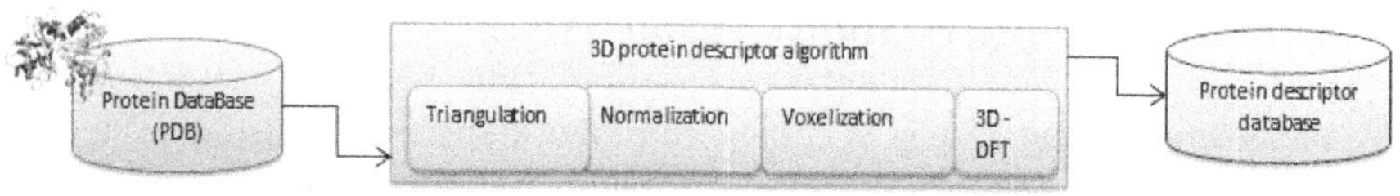

Fig. 2 Voxel protein descriptor generation process

2.1.2 Ray-Based Protein Descriptor

Descriptors consisting of 512 features which describe the protein's backbone folding are generated for each protein from the Protein Data Bank (PDB). Ray-based protein descriptor is based on the protein's backbone structure. First, from the backbone structure, the chain of C_α atoms is extracted. Afterwards, the center of mass of protein's backbone is determined and translated to the center of the coordinate system. After invariance to translation and scaling is gained, the phase of normalization is reached when the protein's backbone is scaled for d_{max}. Because different proteins have different number of C_α atoms and the descriptors have different length, protein's backbone should be uniformly interpolated.

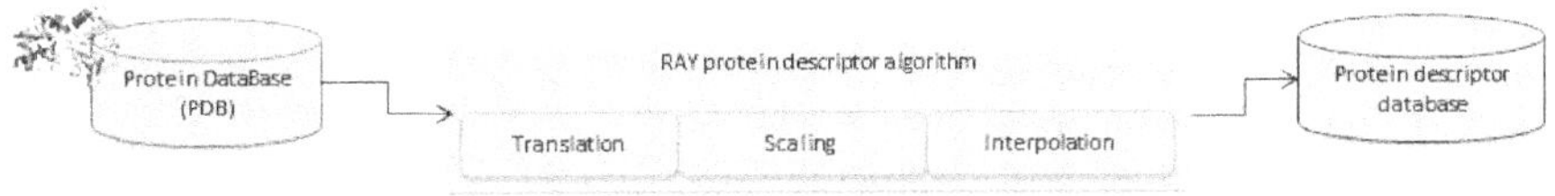

Fig. 3 Ray-based protein descriptor generation process

2.2 Hierarchical Protein Classification Scheme-Gene Ontology

Gene Ontology is a structured dictionary, where the genes products are settled into structured ontology. Gene Ontology defines a set of terms to which any given protein may be annotated and is structured as a directed acyclic graph. Each term has defined relationships to one or more other terms in the same, or sometimes to other domains. GO terms can be connected with three types of relations[16]. The parent–child relation among terms implies that the child term is either a special case of the parent term (IS–A relationship) or describes a process or component that is part of the parent process/component (PART-OF relationship). The REGULATE relation defines that one process regulates the other, while the opposite is not necessarily true. The Gene Ontology project covers three domains: cellular component, molecular function, and biological process. These areas are mutually independent. A cellular component defines an ontology that describes the components of a cell, which are part of some bigger structure or macromolecular complex. Molecular functions describe activities, catalytic or binding processes which occur on molecular level. These molecular functions refer to activities done by one gene product. A biological process is a series of events accomplished by one or more ordered assemblies of molecular functions. Sometimes, it can be difficult to distinguish between a biological process and a molecular function, but the general rule is that a process must have more than one distinct step during its execution. Ontology for the genes products in GO is kept into OBO data format. In the GO annotation is defined that the gene product has some molecular function, is included in some biological process, or is found in some cell component. Each line from the annotation file is one annotation between the gene product and the GO term.

2.3 CLUS–HMC Algorithm for Hierarchical Multi-label Classification

Hierarchical multi-label classification (HMC) is a classification where instances may belong to multiple classes at the same time and these classes are organized in a hierarchy. This hierarchical organization can be tree structure (each class has one parent) or DAG structure (each class can have more parents). The hierarchical organization and classification of proteins can be realized with Gene Ontology, which is structured as DAG structure. In this paper, the CLUS–HMC algorithm implemented in the CLUS system for machine learning is used[20]. When the classification scheme is structured as DAG, it can have two interpretations: if an instance belongs to a class c, then it either belongs to all super-classes of c, or it belongs to at least one superclass of c. CLUS-HMC computes the variance based on the weighted Euclidean distance between class vectors. Therefore, a class c's weight $w(c)$ depends on the depth of c in the class hierarchy. When the classes are structured as a DAG, however, the depth of a class is no longer unique because a class may have several depths. As a result, the class weights are no longer properly defined. Therefore, the following approach is proposed: Observe that $w(c)=w_0^{depth(c)}$ can be rewritten as $w(c)=w_0*w(par(c))$, with $par(c)$ the parent class of c, and the weights of the top-level classes equal to w_0. This recurrence relation naturally generalizes to hierarchies where classes may have multiple parents by replacing $w(par(c))$ by an aggregation function computed over the weights of c's parents. Depending on the aggregation function used (sum, min, max, average), different approaches are obtained:

- $w(c)=w_0\sum_j w(par_j(c))$ is equivalent to flattening the DAG into a tree and then using $w(c)=w_0^{depth(c)}$. The more paths in the DAG lead to a class, the more important this class is considered by this method.
- $w(c)=w_0*min_j w(par_j(c))$ has the advantage that it guarantees $\forall c, j$: $w(c)<w(par_j(c))$. A drawback is that it assigns a small weight to a class that has multiple parents and that appears both close to the top-level and deep in the hierarchy.
- $w(c)=w_0*max_j w(par_j(c))$ guarantees a high weight for classes that appear close to the top-level of the hierarchy. It does not satisfy $w(c)<w(par_j(c))$, but still yields smaller weights than $w(c)=w_0\sum_j w(par_j(c))$.
- $w(c)=w_0*avg_j w(par_j(c))$ can be considered a compromise in between the min and max approaches.

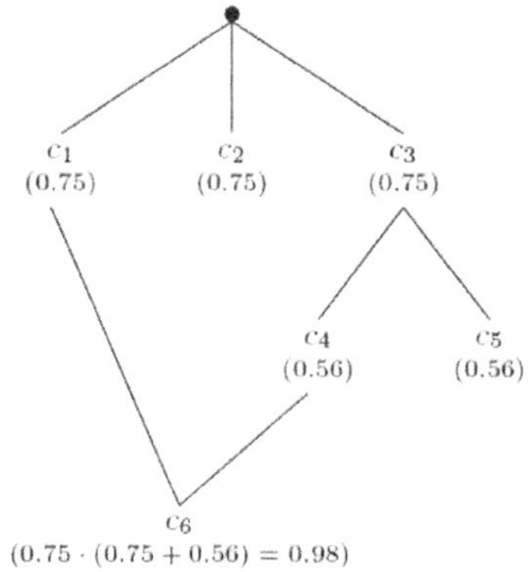

Fig. 4 The class–wise weights computed for CLUS–HMC with the weighted scheme $w(c)=w_0\sum_j w(par_j(c))$ and $w_0=0.75$ are indicated below each class

According to[21] the avg aggregate function gives the best results when Gene Ontology is used.

3 Experimental Results

All of the experiments were conducted on a Mac OS X with a 2.3 GHz Intel Core i5 and 4GB RAM. For generating the GO dependencies OBO v1.2 data file is downloaded and used, while the annotation data is read from Protein Data Bank [multispecies] GO Annotations @ EBI (*gene_association.goa_pdb*) file. This GO hierarchy is parsed for the molecular function purposes. The Voxel and Ray-based protein descriptor extractors were implemented in C++. As a system for predicting protein function, the CLUS system is used. The CLUS system created the model (decision tree) trained by using the protein descriptor datasets. The datasets for both the Voxel and Ray-based protein descriptors were consisted of 24.502 proteins. When generating the model for protein function prediction, 10 fold cross validation was used. For this validation, the original dataset was randomly divided on 10 equal subsets consisted of 2.450 data. The distribution of proteins in GO is not uniform as can be seen in the *hierarchy.txt* file generated by the CLUS system[21].

CLUS is a decision tree and rule learning system that works in predictive clustering framework. CLUS uses different algorithms and heuristics for decision trees training. Also, CLUS system uses parameters that should be defined for generating predictive decision trees. Such parameters are the initial relation's weights, F–test criteria for algorithm stopping, and the minimal number of nodes. While most decision tree learners induce classification or regression trees, CLUS generalizes this approach by learning trees that are interpreted as cluster hierarchies. These trees are called predictive clustering trees or PCTs. Classification and regression trees are special cases of PCTs, and by choosing the right parameter settings CLUS can closely mimic the behavior of tree learners such as CART or C4.5.

A precision–recall curve (PR) plots the precision of the model as a function of its recall. Although a PR curve helps in understanding the predictive behavior of the model, a single performance score is more useful to compare models. A score often used to this end is the area between PR curve and the recall axis (AUPRC). The close the AUPRC is to 1.0, the better the model is.

3.1 Results for Voxel Protein Descriptor

On Fig. 5. is shown the PR curve which is gained from testing the dataset generated from the Voxel protein descriptor.

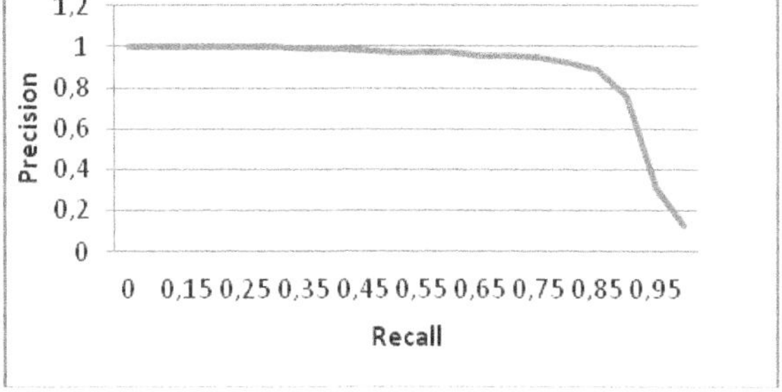

Fig. 5 PR curve for Voxel protein descriptor dataset

3.2 Results for Ray-Based Protein Descriptor

On Fig. 6. is shown the PR curve which is gained from testing the dataset generated from the Ray-based protein descriptor.

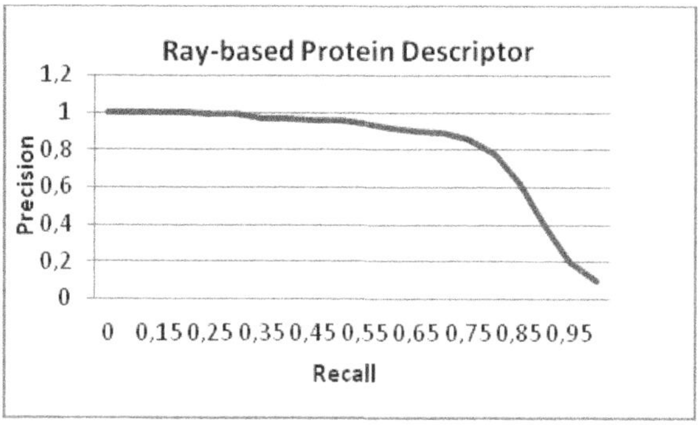

Fig. 6 PR curve for Ray-based protein descriptor dataset

3.3 Comparison of the Results Generated from Voxel and Ray-Based Protein Descriptors

The CLUS system calculates the error measures that appear in the hierarchy. There are two approaches of calculating and comparing the error measures: $AU(\overline{PRC})$ - the area under the average PR curve, and $\overline{AUPRC}$, the average over all areas under the PR curve. $AU(\overline{PRC})$ gives more weight to more frequent functions, while $\overline{AUPRC}$ considers the importance of every function to be equal. $\overline{AUPRC}$ can be calculated either by dividing the sum of all AUPRC scores by the number of classes ($\overline{AUPRC}$), or by weighing each class' area under PR curve by the frequency of instances it contains ($\overline{AUPRC_w}$). In[20] is shown that the best performance is obtained when average weight of all parents' weights is used. The hierarchical error measures gained from CLUS system for our model for prediction of protein's function are shown on Table 1 and Table 2.

Table 1 Hierarchical error measure for Voxel protein descriptor

Hierarchical error measures	Value
Average AUROC	0.8199180828118348
Average AUPRC	0.4900752637180374
Average AUPRC (weighted)	0.8812555970129666
Pooled AUPRC	0.9053965170972142

Table 2 Hierarchical error measures for Ray protein descriptor

Hierarchical error measures	Value
Average AUROC	0.7754675692742395
Average AUPRC	0.3889052049794221
Average AUPRC (weighted)	0.8131988427964493
Pooled AUPRC	0.8443881884132263

If multiple algorithms or different datasets are used, by comparing the AUPRC score, can be seen which algorithms or access is better. In this paper, when Voxel protein descriptor is used, Average AUPRC (weighted) measure = 0.88, while when Ray-based descriptor is used, Average AUPRC (weighted) measure = 0.81. If these values are compared (0.88>0.81), can be concluded that the PR curve gained from the Voxel protein descriptor dataset covers bigger area with the PR axis. That means that by using the Voxel descriptor in generating a dataset used for prediction of protein's function, better and more accurate results are gained than using the Ray-based descriptor (as can be seen on the Fig. 7).

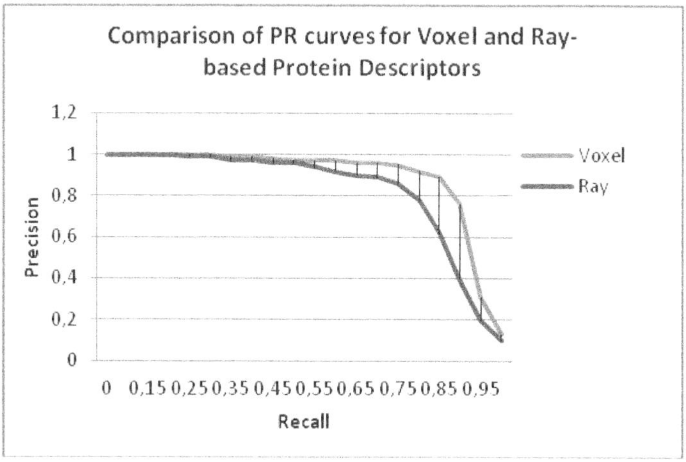

Fig. 7 Comparison of PR curves for Voxel and Ray-based protein descriptors

4 Conclusion

The aim of this paper was to compare the efficiency of Voxel and Ray-based protein descriptors when used for prediction of the protein's function. The prediction is based on the hierarchical structure of Gene Ontology and implemented by a hierarchical multi-label classifier generated by the CLUS–HMC algorithm. The CLUS system was suitable for this purpose because it provides hierarchical, multi-label classification that the GO has. According to the gained results, with predicting a protein's function with CLUS–HMC algorithm, better results are generated with Voxel protein descriptor. The advantage of CLUS system is in its efficiency and predicting precision. Also, this system provides detailed output files. The disadvantages of the CLUS system are in the huge memory complexity, and in the long generating time. Also, the output files are huge and a lot of memory space is needed.

As a future work, the gained results can be compared with other protein descriptors in order to find the best way for protein's function prediction. Also, a comparison between different algorithms for hierarchical classification is possible. In the future, the CLUS system can be modified and optimized. By implementing some of these possibilities, a system can be built for fast and accurate prediction of protein's function with using the Gene Ontology hierarchy.

References

1. Jensen, L.J., Gupta, R., Staerfeldt, H.H., Brunak, S.: Prediction of human protein function according to Gene Ontology categories. Bioinformatics 19(5), 635–642 (2003)
2. Protein Data Bank, `http://www.pdb.org`
3. Protein Databases–Genpept, `http://www.renabi.fr/article202.html`
4. The Reference Sequence-RefSeq Database, `http://www.ncbi.nlm.nih.gov/books/NBK21091`
5. SWISS-PROTprotein sequence data bank, `http://arep.med.harvard.edu/labgc/jong/Fetch/SwissProtAll.html`
6. Protein Information Resource, `http://pir.georgetown.edu`
7. Needleman, S.B., Wunsch, C.D.: A general method applicable to the search for similarities in the amino acid sequence of two proteins. J. Mol. Biol. 48(3), 443–453 (1970)
8. Altschul, S.F., Madden, T.L., Schaffer, A.A., Zhang, J., Zhang, Z., Miller, W., Lipman, D.J.: Gapped blast and psi-blast: a new generation of protein database search programs. Nucleic Acids Research 25, 3389–3402 (1997)
9. Taylor, W.R., Orengo, C.A.: Protein structure alignment. Journal of Molecular Biology 208, 1–22 (1989)
10. Holm, L., Sander, C.: Protein structure comparison by alignment of distance matrices. Journal of Molecular Biology 233, 123–138 (1993)
11. Shindyalov, H.N., Bourne, P.E.: Protein structure alignment by incremental combinatorial extension (ce) of the optimal path. Protein Eng. 9, 739–747 (1998)
12. Madej, T., Gibrat, J.F., Bryant, S.H.: Threading a database of protein cores. Proteins 23(3), 356–369 (1995)
13. SCOP: Structural Classification of Proteins, `http://scop.mrc-lmb.cam.ac.uk/scop`
14. CATH: Protein Structure Classification, `http://www.cathdb.info`
15. FSSP: Families of Structurally Similar Proteins, `http://srs.ebi.ac.uk/srsbin`
16. Gene Ontology, `http://www.geneontology.org`
17. CLUS system, `http://dtai.cs.kuleuven.be/clus`

18. Kalajdziski, S., Mirceva, G., Trivodaliev, K., Davcev, D.: Protein Classification by Matching Voxel Structures. In: IFMBE Proceedings 13th International Conference on Biomedical Engineering, vol. 23, Track 1, pp. 174–178 (2009)
19. Mirceva, G., Davcev, D.: Incorporating several features in the protein ray descriptor for more accurate protein Voxel structure retrieval. In: ACM Workshop on Voxel Object Retrieval (VoxelOR 2010), ACM Multimedia 2010, pp. 51–56 (2010)
20. Vens, C., Struyf, J., Schietgat, L., Dzeroski, S., Blockeel, H.: Decision Trees for Hierarchical Multi-label Classification. Machine Learning 73(2), 185–214 (2008)
21. Struyf, J., Zenko, B., Blockeel, H., Vens, C., Dzeroski, S.: CLUS: User's Manual (2010)

Improving Data Integrity and Performance of Cryptographic Structured Log File Systems

Genti Daci and Megi Shyle

Abstract. Modern File systems like CLFS (Cryptographic Log Structured File System) are aimed to provide security and confidentiality. Current deployments of such File Systems do not ensure data integrity of the encrypted data that is stored on disk. Due to Kernel bugs, racing conditions and arbitrary dead-locks, CLFS data on the disc can be damaged and also there is always the possibility that system users can modify the encrypted data. That's why, we considered essential to modify the way keys are stored in the system, as their safe storage is a clue point to the whole protection this system assures. Implementing a Trusted Platform Module is our suggestion to the case. So in this secure environment, our aim lies towards ensuring data integrity on CLFS without compromising the overall performance. This paper considers the standard data verification methods, with the main goal to overcome one of its major limitations, low performance of File System check-summing. CLFS matches our performance expectations, as it performs close enough to non-cryptographic file systems. To improve the performance of the check-summing process we try to study and examine various design choices and propose metadata check-summing. Several tests are made to prove that this added functionality does not significantly affect performance.

Keywords: cryptography, data security, Log Structured File System, metadata check-summing, Trusted Platform Module.

1 Introduction

A file system is meant to store typically large amounts of data, which may be either critical or sensitive, so they need to be protected. Modern File Systems, such as CLFS [1] ensure confidentiality of data, encrypting them. Usually File Systems that comprise data encryption result to perform considerably slower than

Genti Daci · Megi Shyle
Polytechnic University of Tirana, Department of Information Technology
Tirana, Albania
e-mail: gdaci@icc-al.org, megishyle@yahoo.com

L. Kocarev (Ed.): ICT Innovations 2011, AISC 150, pp. 73–83.
springerlink.com

non-cryptographic File Systems. New techniques implemented in CLFS manage to reach its performance goals, as they go close to fast local file systems. Encryption is a native characteristic of this file system and it ensures that information is accessible only to those authorized to have access, while not being affected by the overhead of encryption with more than an order of magnitude. Being native means including the data encryption code into the file system code. We are generally used to systems, where encryption is added as a layer, thus allowing it to be bypassed, but this is not our case, as Fig. 1 demonstrates.

We consider CLFS an efficient cryptographic file system, one worth working on and trying to improve. Further analysis has spotted its main drawbacks, where we will concentrate. These actually reside in two main points, the lack of data integrity and the lack of secure techniques implementation during the key storing processes.

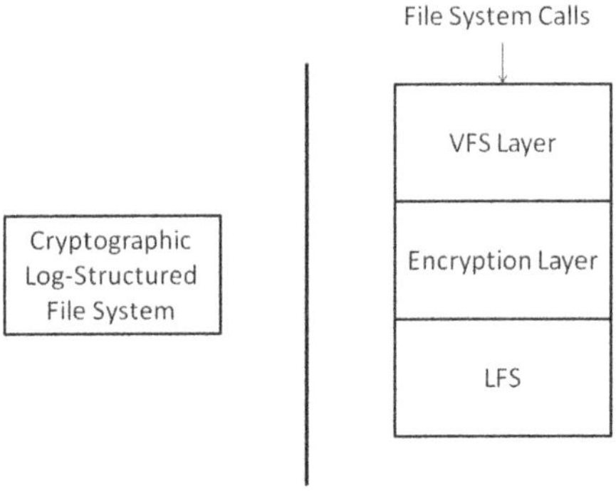

Fig. 1 The difference between the CLFS, which implements encryption as a native characteristic and a Layered File System organization

Although confidentiality is an important part of the overall system storage security, it is not everything. Considering hard disks malfunctioning, data on it can be damaged even though it is encrypted. It can also be affected by attackers, both physically or while communicating over an insecure network. Physical access on disk allows the attacker to change the unencrypted part, which in our case comprises the ifile, without the system knowing it. Thus, suggesting a way to make the file system immune from such data corruption, either as a result of a malicious attack or hardware failure, is our scope. We aim to protect our sensitive data, checking for inconsistencies, to obtain data integrity.

The second important drawback resides on the lack of security in the storage of cryptographic keys. Currently the system is based on software solutions, as it uses the common passphrase to protect the key. We implement a solution, which uses hardware aid, the trusted platform model [2] to securely store keys. Being TMP [3] its main component, that's where we will focus. TCP has been proposed and successfully adopted in various analog scenarios [4]. We believe it will be efficient even in our system, as it resulted with an overhead of only 0.3% in Jin et al. [5], which discusses a similar encrypted storage system. The difference between the above systems and ours is that CLFS overweighs them in performance. However, our objective will be reached if the overhead's percentage is lower than 10%, as this will not be absolutely detectable by the user, so it is more than acceptable.

The following paragraphs of this paper will be organized as follows. First we will describe the context of our working environment, Cryptographic Log Structured File System, which is in turn based on LFS [6]. This work comes to life to improve this file system, thus a detailed explanation of the techniques it uses to avoid the latency encryption carries, is needed. Further we will go deeper into the trusted platform module, to point out the benefits of its involvement into our system. After that, a list of the techniques used to achieve the integrity of data requirement will be unfolded. They will be examined in a selective prospective, to pick out the most congruent solution to our specific case, as Fig 2 brings to attention.

Finally, we will give our approach towards the solution of the integrity issue and conclude the idea.

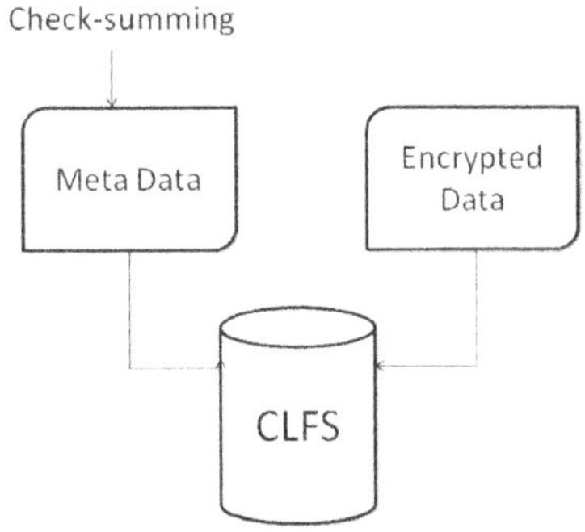

Fig. 2 Overview of one of our scopes, adding integrity checking to CLFS

2 The Background, (C)LFS

In this section, we consider important to unfold the features of the Cryptographic Log-Structured File System, to justify our choice. To make a long story short, the crucial argument that supports CLFS is performance. Previous cryptographic file systems come in different implementation, the encryption can be block based, as well as disk based, on network file systems or stackable ones. Their common characteristic is bumping into the knot of being too CPU sensitive. To provide performance solutions CLFS concentrates on previous cryptographic File Systems weaknesses, being the speed of the encryption algorithm used and the writing latency.

2.1 The Encryption Algorithm

CLFS considered that improving the speed of the algorithm used to encrypt would be useful both ways of read and write operations. Let's take a look at a couple of the most known cryptographic file systems, to check out for their deficiencies.

Blaze (1993) [7] implemented CFS as a network file system and its main drawback resulted the continuous context switching overhead. A completely different approach was treated by Zadok (1998) in CryptFS [8], as it is a stackable file system. This provides it with portability allowing execution above any kind of native file system. It also comes out to be faster than CFS by a factor that fluctuates from 2 to 37 times. The bottleneck of this file system is precisely the encryption algorithm it uses, blowfish.

The algorithm used in CLFS is SEAL 3.0 [9]. It is used exactly to surpass the knot a non efficient algorithm brings to the system. This is achieved by utilizing two major characteristics of the algorithm, namely speed and pre-computation. The tests made by P. Rogaway and D. Coppersmith demonstrated that SEAL uses approximately 4 cycles to process a byte and results up to 10 times faster than DES. SEAL works as a stream cipher, where the encryption depends not only from the plain text and the encryption key, but also from the position of the data. The key is 160bit long and SEAL uses it to map a 32bit string to one composed of L bit, where L is less than 64KB. The trick that makes the work with SEAL that fast is pre-computation. Being LFS the ground upon which CLFS was built, allows us to know precisely the position of the next write and that's the reason why we can pre-compute the whole key stream, so that the encryption process is reduced to a simple XOR operation between the key and the plain text.

2.2 The Underlying File System, LFS

The speed of I/O bound applications is limited by the write performance of the file system. A log-structured file system is designed for high write throughput. Rosenblum and Ousterhout designed e file system with a relatively different behavior from its predecessors. They introduced the idea of a continuous log, where data and metadata are written. The majority of file systems put a great emphasis on spatial locality and transform their data without changing its location, which leads to slow seeks. LFS assumes that taking care of locality will no longer be effective, as cache memory size is increasing to the extends of satisfying all read operations. Storing data in a log avoids seeks, therefore minimizes the movement of disk's head and maximizes write throughput.

The benefits of using it as a base for CLFS is that we know where the position of the next write is located. Besides, file system writes are only performed at the end of the log which means that inconsistencies can only be located in the last segment of the log, which greatly speeds up crash recovery.

2.3 (C)LFS Organization

The Cryptographic Log Structured File System is very similar to LFS, as much as concerns its overall format. Its building blocks are divided into segments, which in turn become active, as shown in Fig 3. The initial part of each of these segments consists of the so called summary block. This structure is composed of a set of pointers that indicate the position of the successive summary block. That is how the idea of a continuous log is implemented in LFS.

Upon creation of the file system a random key is automatically generated for the first segment. Afterwards the segment key and the key stream offset are stored encrypted in the corresponding segment summaries. The key stream is calculated increasing it with the segment size. To increase performance CLFS uses a cache to store the expanded keys and their positions.

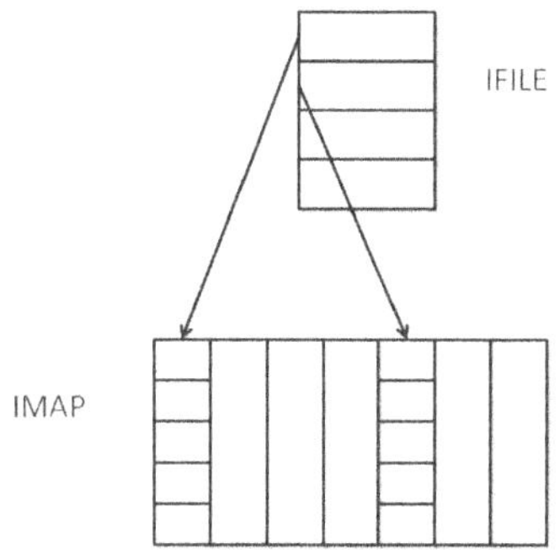

Fig. 3 LFS segment organization

In a read operation CLFS would initially act the same as LFS, by searching for the desired block to read. The search would start by the index file's inode stored in the superblock, where we would find the inode map, followed from the inode. These operations are the majority of times bypassed, since this data are stored in the cache. Obviously, CLFS to ensure effective encryption cannot stop at the point of LFS. It has to read the exact position of the segment summary from the ifile. After reading it, decrypt the segment key and the key offset, to generate the key stream. Finally the read data has to be XOR-ed with the key stream to decrypt them.

The write operation differs from LFS only in a couple of steps. The first chunk of data is removed from the list and its key is already known, so we can perform the XOR operation. This is done with all the chunks of data afterwards.

3 Key Management Disposal

As previously mentioned, our goal is building security facilities on a fast Cryptographic Log-Structured File System. The second step to increased security is dealing with key management mechanisms. This issue is one of the most challenging, but a successful process is crucial and of course shouldn't be neglected. This is due to the fact that encryption is not an undefeatable solution to security. To mention here that it doesn't solve access problems, neither data corruption issues. These problems can be caused by an outrageous number of factors, considering the increasing scale of security threats, coming from malicious attacks.

3.1 Trusted Computing Technology

The current deployment of CLFS holds up on a user-supplied passphrase to protect the encryption keys, but research has concluded that a merely software approach is not enough to security. A way out of this has bloomed in the recent years and is gaining market access. It consists on searching for reliability into a hardware, which is equipped with trusted components, called a trusted platform.

Building a trusted platform certainly requires adding extra hardware modules, if they already aren't part of the architecture. Whereas, logically these platforms are constructed in various layers and components. The main one is TPM (Trusted Platform Module), which is intended to provide integrity, i.e. the system acting as it is expected. To achieve this the platform has the so called, roots of trust. Their integrity cannot be proven in any way, that's the reason why they have to be assumed as trusted modules. The roots of trust comprise the RTM (Root of Trust for Measurements), RTS (Root of Trust for Storage) and RTR (Root of Trust for Reporting). These are the base components upon which a whole secure platform is set.

3.2 Managing Keys through Trusted Platforms

We overviewed Trusted Computer Platforms because that is what we are going to use to manage the storage of encryption keys. The trusted platform module, is created to safely save cryptographic keys. This means that attempting to access the key must pass through logging into TMP.

Having a trusting hardware environment, will give us two main advantages: protected capabilities and shielded memory locations. This means we are equipped with memory locations, where storing is considered to be secure and we can utilize capabilities like cryptographic key management and random numbers generation. This is shown in Fig 4.

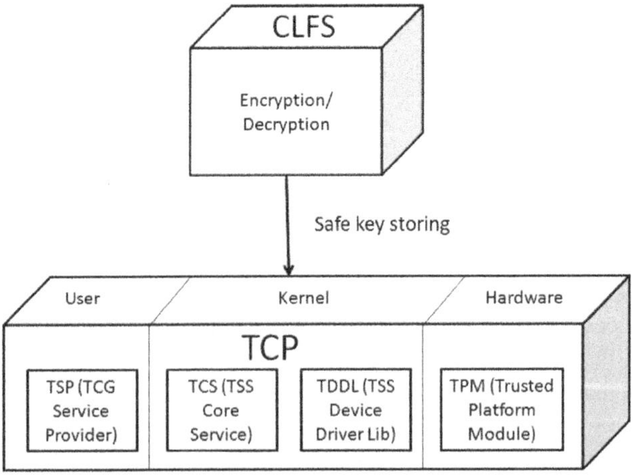

Fig. 4 CLFS uses a trusted computer platform so it can save its keys reliably

4 Data Verification Techniques

Ensuring data integrity is fundamental to computer systems. Several factors may induce to data errors, to mention media failures, kernel bugs and racing conditions. Even an attacker who has reached to gain administrator privileges can modify the data. Threats are multidirectional and our system cannot be left unprotected, so our approach is based on the existing cryptographic file system to which the integrity checking capability is added.

4.1 Mirroring

Making exact copies of our data, i.e. mirroring, can be a way of managing it reliably. The process would comprise the comparison of our data with the mirrored one, before operating with it, as in Fig 5. This method would easily detect changes in one of the replicas, providing integrity in case the changes occur because of hardware malfunctioning.

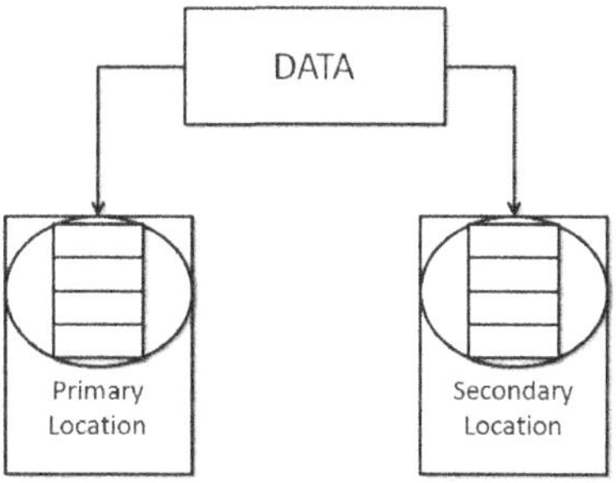

Fig. 5 The mirroring technique, data are stored in two different locations

However, it doesn't tell which of the copies is the genuine one. It is not able to perform correctly in the case of an intruder, either. This is due to the fact that both of the replicas can be changed, so that the system lacks the tools to detect intrusion and integrity is not obtained. The inefficiency of this method also arises when we consider the storage space it requires and the time we need to spend checking both replicas.

4.2 Parity

Parity is a simple yet effective method to assure integrity protection. The procedure is as follows: one bit is added to the stream, as shown in Fig 6. The condition on this added bit is to be the equal to the binary sum of the bits. These parity bits are able to detect only single errors, which makes this technique primitive. We would get a result where a stream with two errors is considered valid. This definitely doesn't fulfill our security requirements.

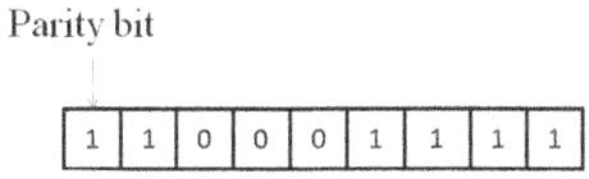

Fig. 6 The parity technique, the parity bit added to a bit pattern

4.3 Check-Summing

Check-sums are exactly the same as parity with two changes: to create the check-sum of a pattern of entities of length n bits, an n-bit entity (the check-sum) is added and the modulo 2 sum of the entities and the check-sum is constrained to be zero. This is shown in Fig 7.

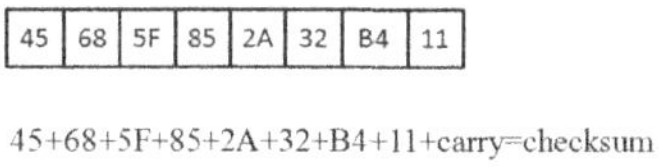

Fig. 7 The check-summing method, the calculated check-sum at the end of the bit pattern

CRC is more secure than check-summing, but it needs more calculation, as it adopts a more complex technique. Check-sums can be implemented in various ways, we can compute a per block check-sum, or a per file check-sum. The latter proposes two alternatives, storing check-sums apart from the data, otherwise interleaving data files and check-sums with the purpose of making more efficient use of data locality.

5 Our Approach

To constitute our design model we have developed our idea in several layers, each of which is built on answering one single question per layer. As we previously explained ensuring data verification in a file system is essential to its integrity. On the other hand we need to preserve as much as possible the major advantage of our native file system CLFS, which despite being cryptographic fully meets performance requirements. So, naturally the decisions we need to make involve which method of verification is more appropriate, what part of the data will be verified unless all of it, where will this extra information be stored and finally how frequently the verification will occur. Let's analyze them further.

5.1 The Method Used to Provide Data Integrity

Check-summing is the most common method to ensure data integrity [10]. If we want our data to be protected from intruders as well as from transient errors, the check-sum need to be protected with a secure hashing scheme such as SHA1 [11].

We considered efficient to compute check-sums for metadata. It comprises all the inodes and the ifile. Checking different fields of the metadata will allow us to find out if any malicious modification has been made to the data, because almost any modification to our data will be reflected to the metadata [12]. Furthermore, this decision is more advantageous as the amount of metadata is considerably less than data, and results to be efficient on timing, hence giving a better performance.

5.2 Benchmarks on Improved LFS under Stress Test Using Tar and Compiling

Knuttson, after implementing his cryptographic File System, made several tests to check its performance affected by the overhead encryption introduced. It resulted to perform close to non-cryptographic File-Systems, being an ideal solution for systems which store large amounts of sensitive data.

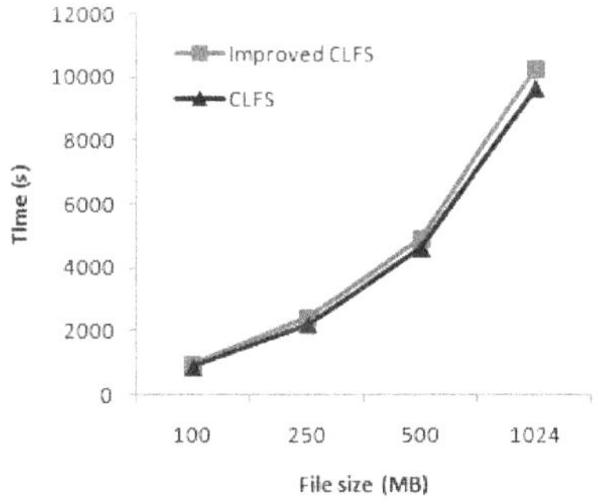

Fig. 8 Results of the tar test, the results of four different sized tar files. They are all composed of a great number of small files and this test was chosen considering it is a very frequent operation in a file system and its overhead reached a 6%.

Now, apart from encryption we have added a new feature, file integrity checking, through metadata check-summing. We need to test our system to evaluate the latency, this further overhead cause. The tests made are on a .tar file and also tying a kernel compilation. We have submitted four different file sizes, respectively 100MB, 250MB, 500MB and 1GB.

Knuttson tar tests completed in approximately 562s for a 64MB file. Our 100MB file completed in 927s, as shown in Fig 8. Whereas, Fig 9 shows the comparison of the two systems during a compilation test.

After a comparative analysis of the tests, we notice that the overhead reaches the margins of 4% on compilation tests and 6% on tar tests. The first is due to the fact that compilation is a CPU bound application and is not particularly affected by the File System, whereas the latter is slightly affected. We also have to emphasize that the gap between the two graphs is due to the longer time this test needs to complete.

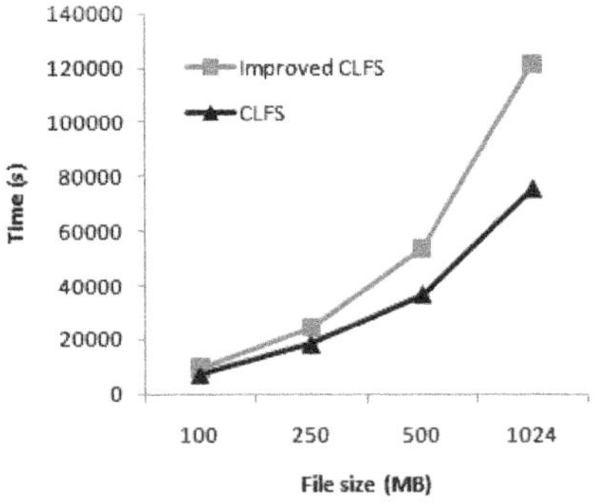

Fig. 9 Results of the compilation test, the time of compilation of four different sized files, after adding integrity checking to metadata and after calculation the overhead results a mere 4%

6 Conclusions

Trying to add integrity to a cryptographic File System, like CLFS, initially sounds risky from the performance point of view, but choosing the most appropriate method and a reduced amount of data to check is the clue to success.

We also managed to address the key storing security issue, by adding a secure platform in which we trust. This hardware solution overweighs previous exclusively software ones.

Our benchmark showed an overhead that varies from 4-6%. This is an appealing result, considering the systems where CLFS is intended to work. We provided secure key storage and highly secure metadata integrity checking, i.e. the data stored in our system are correct, or at least cannot be modified undetectably.

We assumed that checking the integrity of both data and metadata would considerably affect performance, while not evidently improving integrity. Metadata check-summing was considered sufficient to meet our requirements. But, in spite of this, further work can be done on testing what this overhead would exactly be.

As a conclusion, we successfully managed to add a reliable hardware environment and to improve a cryptographic file system, enabling it with integrity without sacrificing performance.

References

1. Knutsson, K.: Security without cost: A cryptographic log-structured file system. Department of Software Engineering and Computer Science Blekinge Institute of Technology, 1–26 (2002)
2. Pearson, S.: Trusted Computing platform, the next security solution. HP Laboratories, Bristol (2002)

3. Lioy, A., Ramunno, G., Vernizzi, D.: Trusted-Computing Technologies for the Protection of Critical Information Systems. Journal of Information Assurance and Security 4, 449–457 (2009)
4. Chen, J., Liu, J., Feng, D.: Design and Implementation of Object-oriented Encryption Storage System Based on Trusted Computing Platform. In: International Conference on Computational Intelligence and Security, pp. 546–550 (2010)
5. Jin, G., Bo, L.: Design and Implementation of a Cryptographic File System for Linux Based on Trusted Computing Platform. In: Fourth International Conference on Intelligent Computation Technology and Automation, pp. 102–104 (2011)
6. Rosenblum, M., Ousterhout, J.: The design and implementation of a log-structured file system. In: Symposium on Operating System Principles Proceedings, pp. 1–15 (1991)
7. Blaze, M.: A cryptographic file system for unix. In: Proceedings of the 1st ACM Conference on Computer and Communication Security, pp. 9–16 (1993)
8. Zadok, E.: Cryptfs: A stackable vnode level encryption file system. Technical report, CUCS-021-98, pp. 1–14 (1998)
9. Rogaway, P., Coppersmith, D.: A software-optimized encryption algorithm. Journal of Cryptology: The Journal of the International Association for Cryptologic Research, 1–14 (1997)
10. Sivathanu, G., Wright, C., Zadok, E.: Enhancing file system integrity through checksums. Stony Brook University Technical Report FSL-04-04, pp. 1–5 (2004)
11. SHA1: Secure hash standard, Federal Information Processing Standards Publication 180-1, pp. 1–11 (1997)
12. Patil, S., Kashyap, A., Sivathanu, G., Zadok, E.: I3FS: An in-kernel integrity checker and intrusion detection file. Stony Brook University, pp. 67-77 (2004)

Effects of Terrain Details on Wireless Network Simulation Macro Results Accuracy

Sonja Filiposka, Igor Mishkovski, and Dimitar Trajanov

Abstract. In this paper we investigate the use of terrains with various level of detail and its impact on the simulation results on the macro level. We argue to which extent this problem has to be taken into account and what is the appropriate level of terrain accurate representation in order to achieve results that are valid and can be used as a simulation projection for a given real life scenario.

1 Introduction

Network simulators are used by researchers in order to test new scenarios and protocols in a controlled and reproducible environment, allowing the user to represent various topologies, simulate network traffic using different protocols, visualize the network and measure the performances. There are many different network simulators used by the research and the educative and corporative communities today, varying in the type of implementation, as to whether they are open for custom changes, as well as in the number of protocols implemented, the ability to simulate wired and wireless scenarios, the level of detail when describing the network scenario (both traffic pattern and node positioning) and more. In this paper we are mostly concerned with one specific feature of the network simulator, that is, the ability to simulate a wireless ad hoc network that is deployed over a 3D terrain.

The field of terrain aware radio propagation models for wireless networks is a vibrant and popular field especially in the recent years. While the simulations conducted in the past where mainly occupied with statistically based propagation

Sonja Filiposka · Dimitar Trajanov
Faculty of Computer Science and Engineering, Ss. Cyril and Methodius University, Skopje, R. Macedonia
e-mail: {sonja.filiposka,dimitar.trajanov}@finki.ukim.mk

Igor Mishkovski
Politecnico di Torino, Turin, Italy
e-mail: igor.mishkovski@polito.it

L. Kocarev (Ed.): ICT Innovations 2011, AISC 150, pp. 85–94.
springerlink.com

models that described the simulation environment on a large scale level with or without shadowing effects, lately the research community has shifted its interest to creation of a simulation scenario that will more closely reflect real life conditions.

Propagation models for wireless networks have traditionally focused on predicting the average received signal strength at a given distance from the transmitter, as well as the variability of the signal strength in close spatial proximity to a particular location. Typically when estimating the ad hoc network performances very little attention is given to the terrain [1], and thus, to the propagation model used for the estimation [2][3][4]. The nodes are assumed to be scattered on a flat, regular terrain and a simple ground reflection propagation model is used.

However, radio transmission in a mobile communications system often takes place over irregular terrain. Since terrain irregularities can greatly affect and distort the expected network performances, great care must be taken when trying to picture the real-life use of the modeled network. Therefore, the terrain profile of a particular area needs to be taken into account when estimating the path loss since the transmission path between the transmitter and the receiver can vary from simple line-of-sight to one that is severely obstructed by hillsides or foliage.

There have been a number of different implementations of terrain aware radio propagation models, mostly in the form of a new framework for simulations. Examples typically include introducing a new realistic simulation framework, like the case for urban vehicle to vehicle networks [5] wherein the authors developed this new simulation tool to evaluate the performance of event dissemination protocol in realistic city traffic environment. Another interesting implementation is the AMADEOS extension [6] to NS-2 and Glomosim [7]. In addition to the developed mobility models, it provides a ray-tracing routine combined with a geometry DXF file that describes environment characteristics. The algorithm is based on a 2D ray tracing technique and is more than seven times slower than two-ray ground.

The OPNET's Terrain Modeling Module (TMM) [8] also adds Earth topology, such as mountains, forests and valleys, as well as user-selectable environmental conditions to the simulated network model. There are several terrain aware propagation models that are included: the Longley-Rice model [9], TIREM [10] and the Walfisch-Ikegami model [11]. TMM imports and graphically displays elevation information from Digital Terrain Elevation Data (DTED) and the US Geological Survey Digital Elevation Maps (USGS DEM) formats, enabling accurate calculation of signal strength.

We decided to follow the footsteps of OPNET and create an extension for the free open source NS-2 [12] that will provide the users with the ability to add terrains into their wireless scenarios. The Durkin's model, that we have chosen for our implementation, works with DEM files in order to read in the profile data and is based on edge diffraction in a very similar approach to the one used by TIREM.

After our successful DEM based implementation [13], in order to make our simulations run faster, we decided to use a different file format to describe the terrain, and we chose the triangular irregular network (TIN) format for this task. As it is shown in [14], our new implementation gives as accurate results as the DEM implementation, but in the same time makes the simulations run faster and allow the user to choose the detail level of terrain with a wide gradation.

Since the TIN format needs less data to represent the terrain the simulations are a lot faster, but they also offer another opportunity for the user. When creating a TIN file, which can be done using different free software packages, the user can define the detail of the terrain description by giving the acceptable level of mean or max error while creating the file. In this way, a number of TIN files can be obtained while all can represent the same real terrain but with a different detail level.

In this paper we investigate the influence of the detail level on the macro results from a typical ad hoc network simulation. The results from this paper are intended to help the NS-2 users that create 3D terrain wireless simulations with guidance as to how to choose the appropriate level of detail for their chosen terrain and what is the confidence interval of the simulation for terrains with less detail, which on the other hand are leading to much faster simulations.

The rest of this paper is organized as follows: Section 2 is an introduction to the triangular irregular network file format is given as well as the description of the creation of terrains with different detail level that are described in this file format. In Section 3 case study simulation scenarios are shown and the analyses of the performances of the case study ad hoc network over 3D terrains with different detail level are presented. Finally, Section 4 concludes this paper.

2 Terrain Model

The Geographic Information System (GIS) [16] integrates hardware, software and data for storing, management, analysis and representation of all forms of geographically referenced data. GIS technology [17] can be used for science purposes, cartography, urban planning, rescue missions and everywhere where need for geospatial parameters of the earth's surface arises. GIS files are defined using very strict standards for keeping record of digitalized data. GIS files contain elevation data, terrain shape, coordinate systems and projections and data that describe the precise shape of the earth's surface according to the given coordinates.

The geographic systems are differentiated according to the way they store the elevation data. There are two formats for data storing: raster and vector. The raster type of data consists of rows and columns of cells wherein a unique value is stored. Each cell gets a numeric value that can be represented by a unique identifier. The resolution of the raster data is the cell width and length in earth units. Usually the cells are square terrain areas, but other shapes can also be used. The USGS DEM standard [18] [19] is a geospatial file format developed by United States Geological Survey (USGS) for storing raster based DEMs.

Contours and TINs are the most common vector-based models of surfaces. Source data may be provided in either of these formats, or may be converted to these formats from DEM files. Typically they are far more compact than DEMs.

The TIN model represents a surface as a set of contiguous, non-overlapping triangles. Within each triangle the surface is represented by a plane. The triangles are made from a set of points called mass points. These mass points can occur at any location, the more carefully selected, the more accurate the model of the surface. TIN-based model is able to describe features very well. Although grid DEMs are easy to handle, triangulated irregular network models (TINs) have become increasingly popular for the representation of terrain surfaces because of their ability to accommodate irregularly spaced elevation data. It enables them to adapt to the variable complexity of the relief and to integrate structural features (like breaklines) in the terrain model [20]. A diversity of TIN-based interpretation procedures is already available today. It has been shown that many of these procedures have clear advantages over their grid-based counterparts [21] [22].

The roughness of the terrain surfaces determines the difficulty of digital terrain model (DTM) representation of terrain. Only few points need to be sampled if the terrain is simple. Alternatively, if the terrain is complex, more points need to be measured. Hence, DTM resolution plays an important role in determining its detail. It depends on the intended application of DTM. Lower horizontal resolution DTMs produce lower slope gradients on steeper slopes and steeper slope gradients on flatter slopes [23]. However, higher-resolution DTMs may not be necessary for generating useful soil-landscape models. Therefore, an optimal DTM resolution is needs to be determined.

In order to visualize and statistically present the broadness of the detail level of a TIN terrain we investigate the results from a case study using the Landserf software package [24]. LandSerf is free software that allows you to visualize and perform analysis on spatial data.

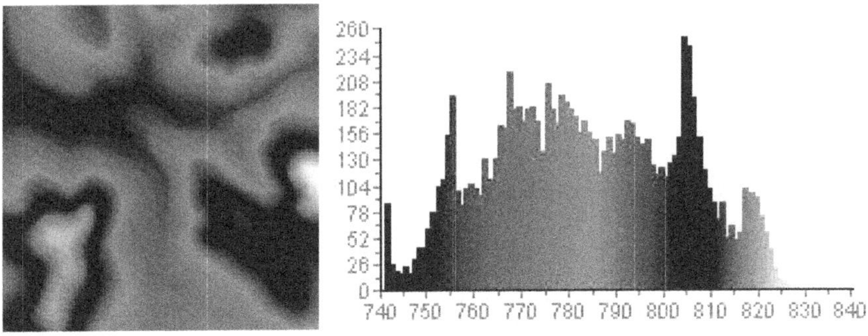

Fig. 1 Source DEM file and its height frequency distribution

Our case study involved a number of source DEM files with different test areas on which we used the convert to TIN function of the software. To control the number of triangles produced by the process, one needs to either specify the number of sample vertices explicitly or specify some error criterion. The error (either average as specified by the Root Mean Square Error, or the maximum error) represents the difference in elevation between any point on the original DEM and its elevation in the TIN. The smaller the error specified, the greater the number of triangles required in the network. The triangulation process works by successively adding triangles to the network until either any of the selected error criteria are met, or the maximum number of triangles is reached (if specified).

Figure 1 shows the test area source DEM together with the histogram that represents the frequency distribution of the heights in the file. The statistical info on the source DEM file is as follows: the terrain size is 1000 to 1000 m, with

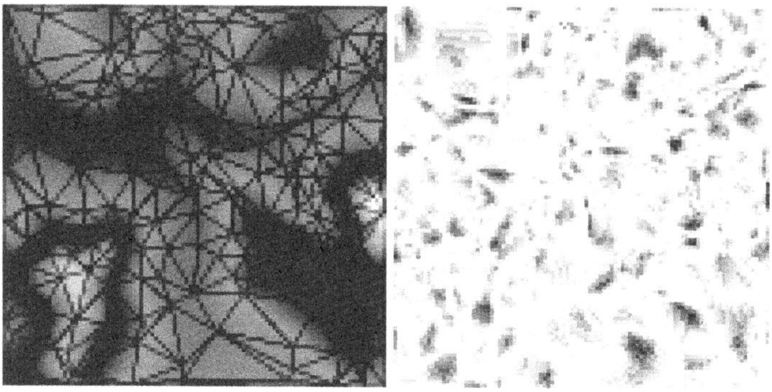

Fig. 2. Derived TIN representation and error surface using 400 triangles

minimum height of 741 m, and maximum height of 830.33 m, mean height is 783.20 m and the standard deviation to this value is 19.93 m.

The derived TIN representation, created using the Landserf package is shown on Figure 2. The TIN example is based on a program selected set of 200 vertices, resulting in approximately 400 triangular regions. The statistical information on the obtained TIN is as follows: 376 triangles, minimum height 741.33 and maximum height 826.33.

The differences between the TIN-generated DEM and the original DEM can be plotted as a grid, as shown on the right in Figure 2. The darkest reds (grays) are the areas of maximum error – in this case up to 4.12 m (as you can see from the statistical information about the two terrains, this is the case for the points with maximum height). The presented example was chosen due to the fact that this is the minimum number of triangles needed in order to get mean error of less than 1m, while the median error equals 0.8 m, and the root mean square error (RMSE) calculated from a number of control points for which exact elevation values are known equals 1.26 m. The error map can be improved (the quality of the TIN representation enhanced) by increasing the number of vertices.

Please note, however, that the representation is not unique, and a similar computation using another package (e.g. TNTMips, which uses a similar adaptive densification) may result in a different pattern. Also, one must bear in mind that the representation of flat regions with the TIN model can be problematic.

In order to get a more clear picture how the RSME and max error change with the increasing number of triangles we generated a great deal of TIN terrains from a given source DEM. Table 1 summarizes our findings.

As it is clearly shown in Table 1 the error is rapidly decreasing with the increased number of triangles up to the point when we start using more than 2000 sampling points in which case we are sure that the max error is less than 1 m which is quite acceptable for our network simulations. However further increase in the accuracy of the terrain details implies involving a great deal more triangles eventually reaching 12456 for which value we obtain a TIN representation that is an exact match to the DEM source. Please note that the results presented on Figure 2 are for the first case when the mean error is less than 1, but, as it can be seen from the table, this parameter is not sufficient constraint since the values for the max error are still quite high.

Table 1 Statistical data on TIN terrain accuracy compared to a DEM source when representing with a different number of sampling points / number of triangles

No. sampling cells	No. of triangles	RSME	max error	mean error	median error	min height	max height
/	12456	0	0	0	0	741	830.33
6200	12210	0.3	0.33	0	0	741	830.33
6000	11812	0.05	0.5	0.01	0	741	830.33
5400	10649	0.1	0.5	0.03	0	741	830.33
3800	7460	0.2	0.5	0.1	0	741	830
2000	3902	0.3	0.75	0.21	0.1667	741	830
1200	2327	0.4	1.25	0.3	0.25	741.33	830
700	1346	0.5	1.59	0.39	0.333	741.33	830
500	953	0.62	1.85	0.48	0.4	741.33	830
400	759	0,72	2,93	0,56	0,4681	741,333	827,667
300	569	0,94	3,77	0,73	0,6	741,333	827,667
200	376	1,26	4,12	0,97	0,7969	741,333	826,33
100	186	2,18	7,31	1,69	1,356	741,667	821,667
50	92	3,67	11,11	2,85	2,333	741,6667	818,333
25	47	6,47	27,2	4,95	3,95	747,333	816,6667
10	20	13,39	42,81	10,09	7,625	748	810
5	11	16,68	46,5	13,15	10,82	767,667	809,667

Having done this analysis as a first step or our research, we concluded that while creating terrains for our wireless ad hoc networks scenarios it would be reasonably sufficient if we are careful to use a TIN terrain that has accuracy with a max error less than 1 (see Figure 3) while keeping the number of triangles as low as possible in order to get the simulations run faster. In the remainder of this paper we have conducted a series of simulations in order to prove this hypothesis.

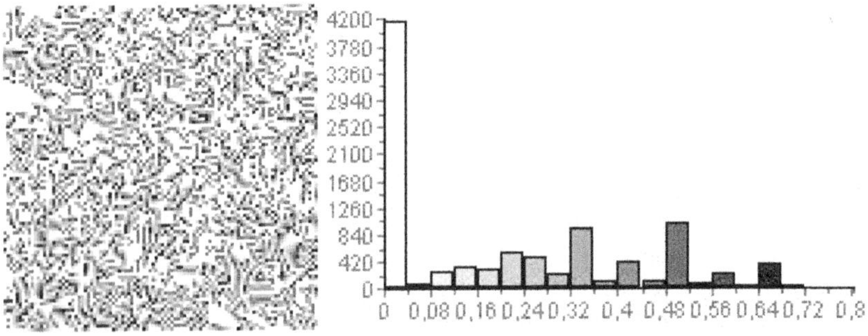

Fig. 3 TIN error surface and its histogram for TIN representation using approx. 4000 triangles

3 Terrain Details Effect on Network Simulation Results Analysis

The primary objective of our case study simulations is to understand the impact of the amount of details in the terrain presence in a simulation environment for ad hoc networks. To this end, we evaluate the aspects of traffic throughput of the Durkin's propagation model using TIN files with different level of details. For the case study evaluations made in this paper, we decided to work with a number of real terrains, but since the obtained results are statistically equal, here we present the results for the real terrain shown in Figure 1.

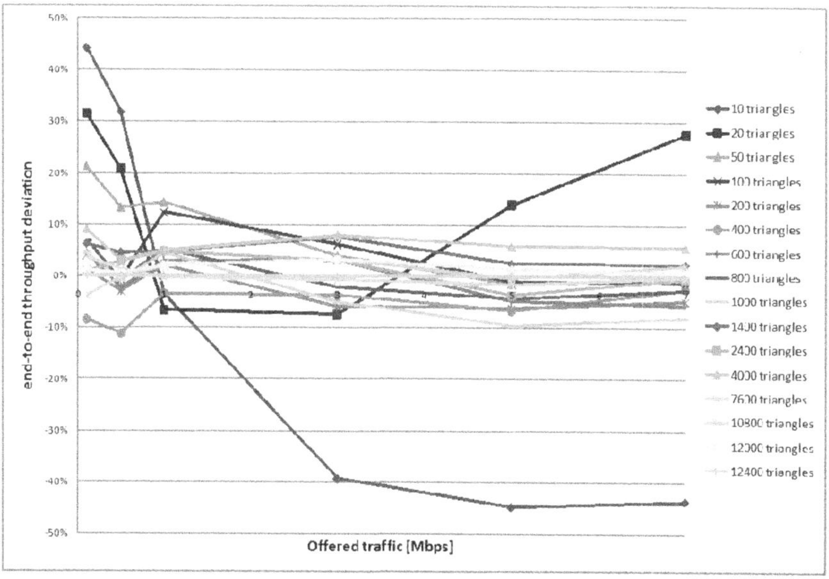

Fig. 4 End-to-end throughput deviation for given offered traffic when the simulation is run with a terrain described using a different number of triangles

The created DEM and TIN files we used for the simulations were made with dimensions of 1.000m x 1.000m. The complete ad hoc network of 100 nodes all equipped with an IEEE 802.11b radio and a transmission range of 250 m. The nodes are uniformly scattered in the terrain area. The traffic pattern is a completely random traffic between the nodes with the offered traffic amount ranging from 0.1 Mbps up to 7 Mbps. We created a one scenario set for each offered traffic which we then run on the converted TIN terrains with a different detail level based on the number of triangles. It was crucial that the traffic pattern and positioning patter scenarios are defined only once and are used for all the terrains so that the differences in the results are purely due to the terrain discrepancies. The results shown here are averaged over the results obtained for 10 different generated scenarios. As a basis for comparison we took the TIN file with highest level of details

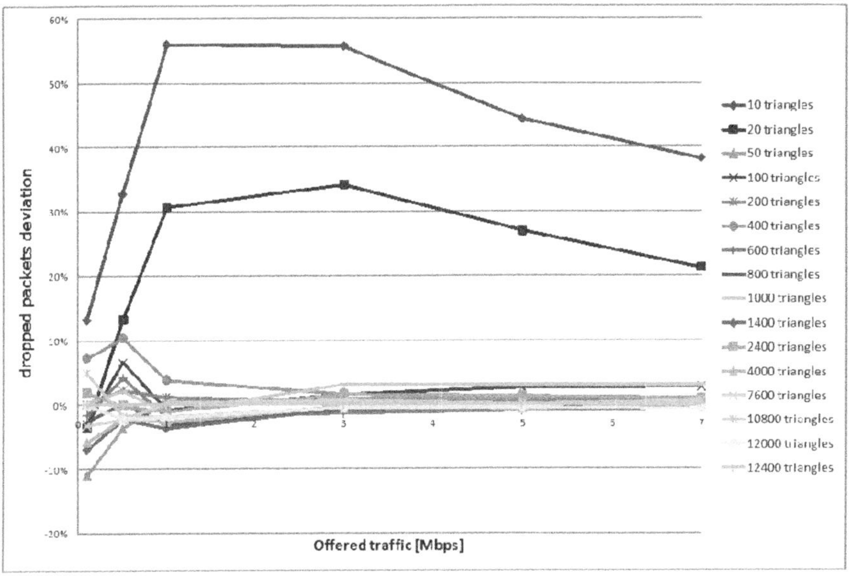

Fig. 5 Dropped packets deviation for given offered traffic when the simulation is run with a terrain described using a different number of triangles

which is statistically equal to the information contained in the source DEM file with maximum and mean and median errors as well as RSME equal to zero.

On Figure 4 the obtained results for the end-to-end throughput of the simulated network are shown when simulating using the terrains with the properties as given in Table 1. It can be seen that our first estimation of the hypothesis given in the previous section is verified. When using terrains that are described using 4000 or more triangles (for which we are sure that the maximum error is less than 1 m), the obtained results are deviating from the ideal values (obtained for the terrain with 12456 triangles) for maximum of 9% (this is the bordering case of 4000 triangles). For all other samples with more than 4000 triangles the results are different by no more than 4% (2% on average), which is quite acceptable. It is also notable that the worst results are obtained for the cases when the offered traffic is very low (0.1 Mbps) due to the fact that in this case very small number of packets are being sent and the results are very sensitive to a loss of even few packets. However, it is more interesting to conclude that even for a lesser number of triangles the results can still be taken into consideration with some skeptics in mind. As it is shown, for terrains with more than 400 triangles the results fit within the +/-10%. Given that the simulations are many times faster when the number of triangles is smaller, with regards to Table 1, we can conclude that for times when we need fast first order approximation we can use terrains that are generated with a mean and median error of less than 0.5 m.

On Figure 5 the dropped packets deviation for the given simulations is shown. As it can be seen, the results are not just confirming our conclusions given with the previous results, but are even much more relaxing. We can still confirm the problem with the low offered traffic where we find the highest discrepancies. However, even for a low number of triangles the results fit well into the 10% deviation region. It is interesting to note that in this case the deviation is not symmetrical; the number of dropped packets in the simulation is almost always higher when compared to the ideal case.

4 Conclusion

In this paper the effect of the terrain detail level on the macro results obtained from wireless network simulations is investigated. The aim of the paper is to investigate the size of the effect and to recommend some guidelines as to what level of terrain detail is considered acceptable while bearing in mind that the running time of the network simulation is considerably under the direct influence of the number of triangles needed to describe a certain terrain.

The paper followed two lines of investigation: first, how does the statistical data that describes one terrain change with the level of details that the terrain has incorporated in it, and second, how does the terrain detail level influence the macro results from a typical wireless ad hoc network simulation, while trying to find a correlation between these two results and thus make a recommendation on choosing terrains for the simulations.

The results have clearly shown that as long as the maximum error of the terrain data is below 1m, the obtained results there will be no more than 5% discrepancies in the obtained results compared to the ideal case of a perfectly described terrain with zero errors. Furthermore, for the purposes of obtaining results that will act as a first approximation and be used on a global scale, the conditions can be relaxed so as to use terrains that have median or mean error of less than 0.5 m.

References

1. Lee, C.M., Pappas, V., Sahu, S., Seshan, S.: Impact of Coverage on the Performance of Wireless Ad Hoc Networks. In: Proceedings of the Second Annual Conference of the International Technology Alliance, UK (September 2008)
2. Ozan, K., Tonguz, G., Ferrari: Ad Hoc Wireless Networks: A Communication-Theoretic Perspective. John Wiley & Sons (2006)
3. Shan, W., Jian-xin, W., Xu-dong, Z., Ji-bo, W.: Performance of anti-jamming ad hoc networks using directional beams with group mobility. In: IFIP International Conference on Wireless and Optical Communications Networks, April 11-13 (2006)
4. Baumann, R., Heimlicher, S., May, M.: Towards Realistic Mobility Models for Vehicular Ad-hoc Networks. In: 26th Annual IEEE Conference on Computer Communications IEEE INFOCOM 2007, Alaska, USA, May 6-12 (2007)
5. Vuyyuru, R., Oguchi, K.: Vehicle-to- vehicle ad hoc communication protocol evaluation using realistic simulation framework. In: Fourth Annual Conference on Wireless on Demand Network Systems and Services, WONS 2007 (2007)
6. Harouna Souley, A.-K., Cherkaoui, S.: Realistic Urban Scenarios Simulation for Ad Hoc Networks. In: The Second International Conference on Innovations in Information Technology, IIT 2005 (2005)

7. Glomosim, Global Mobile Information Systems Simulation Library, http://pcl.cs.ucla.edu/projects/glomosim/
8. OPNET Modeler Wireless Suite, http://www.opnet.com/solutions/network_rd/modeler_wireless.html
9. Hufford, G.: The ITS Irregular Terrain Model, http://flattop.its.bldrdoc.gov/itm.html
10. TIREM specifications, http://www.alionscience.com/Technologies/Wireless-Spectrum-Management/TIREM/Specifications
11. Cost 231 Walfisch- Ikegami, Cost Final Report, http://www.lx.it.pt/cost231/
12. NS-2 network simulator, http://nsnam.isi.edu/nsnam/index.php
13. Vuckovik, M., Filiposka, S., Trajanov, D.: Performances of Clustered Ad Hoc Networks on 3D Terrains. In: SIMUTools 2009, Rome, Italy (2009)
14. Filiposka, S., Trajanov, D.: Terrain Aware 3D Radio Propagation Model Extension for NS-2. Simulation Journal (January 2011)
15. Vuckovik, M., Filiposka, S., Trajanov, D.: Durkins Propagation Model Based on Triangular Irregular Network Terrain. In: Springer Book of Proceedings for ICT Innovations 2010, Ohrid, Macedonia (2010)
16. The Guide to Geographic Information System, http://www.gis.com
17. Harmon, J.E., Anderson, S.J.: The Design and Implementation of Geographic Information Systems. John Wiley & Sons (2003)
18. U.S. Geological Survey National Mapping Division: Part 1 General, Standards for Digital Elevation Models
19. U.S. Geological Survey National Mapping Division: Part 2 Specifications, Standards for Digital Elevation Models
20. Weibel, R., Heller, M.: Digital Terrain Modeling. In: Geographical Information Systems: Principles and Applications, pp. 269–297. Longman, London (1991)
21. Goodchild, M.F., Lee, J.: Coverage problems and visibility regions on topographic surfaces. Annals of Operations Research 18, 175–186 (1989)
22. Falcidieno, B., Spagnuolo, M.: A new method for the characterization of topographic surfaces. International Journal of Geographical Information Systems 5(4), 397–412 (1991)
23. Matori, A.N.: Low Cost DTM for Certain Engineering Purposes. In: Map Asia 2010 & ISG 2010, Kuala Lumpur, Malaysia, July 26-28 (2010)
24. Landserf, http://www.landserf.org/

Concept for Deploying Wireless in the Enterprise Infrastructure: Balancing Security and Positive QoE for the End-Users

Toni Malinovski and Vladimir Trajkovik

Abstract. This paper presents a concept for deploying wireless in the enterprise infrastructure, especially intended for companies that are usually considered conservative and have strict information security policies. These companies are usually careful while introducing new technologies, so our discussion presents a wireless concept that we believe can satisfy such requirements. The network will run business-critical application and data, voice and video applications, which will extend the wired network. The paper presents comparative analysis for the options that are currently available for wireless networks, while the claims are discussed through live implementation of the proposed concept in a financial institution with commercially available wireless products, produced results from technical point of view and quality of experience (QoE) for the end-users.

Keywords: Wireless, Enterprise Infrastructure, Security, 802.11i, Centralized Management, QoE, Concept.

1 Introduction

Wireless LAN technologies have become increasing popular while introducing mobility, simple deployment methods and easy access to the end-users to networking resources. Due to ease of installation in a home environment, possibility to offer internet access at airports, coffee shops and malls, this technology has spread as type of computer network, often seen as "LAN without cables", which offers most of the

Toni Malinovski · Vladimir Trajkovik
Faculty of Information Science and Computer Engineering,
Ss. Cyril and Methodius University, 1000 Skopje, Republic of Macedonia
e-mail: tmalin@nbrm.mk
trvlado@feit.ukim.edu.mk

L. Kocarev (Ed.): ICT Innovations 2011, AISC 150, pp. 95–105.
springerlink.com

possibilities of the wired network. The new-age wireless solutions focus on challenges that require complete extension of wired infrastructure, while continuing to offer mobility and flexibility to the end-users. The IEEE 802.11 standard, variants and alternatives, have played a major role in wireless popularity for nomadic access, point-to-point LAN bridges, ad-hoc networking, and even larger applications through internetworking.

Besides being present at home or public businesses, which offer wireless services to their customers, wireless LAN has extended the enterprise infrastructure of different companies, to avoid costly process of introducing cables into to a building, or as a connection between various equipment locations and mobility to their end-users.

Wireless network are more vulnerable and susceptible to outside attacks since anyone can intercept or divert the signal. Due to this nature of wireless, this technology has gone under a negative publicity when security is concerned and this notion has not changed dramatically in the past years. Home and public installation are not really much affected by this notion, but the enterprises, especially the conservative ones, that value their reputation and information security, can be skeptic to wireless deployment even these days.

In the early days, as main problem in wireless security, very few wireless LANs were implementing any form of security at all. Even more, most of the devices and solutions were supporting Wired Equivalent Privacy (WEP) [4] protection only, that was vulnerable to attack and intrusion, so more suitable security measures had to be introduced. Properly placed firewall in the whole infrastructure could mitigate the risk, but still this was not good enough to overcome the problem. An interim solution, WPA standard, provided improved encryption with stronger user-base authentication. The latest standard, WPA2 or IEEE 802.11i [2], ratified in 2004, encompasses the prior features, plus a number of security improvements [3] while introducing secure wireless possibilities to enterprise infrastructure. This standard is still not widely utilized, since its security mechanisms require new hardware resources and additional enhancements to the existing infrastructure. If not deployed properly, wireless will bring additional risk in the environment [1, 5, 6] that can compromise the security of the whole infrastructure. So new approaches and schemes have to be developed to provide guidelines, concepts and models for secure implementation of wireless in enterprise environments, while being able to satisfy the business requirements and provide positive quality of experience (QoE).

The main contribution of this paper is the presentation of concept for deploying a wireless in the enterprise infrastructure, by utilizing the latest standards and possibilities, while successfully reaching business objectives with satisfactory QoE for the end-user. The concept's claims will be discussed through gathered data and information from a live implementation of the proposed concept, in a financial institution with commercially available wireless products.

In section 2, the system infrastructure and concept for wireless deployment in enterprise infrastructure is presented, proof of concept and QoE measurements are given in section 3, discussion in section 4, while section 5 concludes the paper.

2 Concept for Wireless Deployment into Enterprise Infrastructure

Enterprise infrastructure consists of different layers that further divide the enterprise network into different physical, logical and functional parts, which should provide performance, availability and scalability to meet the business needs at the end. Modern conceptual models [7, 11] of the enterprise network integrate network resources and informational assets connect local and remote networks, branch offices and business partners, in a secure and productive manner. The wireless is extension of the wired network within the enterprise and should extend the intelligence built into each component of the network, to deliver services and applications to the end-users. Thus the whole network should be centrally managed, monitored and optimized, so it can promote true business transparency and agility.

This paper reviews the options that an enterprise may face in wireless deployment and provides a concept to securely deploy such solution. Such infrastructure will be position to meet the business objective and will run business-critical application and data, voice and video applications, as an addition to the wired network. The claims will be presented through findings in a real production environment. The QoE of the users will be evaluated and quantitative measures will be produced.

2.1 *The Concept and the Design of the System Infrastructure*

In order to develop a concept that can successfully guide the enterprise organizations into secure wireless deployment different aspect have to be considered. This paper proposes concept that balances the security measures, creates an abstraction from the transport infrastructure while fulfilling the business needs for productivity, efficiency and flexibility. The concept we propose consists of the following three components: end-users, wireless platform and enterprise infrastructure; and reviews the following aspects:

- Placement of each of the components within the infrastructure;
- Interaction between the components;
- Security measurements that need to be implemented;
- Provided services and applications.

The foundation of the wireless concept deployment begins while choosing the nature of the wireless access-point (AP) devices, which can help to decide how the elements will be placed between each other. These options can be generally sensitized in two approaches: infrastructure with individual APs or with depended APs from a centralized controller device. It is important to have scalable and next-generation architecture, so it is our impression that enterprise infrastructure solutions are definitely shifting towards centralized architectures that utilize wireless controllers. It has become clearly obvious that independent solutions, even though provide high speed connectivity, introduce problems in management, scalability and difficulties in overall support and integration. Our concept follows the depended AP architecture and presents the additional advantages that are introduced to the enterprise in general.

To ensure that wireless will extend the infrastructure and not compromise the existing security mechanisms, the correct placement of the controller and the AP within the infrastructure should be decided. While developing the concept, we have tested several simulation scenarios, which had prime objective to meet the business needs in a balance with security measures. The devices were placed in different points of the infrastructure: deep inside of the local network, within a separate VLAN, within separated zone and even outside of the local network. These simulations have shown us the advantages and the disadvantages of such approaches. Even though the security measures are important and have to meet the regulatory and compliance pressure, we believe that the placement is also decisive in correct concept for an enterprise infrastructure.

After evaluation of all the possible scenarios, we have come to conclusion that the wireless addition to the local network in an enterprise infrastructure should be placed within a secure zone, if possible protected by specific firewall, so additional measures can be implemented on the top of the wireless security mechanisms. This way the enterprise will minimize the risk that the wireless will introduce and mitigate the threats of the additional technology. Some may argue that this might not be necessary, but still we believe that the overall concept will benefit at the end and the QoE of the end-users won't be really degraded. A potential bridge in the security zone over the wireless platform will be localized, and the enterprise won't be fully compromised.

To meet the end-users requirements regarding business processes and to provide possible ideal transparency of the transport media, but still keep the current position, the concept should have the option for connection from the secured zone towards the main part of the enterprise network. So when the end-users gain access to the wireless, they should have means to reach enterprise resources in same manner as being in the wired network. By reviewing the interaction between the three mentioned components, we have come to the conclusion that this can be easily provided with controlled approach. The end-users can create a VPN tunnel towards the inside of the network, or transparent connection through gateways for special devices can be applied. Most of the companies may already have a remote access solution through VPN, so it can be easily reflected on the wireless region. Even more, after the association to the wireless infrastructure, the end-user may gain internet access and through VPN tunnel can reach all the necessary resources as being in the wired network. The standard network admission controls can be utilized while following such concept and the enterprise overall security policy can be met without introducing additional risk.

Having all this in consideration, Figure 1 shows a design that follows our proposed concept for deployment of the wireless network in the enterprise infrastructure.

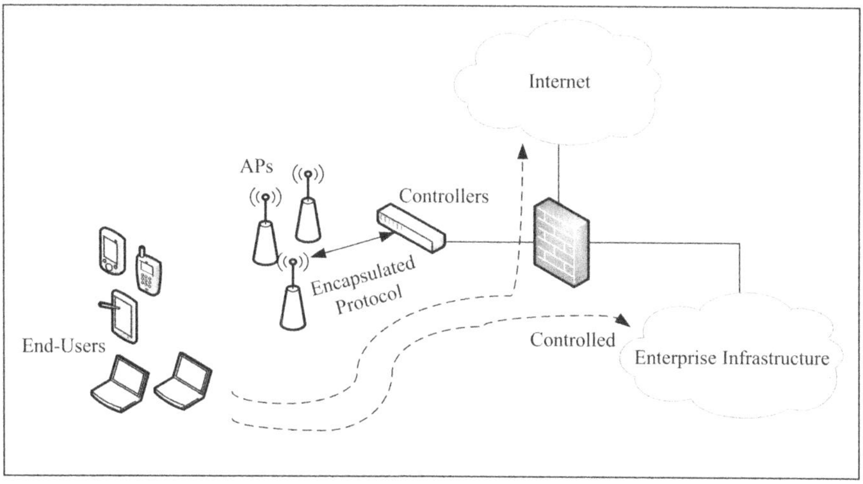

Fig. 1 Design of the system infrastructure following the proposed concept for wireless deployment

The centralized approach offers predictive and controllable behavior of the infrastructure, which splits the processing of the 802.11 protocol between two devices: the APs and the centralized controller. These devices can be in different domains, so this approach can introduce even stronger security for the centralized device. The concept provides high availability and redundancy since the centralized controller device can be easily doubled, while the APs provide self-healing network themselves. AP failure will be detected by the controller, which will boost the signal of the neighboring AP to provide additional coverage. Different SSIDs for corporate users with VPN access, guest users, VoIP and video devices can be deployed in a centralized manner, so the whole infrastructure can be aligned for traffic segmentation. Because the whole communication goes from the APs to the controller, rogue (unauthorized) AP and rogue clients can be easily detected and automatically blocked from the enterprise infrastructure.

Security measures should be part of every concept, so we follow the 802.11i standard [2] for data protection and strong user authentication. Through several simulations, different security methods were evaluated and tested. Following the best practices on this topic and results from our tests, we have come to conclusion that the concept for wireless deployment in modern enterprise infrastructure should always utilize the latest protocols for data protection and encryption (such as AES-CCM) [8] and the more advanced variance of 802.1x user authentication protocol [9] for digital identity. The enterprise that considers wireless deployment should have centralized authentication server for user authentication and Public Key Infrastructure (PKI). We suggest using certificates for enterprise users and devices (EAP-TLS) [10], while allowing access to the wireless infrastructure. By deploying certificate to company assets and devices, the enterprise will distinguish its property and provide access accordingly. Furthermore the same digital identifications can be used in VPN tunnel creation and end-to-end connectivity. Guest and visitors can be supplied with certificates too, or PEAP authentication (server-side certificate only) with usernames and passwords can be introduced for this purpose. Either way, the enterprise will have centralized solution, controlled, logged and monitored, that can be easily integrated with the existing monitoring and logging solutions.

Figure 2 represents the chosen security authentication and data protection mechanisms in our concept.

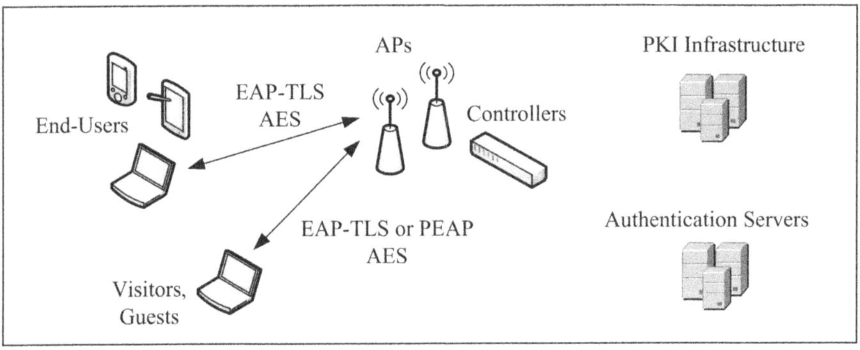

Fig. 2 Security mechanisms for authentication and data protection

The proposed concept can be realized through various different approaches, but its final goal is to provide possible transparency from the transport infrastructure and deliver all the necessary services and applications to the end-users.

2.2 Wireless Platform as a Service in the Proposed Concept

The end-users expectations from the IT technology in every enterprise organization can be generalized and simplified in few statements: applications and services should be available as need, security should be incorporated into the system so the risks are always minimized and business objectives and goals should always be meet at the end. The introduction of the wireless platform in the infrastructure will bring additional mobility, flexibility and efficiency, but the current expectations that came from the wired infrastructure, should not be compromised. We believe that the proposed concept could meet these expectations. By carefully choosing its main components and their interaction, the concept brings the wireless platform securely to the users.

The proposed concept provides secure integration, while the end-users get complete on-demand services from the infrastructure. The end-users can connect through the wireless network and the VPN tunnel using the same digital identity, and have access to all the resources in the system. Even though VPN is proposed, while working in the office environment, the process can be fully automated, so the end-users won't notice any difference in acquired services, while connecting through the wired/wireless network. In [1, 17, 18] different measures are presented towards tighter security for wireless, but the necessity for VPN tunnel towards enterprise resources in this case, provides different approach, with already proven security measures. The centralized concept for wireless architecture based on the new 802.11n standard provides broad access, while the infrastructure maintenance is minimized.

This concept is positively positioned towards the integration of different media - data, video, and voice, which are already widely consolidated into a converged network creating single, standard base, modular infrastructure. The split design of the wireless depended platform and constant AP and controller communication, allows extension of the current QoS schemes into this infrastructure. Most of the commercially available wireless products with centralized approach are able to follow the marked (colored) packets from the wired network in the wireless platform, thus providing desired QoS for the wireless users and devices. We believe that this concept fully supports the evolution towards new multimedia systems and services, where user requirements are not limited to requirements on connectivity.

Furthermore, the proposed concept provides possibilities for development of stronger security measures, which traditional enterprise organizations may require. The centralized controller collects all the information from every transmission, so additional location base restrictions can be introduced. Most of the vendors that deliver such solutions have interface through SOAP or plain XML, for third party applications, which can introduce location based restrictions [12] of the allowed users in the enterprise infrastructure. This location based restriction can be measured within few meters from the enterprise physical perimeters, so the threats can be really minimized.

3 Proof of Concept and QoE for the End-Users

To support our claims, the proposed concept is used in a deployment of wireless platform into an enterprise organization. Information and data gathered through this deployment will be discussed and results will be evaluated. The proposed enterprise organization is a financial institution, which has solid infrastructure, information security policy, compliance with international standard and best practices, and structured and developed IT governance. This institution has around 400 employees which are situated in several buildings with close approximate to each other. These employees represent office workers mostly, which are performing their business activities from working desks, but need certain mobility when attending meetings, move around the premises etc. This enterprise needs to introduce flexibility, increase productivity and mobility by deploying the wireless platform, so the employees may use their laptops, mobile devices and phones, voice and video services, for their business processes over the new infrastructure in same manner as they are working in the wired network. Being an important financial institution, security and proper deployed new technologies, are at utmost importance to this enterprise organization.

The wireless was deployed in this enterprise infrastructure, following the proposed concept and the interaction between its three components: end-users, wireless platform and enterprise infrastructure. The wireless platform was placed accordingly and the proposed security measures were adopted and implemented.

We used descriptive analysis to evaluate the results and produce quantitative measures for this proof of concept. The data was gathered through a three months period, after the wireless platform was put in production. One of the sources for information was the enterprise's service desk, which could give us actual numbers for this time period, regarding the service requests for requesting access and usage of the wireless platform; the end-users' reported problems that were recorded while using the wireless platform; and the number of security incidents that were related to this new technology. Figure 3 shows gathered results for this time period.

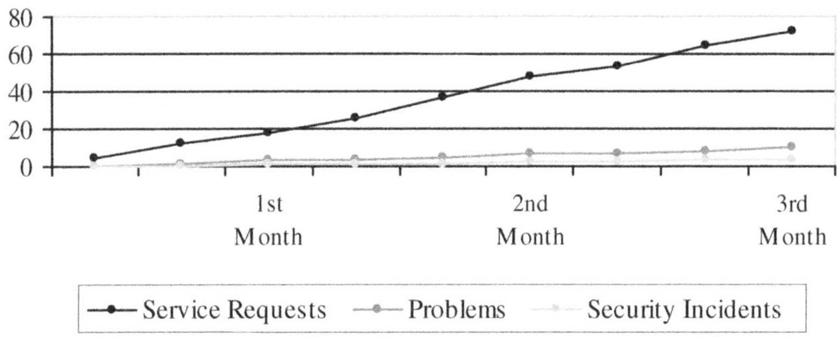

Fig. 3 Results gathered from the service desk regarding the wireless platform

The results show that employees were interested in this new technology and have made requests, which are recorded as a service request, to gain access to the wireless platform. The interest was constant, even slightly increased of the evaluation period. This enterprise has proper procedures and policies for approval, access and rights management, which leads us to positive conclusion that this technology was quickly spread to the company as a result of actual business needs, and not just an employees' wish to try something new. Through the results, the concept can be also evaluated for user friendliness and possibilities to deliver business applications and services to the end-users. We can note that only few problems were recorded, which mostly came from the adaptation of the end-users to the introduced technology. The security incidents number was low, mostly from surrounding rogues access-point and clients, who were testing an access to this newly introduced wireless platform in the area. There weren't any security incidents which were reported as potential bridge of security or unauthorized access to the infrastructure. Therefore, these numbers show that the proposed concept provided secure wireless deployment in the enterprise infrastructure.

To further study the QoE for the end-users, while using the proposed concept wireless platform, additional quantitative path analysis was taken. The end-users were educated through presentations and prepared user-guides, how to use the new technology in their business activities, while utilizing the standard services and applications through the new mobile devices and laptops. Throughout this process of education and training, end-users discussions were initiated and feedback was provided. To provide quantitative measures, that can specify if this specific system can reach the business objective, and will take into account the experience and the expectation of the end-user, a survey that measures QoE using 6-item scale was conducted among all wireless end-users after the three months period. Graded by the end-users and therefore, somewhat subjective, each of the items was measured using five-point scale, ranging from 1 as "strongly disagree" to 5 as "strongly agree". Table 1 shows the items and mean score from the obtained results.

Table 1. QoE grades from the wireless survey.

Study Variable	Mean Score
The connection to the wireless network was easy	4.74
I was able to use additional mobile devices	4.47
The wireless platform increased productivity and flexibility	4.56
I was able to use all the necessary applications	4.81
Wireless service was always available	4.70
In general, how would you rate your experience	4,67

These results show different aspect of the wireless deployment concept such as quality of effectiveness, usability, availability and expectations of the end-users. These measurements show positive QoE from the end-users in general. This type of analysis could be used as input for optimization purposes of the proposed concept, based on the predicted impact of specific QoE parameters on an end-users' experience.

4 Discussion

Efficient and on time delivery of the information to the relevant recipients is decisive competitive success factor for every enterprise organization. Introducing mobility within enterprise organization is no longer at the fringe of discussions and is a part of strategic agendas of most of the companies. But mobility projects raise concerns in enterprises about costs, security and reliability, especially as they start to enhance their enterprise applications and infrastructure.

By covering different aspect and reviewing possible solutions, we have designed a concept, which emphasizes the components and their interaction, while deploying wireless infrastructure into an enterprise environment. The concept was designed to provide necessary security measures, deliver application and services, but also positive QoE for the end-users. In our opinion, this research has shown that the proposed concept provides desired balance between the security measures and positive experience of the end-user, while supporting business processes which relay on different applications running within the network infrastructure.

The centralized approach of the wireless concept provides field for research of models and solutions, for location based restrictions [12] within the wireless platform integrated in the enterprise infrastructure, which will increase the security and minimize the risk of potential threats and attacks.

The results of statistics indicate that there was moderate to high level of satisfaction and effectiveness as perceived by the end-users. Through usage of statistical tools of principal component analysis [13, 14] and the discussed parameters additional results can be produced for end-to-end perceived QoE.

Further research based on this concept can be conducted. Larger mesh networks positioned following the concept can be reviewed, and key technical aspects such as topology creation, routing, medium access control, security, quality of service, and power efficiency can be evaluated [15]. In our opinion this concept is also well positioned for emerging applications like audio/videoconferencing and collaboration, which require simultaneous communication between groups of users [16] and can provide advantages in solving major challenges for a multicast client in a wireless environment to maintain its multicast group membership when moved in different locations.

5 Conclusion

This paper presents concept that can help enterprise organizations when planning and deploying wireless platform into their infrastructure. It distinguishes the main components, their interaction and behavior. Following this concept a wireless solution is deployed in a financial enterprise institution, while claims and results are being discussed. Quantitative measures are provided, which take into account the experience and the expectation of the end-users (QoE), their ability to utilize the new technology which was deployed following the concept, and overall positive experience of the consumers accessing and using the provided service.

In our future work, we plan to research the behavior of the proposed concept for new multimedia systems and services within the wireless platform, the possible limitations due to the characteristics of the transmission channel and QoE evaluation that enables user centric design of novel multimedia wireless systems.

References

1. Martinovic, I., Zdarsky, F., Bachorek, A., Jung, C., Schmitt, J.: Phishing in the Wireless: Implementation and Analysis. In: Venter, H., Eior, Y.M., Labuschagne, L., Eloff, J., von Solms, R. (eds.) New Approaches for Security, Privacy and Trust in Complex Environments. IFIP International Federation for Information Processing, vol. 232, pp. 145–156. Springer, Boston (2007)
2. IEEE P802.11i/D10.0, Medium Access Control (MAC) Security Enhancements, Amendment 6 to IEEE Standard for Local and Metropolitan Area Networks part 11: Wireless Medium Access Control (MAC) and Physical Layer (PHY) Specifications (2004)
3. Lashkari, A., Danesh, M., Samadi, B.: A Survey on Wireless Security Protocols (WEP, WPA and WPA2/802.11i). In: 2nd IEEE International Conference on Computer Science and Information Technology, pp. 48–52 (2009)

4. IEEE P802.11 D5.0, IEEE Standard for Local and Metropolitan Area Networks: Wireless LAN Medium Access Control (MAC) and Physical Layer (PHY) Specifications. IEEE Standards Department (1996)
5. Bicakc, K., Tavli, B.: Denial-of-Service Attacks and Countermeasures in IEEE 802.11 Wireless Networks. Computer Standards & Interfaces 31(5), 931–941 (2009), Specification, Standards and Information Management for Distributed Systems
6. He, C., Mitchell, J.: Analysis of the 802.11i 4-way Handshake. In: Proceedings of the 2004 ACM Workshop on Wireless Security., ACM Press, Phil. (2004)
7. Frank, U.: Multi-perspective Enterprise Modeling (memo) – Conceptual Framework and Modeling Languages. In: Proceedings of 35th Hawai International Conference on System Sciences (2002)
8. Advanced Encryption Standard (AES), FIPS PUB 197 (November 2001)
9. Chen, J.C., Wang, Y.P.: Extensible Authentication Protocol (EAP) and IEEE 802.1x: Tutorial and Empirical Experience. IEEE Communications Magazine 43(12), 26–32 (2005)
10. The EAP-TLS Authentication Protocol. IETF RFC 5216 (March 2008), `http://tools.ietf.org/html/rfc5216`
11. Närman, P., Schönherr, M., Johnson, P., Ekstedt, M., Chenine, M.: Using Enterprise Architecture Models for System Quality Analysis. In: 12th International IEEE Enterprise Distributed Object Computing Conference, pp. 14–23 (2008)
12. Chen, M.X., Wang, Y.D.: An Efficient Location Tracking Structure for Wireless Sensor Networks. Computer Communications 32(13-14), 1495–1504 (2009)
13. Jolliffe, I.T.: Principal Component Analysis, p. 487. Springer (1986)
14. Krzanowski, W.J.: Principles of Multivariate Analysis. Clarendon Press, Oxford (1988)
15. Faccin, S.M., Wijting, C., Kenckt, J., Damle, A.: Mesh WLAN networks: concept and system design. IEEE Wireless Communications 13(2), 10–17 (2006)
16. Silva, J.S., Camilo, T., Pinto, P., Ruivo, R., Rodrigues, A., Gaudêncio, F., Boavida, F.: Multicast and IP Multicast Support in Wireless Sensor Networks. Journal of Networks 3(3) (2008)
17. Aswall, M.S., Rawat, P., Kumar, T.: Threats and Vulnerabilities in Wireless Mesh Networks. International Journal of Recent Trends in Engineering 2(4) (2009)
18. Moshaddique, A.A., Jingwei, L., Kyungsup, K.: Security and Privacy Issues in Wireless Sensor Networks for Healthcare Applications. Journal of Medical Systems 0148-5598, 1–9 (2009), doi:10.1007/s10916-010-9449-4

An Energy-Efficient Approach for Time-Space Localization in Wireless Sensor Networks

Ustijana Rechkoska Shikoska and Danco Davcev

Abstract. Space and time play a crucial role in wireless sensor networks, since sensor nodes are used to collaboratively monitor physical phenomena and their space-time properties. Wireless sensor networks (WSN) are envisioned to be used to fulfill complex monitoring tasks. A number of techniques and distributed algorithms for location estimation and time synchronization have been developed specifically for sensor networks. A similarity of time and space affects the location estimation and time synchronization, ranging from applications and requirements to basic approaches and concrete algorithmic techniques. An original approach for space and time localization in WSNs is given in this paper, also an energy-efficient approach is given, by clustering the WSNs. In both algorithms, we use only one mobile beacon for both localization and synchronization, a sensor node which moves around the sensor's field, aware of its time and position, equipped with a GPS receiver. The synchronization component uses the packets required by the positioning component to improve its performance. The positioning component gets use from the communication, required by the synchronization component to decrease errors. A set of simulations are presented to evaluate the performance of our algorithms in order to reduce communication and processing resources and save energy and network resources.

Keywords: wireless sensor networks, space-time localization, a mobile beacon, GPS, energy-efficiency.

1 Introduction

Enabled by technological advancements in wireless communications and embedded computing, wireless sensor networks (WSNs) [1, 2] were first considered for military applications, where large-scale wireless networks of autonomous sensor

---

Ustijana Rechkoska Shikoska · Danco Davcev
University for Information Science & Technology,
"St. Paul The Apostle" – Ohrid, Republic of Macedonia
e-mail: {ustijana.r.sikoska,dancho.davchev}@uist.edu.mk

L. Kocarev (Ed.): ICT Innovations 2011, AISC 150, pp. 107–118.
springerlink.com

nodes would enable the unobtrusive observation of events in the real-world. Since then, the use of sensor networks [4] has also been considered for various civil application domains. The categories time and location [3, 5] are fundamental for many applications of sensor networks, due to the close integration of sensor networks with the real world.

In this paper, new algorithm for solving this problem using a mobile beacon is proposed. A mobile beacon is a node that is aware of its time and position (e.g. equipped with a GPS receiver) and that has the ability to move around the sensor field. This beacon can be a human operator, an unmanned vehicle, an aircraft, or a robot. A mobile beacon has been successfully applied to solve the positioning problem. The proposed time-space localization algorithm can synchronize nodes by using the packet delay measurement [10, 11]. The algorithm is implemented on different network topologies.

In the next section, the related work, considering the research field is presented. Section 3 presents wireless sensor networks, their performances, as well as an overview and definition of the positioning and synchronization problems in WSNs. Also, WSNs topologies are given in this section. In section 4, the proposed energy-efficient algorithm for space-time localization is given, which is evaluated in section 5. Section 6 gives the conclusion and future work.

2 Related Work

There are different ways of classifying space - time localization algorithms, they can be classified according to the measurement assumptions: 1) connectivity-only; 2) range-based; 3) angle-based; 4) hybrid. A number of localization methods rely on connectivity information only. These types of methods are also referred to as "range-free" methods. The Centroid method [4] estimates the location of an unknown node as the average of its neighbors' locations. The DV-Hop method [6] counts the hop numbers to beacons and uses them as crude estimates for distances. Range-free methods require no additional hardware, but they generally only work well when networks are dense. Range-based methods include the Ad Hoc Positioning System (APS) methods such as DV-Distance and Euclidean proposed in [6, 7]. In [9], ranging data are exchanged between the neighbors to refine the initial location guess. While those methods compute the absolute node locations, the GPS-Free method [10] calculates the relative node locations from the distance measurements. Compared to range-free methods, range-based methods give more accurate location estimates when ranging data is reliable.

Some recent researches [15, 16] have proposed the use of mobile beacons to assist the nodes of a WSN in estimating their positions. The localization algorithm proposed by Sichitiu and Ramadurai [16] uses mobile beacons to allow nodes to compute their positions by using range-based distance estimations. In Pathirana et al. [17], the mobile beacon itself computes the positions, instead of the nodes. Mobile beacons have also been used to localize nodes in underwater WSNs.

In the Flooding Time Synchronization Protocol (FTSP) [11], all delays in the packet transfer time are computed to synchronize precisely both sender and receiver using only one broadcast. An accuracy of about *1.5* μs is reported. A similar algorithm is Delay Measurement Time Synchronization (DMTS) [12], which reports a single hop accuracy between *2 μs* and *32 μs*.

GPS (Global Positioning System) is a good example of a time-space localization system that can both localize and synchronize sensor nodes; however, to equip all the sensors in a WSN with a GPS receiver is not a good solution because it increases their cost, size and energy consumption [12]. The Synapse algorithm is proposed [14], a time-space localization algorithm that computes the average multi-hop time of a packet to synchronize the nodes. Also, the Lightness algorithm [15] had been proposed, a novel and lightweight time-space localization algorithm that is able to localize and synchronize all nodes in a WSN with the communication cost of a single flooding.

In this paper, improved, energy-efficient algorithm for localization in time-space, using a mobile beacon, by clustering the WSNs is proposed to solve this problem.

3 Wireless Sensor Networks (WSNs)

Wireless sensor networks consist of sensor nodes, computing devices that include a power source, a transceiver for wireless communication, a processor, memory, sensors, and actuators. Although the exact properties and capabilities of these components may vary, a common property of sensor nodes is their resource scarcity.

The output of a sensor network may be used to control the operations without human intervention by enabling/disabling sensors, or by controlling operation parameters of sensors (sampling rate, sensitivity, orientation, position).

The characteristics of wireless sensor networks can present a number of major challenges to the development of algorithms, protocols and systems [1]. The main technical challenges are resource and energy constraints, network dynamics, network size and density, unattended and un-tethered operation. It is important to ensure that resource usage and energy consumption are equally spread among the nodes of the network.

The Positioning Problem - to define the positioning problem in wireless sensor networks, we take a WSN composed of p nodes, with a communication range of c units, and distributed in a two-dimensional squared sensor field $S = [0, q] \times [0, q]$.The network is presented by a graph $P = (L,M)$ with $L = \{l_1, l_2, . . . ,l_n\}$ like a set of sensor nodes; $h_i, j_i \in M$ *iff* l_i reaches l_j , i.e. the distance between; l_i and l_j is smaller than c; $u(e) \leq c$ is the weight of edge $e = h_i, j_i$, i.e. the distance between l_i and l_j. We also consider only two dimensions for a node's location, but the methods presented herein can be easily extended to accommodate three dimensions. Therefore, P is Euclidean graph in which every node has a coordinate $(x_i, y_i) \in R^2$ in a two-dimensional space, which represents the location of node i in S. The unknown node is a node that doesn't know its position; settled node is a node which was initially an unknown node, but has managed to estimate its position by using a positioning system. Also, there is a beacon node or anchor, which is always aware of its physical position and helps to locate other nodes. Its position is obtained by manual placement or by external means such as a GPS receiver. This node forms the basis for most positioning systems in WSNs. The positioning problem is stated as finding the position of as many unknown nodes as possible, referring to a given multi-hop network, represented by a graph, and a set of beacon nodes and their positions.

An advantage of this mobile beacon approach is that the nodes' positions are computed based on the same node - mobile beacon, keeping the mean localization error low and preventing the propagation of this error. It avoids the use of nodes equipped with GPS, except for the mobile beacon. An important aspect that influences the position estimates is the trajectory of the mobile beacon. The less rectilinear the trajectory, the better the estimates will be. Two possible trajectories are evaluated in this work: sinusoidal and spiral trajectory.

Time synchronization - the significance of physical time for sensor networks has been reflected by the development of a number of time synchronization algorithms in the recent past. Most computer systems in use today are based on clocked circuits and hence contain so-called digital clocks. Such hardware clocks are a valuable tool for time synchronization, since they can be used to maintain synchronization over time [2]. In this work, the actual time of the network is considered, in which the nodes must be synchronized (e.g., UTC), is represented simply by t. The hardware clock of node i is defined as $t_i(t)$, since it is a monotonically non-decreasing function of t. Because no hardware clock is perfect, $t_i(t)$ has two components: $t_i(t) = d_i t + o_i$, where o_i is the offset, i.e., the difference between t and t_i at that instant, and d_i is the drift, i.e., how the local clock gradually deviates from t due to conditions such as temperature or battery voltage. The synchronization problem is stated in finding the offset and drift (o_i, d_i) of as many unsynchronized nodes $i \in N$ as possible, by a given multi-hop network, represented by a graph, a set of beacon nodes and their synchronized clocks. In the packet delay measurement, all delays in the packet transfer time are estimated to synchronize precisely both sender and receiver by using only one packet. The packet delay measurement synchronization technique has been used in a number of synchronization protocols for WSNs [11, 12].

Localization in Time and Space - in a wireless sensor network, most of the applications that require *position information* also require *time information* [3]. Number of similarities can be identified [13, 14]. The need of both time and space information and the similarities between them, have shown the importance of combining these two problems into a single one: *localization in time and space*. Doing so, *energy* and *network resources* could be saved, also the opportunity to improve time and position estimations in contrast to the scenarios in which these problems are solved separately is given. Synchronization algorithms can take advantage of the greater number of beacon nodes required by the positioning algorithms, while positioning algorithms can take advantage of the techniques and additional communication resources used to synchronize nodes.

The time-space localization problem can be stated as finding the position (x_u, y_u) and time $t_u(t)$ for all unsynchronized and unknown nodes, by a given multi-hop network, represented by a graph and a set of beacon nodes, their positions (x_b, y_b) and synchronized clocks $t_b = t$, where u and b belong to the sets of unknown nodes and beacon nodes, respectfully.

Several *wireless sensor networks topologies* are considered in our work and they are presented in Figure *1*. The first network topology presented here is C – random network topology (Figure *1 (1)*). Together with the presented H – random topology (Figure *1 (4)*) are irregular network topologies, as well as the random topology, shown in Figure *1 (5)*.

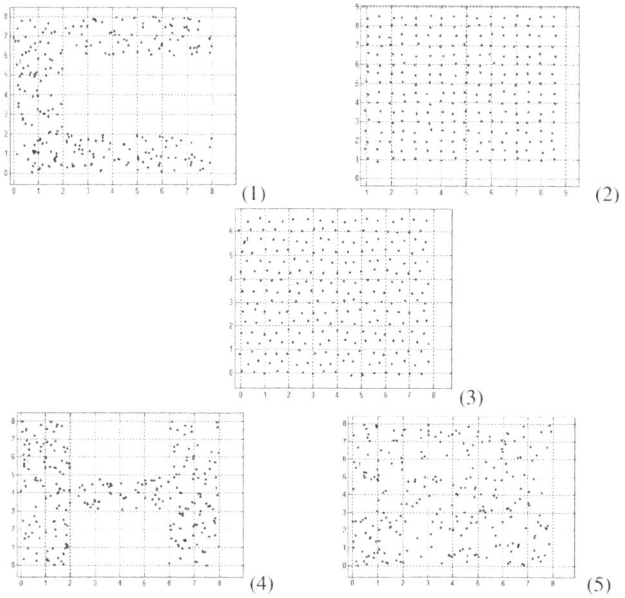

Fig. 1 Network topologies - (1) C – random topology (2) Disturbed grid topology (3) Disturbed hexagonal topology (4) H – random topology (5) Random topology.

C and H – random topologies are random topologies, but irregular once, because only a part of the square of the surface is considered. Disturbed grid topology and disturbed hexagonal topology shown in Figure *1* are more regular network topologies than previous mentioned ones.

4 Energy-Efficient Solution for Time-Space Localization in WSNs

Clustering the WSNs

There are several techniques of clustering in WSNs. Clustering may be done by broadcasting control information or using localization with GPS-enabled nodes. Base station is the part of the network where all source nodes send all the information or data, directly or multi-hop manner, in a particular network. Cluster-head is a node of a particular cluster, where all other nodes of that cluster send all their information when required. Large number of sensor nodes form a sensor network, which works as a source, i.e., senses and sends. Gateway is a node through which one cluster-head interacts with other cluster-head.

In the cluster construction phase, the remaining energy levels of each sensor node and its position i.e., the location of each sensor node are considered. Each sensor node sends messages to its neighbor nodes which includes sensor ID, remaining energy and the location.

Each mobile host (MH) has unique identifier (MH_ID), which is a positive integer. MH also keeps information about the identity of their assigned cluster head (CH_ID). Control information is communicated through packet messages.

Another attribute of MHs is battery power level (percentage of remaining battery power), which is a positive integer. Cluster head consumes more energy than ordinary MHs as they serve a number of MHs for forwarding messages [20].

Both location and power information to partition of wireless sensor network in to separate clusters are considered. Each packet message consists its own identity (MH_ID), i.e., sender identifications, its cluster head ID, (CH_ID), i.e., destination ID. To form the cluster, location information is needed for each node. The location information is measured by different node localization techniques.

There are several range measurement methods. Another technique utilizes *Received Signal Strength (RSS)* measurements, which is available in most of the current transceivers. The transmitted power is included in beacon message while the *RSS* is calculated during the reception of each message. Every unknown node stores this data and uses it to calculate the range/distance from assisting nodes.

In the present condition every node knows their exact location, coordinate and remaining energy of their neighbor nodes also. Using this information clusters with cluster-heads can be formed. In a cluster the radius of a cluster i.e., the farthest a node inside a cluster can be from the cluster-head is fixed and is called the cluster range (*r*). Since cluster-heads forward data collected from nodes inside its cluster to the base station, a cluster-head consumes more energy than any other node. Hence more nodes inside a cluster will consume greater energy of cluster-head, so MHs far away from cluster head consume more energy. The radius of a cluster may be so large that energy consumption becomes an important factor due to the distance between the cluster head and the nodes away from it. So, an optimal distance d between the cluster head and a node is considered, in which energy consumption is optimal, as well as number of clusters is optimal.

To transmit a *k*-bit message to distance *d,* the radio expands the following energy:

$$\mathrm{ETx(k,d) = ETx\text{-}e(k)+ETx\text{-}a(k,d)} \tag{1}$$

ETx-e is the transmitter energy and *ETx-a* is the amplifier energy. The transmitter energy *ETx-e* is assumed to be the same as the receiver energy and its value is *50 nJ/bit*. Amplifier energy *ETx-a* is *100 pJ/bit/m*^{2} [19].

An Energy-Efficient Algorithm for Time-Space Localization, Using a Mobile Beacon

In our work, an original approach for time space localization, for different network topologies, is presented in this section. Similar work has been done in [5]. The positioning component is essentially the Mobile beacon localization algorithm [15]. The synchronization component obeys the same principle as the positioning algorithm, but it is extended to deal with time estimations. Then, these two components are combined into a single algorithm. The delay measurement technique is used to synchronize nodes. Different network topologies are employed, generated by our algorithm.

In this algorithm, once nodes are deployed, the mobile beacon travels through the sensor field broadcasting messages that contain its current coordinates and timestamp. When an unknown node receives a packet from the mobile beacon, it can estimate the packet travel time and, based on the times tamp stored in the packet, its own offset. Also, when an unknown node receives more than three messages from the mobile beacon, it can estimate its position based on the received coordinates and on the *RSSI* distance estimates. Since the nodes will require at least three packets for positioning, synchronization can be improved by computing the average offset of all packets. First, set of position information is given as a variable, later, set of received timestamps follows. Timer to send packets is set; also position and time information through variables is given. After that, the nodes' GPS info is returned, packets' travel distance estimation is given and nodes' position is computed. Later, packets' travel delay estimation is given and also the nodes' offset is computed. Case if this node is a beacon node is explored and number of references is tested, if there are three received messages from mobile beacon, it confirms enough references and proceeds with the algorithm.

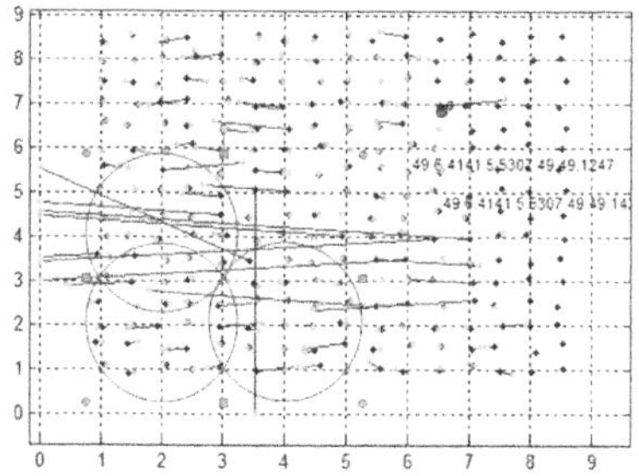

Fig. 2 The running application of the algorithm, for clustered WSNs

The running application of this algorithm is presented in figure 2. The red color point is the mobile beacon which is moving through the sensor network, marked with blue marks here. The options for choosing five types of sensor networks is put here whether the user wants to choose grid oriented network, randomly chosen network, hexagonal network topology, or H and C random network topologies, presented in the previous section. Also, information about the parameters used and display options are given and choice by the user, according to the parameters and network topology, is enabled.

5 Evaluations

The evaluation of our algorithm is done, by performing simulations, parameters used and their values are given in the next table.

Table 1. Simulation parameters and their values

Parameter	Value
Sensor field	92×92 m^2
Number of nodes	256 nodes (different network topologies)
Density	0.03 nodes/m^2
Communication range	15 m
Number of beacons	1 - mobile beacon
RSSI inaccuracy	10 % of communication range

We did an experiment within the sensor field *92 x 92 m*^{2}. *256* nodes are employed, for different network topologies, we have chosen five types of network topology, explained previously, for this evaluation. The density of a sensor network is picked to be *0,03 nodes/m*^{2}, and the communication range is *15 m*.

The evaluation is done by taking two types of trajectories for the mobile beacon: sinusoidal and spiral trajectories, presented in figure *3*.

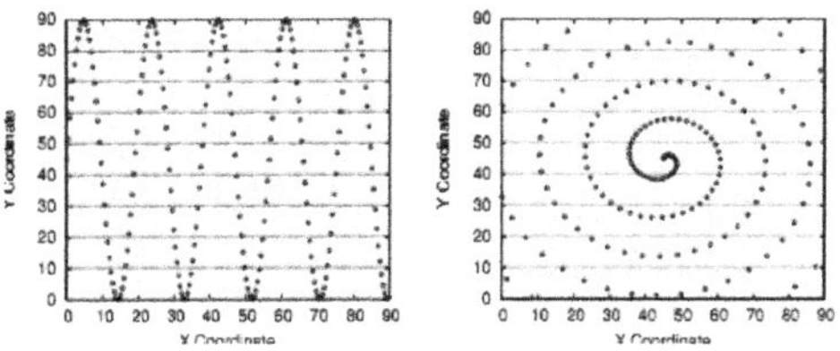

a) b)

Fig. 3 Evaluated trajectories for the mobile beacon: a) sinusoidal; b) spiral trajectory.

In the next figure, localization error, synchronization error and also impact of *RSSI* inaccuracy are presented, for two evaluated trajectories of the beacon, sinusoidal an spiral ones, and for five types of network topologies: C – random topology, disturbed grid topology, disturbed hexagonal topology, H – random topology and random topology.

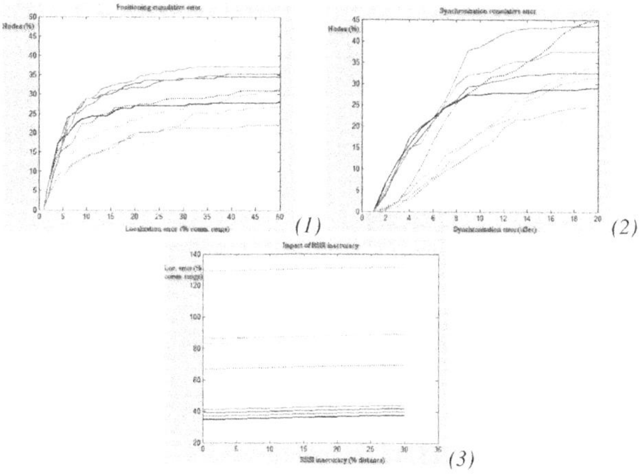

Fig. 4 *(1)* Positioning cumulative error *(2)* Synchronization cumulative error *(3)* Impact of *RSSI* inaccuracy, for clustered WSNs

Localization Error - the distribution of position errors among the sensor nodes is depicted in Figure *4 (1)*. The cumulative error identifies the percentage of nodes (*y*-axis), with a positioning error, smaller than a parameterized value (*x*-axis). A sharp curve means that the majority of nodes have small error. This graph also shows that spiral trajectories result in better positioning than sinusoidal ones, since these trajectories are less rectilinear.

Synchronization Error - the distribution of synchronization errors among the sensor nodes is depicted in Figure *4 (2)*. In this case, the cumulative error identifies the percentage of nodes (*y*-axis) with a synchronization error smaller than a parameterized value (*x*-axis). Again, a sharp curve means that the majority of nodes have a small error.

Impact of *RSSI* Inaccuracy – since the fact that the distance estimations using *RSSI* measurements are not accurate, depending on the environment, such an inaccuracy may lead to greater errors in the estimated positions. The evaluation of this impact is done by adding some noise to the real distances. This noise is generated by a normal distribution, with the actual distance as the mean and a percentage of this distance as the standard deviation. The comparison between the increase of the standard deviation of the normal distribution (used to simulate the noise) and the actual distance for the algorithms is presented on Figure *4(3)*. It can be noticed that the positioning part takes advantage of the synchronization part to improve its performance, which shows the significance of solving both positioning and synchronization problems at the same time.

The impact of clusters in sensor networks over the localization error, as well as on the synchronization error is presented in figure *5 (1)* and *(2)*, respectively.

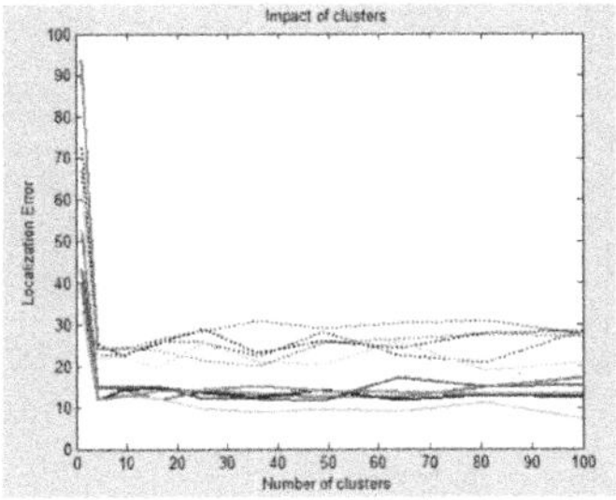

(1)

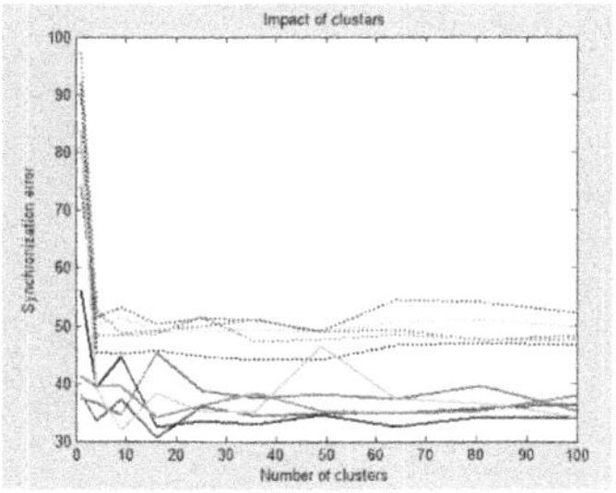

(2)

Fig. 5 (1) Impact of clusters over the localization error; (2) Impact of clusters over the synchronization error

The localization error, while implementing the algorithm, improves with increased number of clusters in the sensor network, which confirms the accuracy improvement here. Different colored curves are shown, for two evaluated mobile beacon trajectories, sinusoidal a spiral one, and for five types of network topologies: *C* – random topology, disturbed grid topology, disturbed hexagonal topology, *H* – random topology and random topology. Better results are gained for *C*-random network topology. Other network topologies follow, and they are improved as it is shown in this figure. It can be noticed that after the *80-th* clustered the results performs better, which again shows the importance of clustering the sensor network.

In figure *5 (2)*, the impact of clusters in the sensor network over the synchronization error is presented.

Different colored curves are shown, for two evaluated mobile beacon trajectories, sinusoidal a spiral one, and for five types of network topologies. The synchronization error improves with the increased number of clusters in the sensor network.

Better results are performed for *H*-random network topology, for spiral mobile beacon trajectory, also other topologies follows, with improved synchronization performing, as it is presented in the figure.

Also, one more comment can be made here which infers the impact of the clusters in the sensor network, and maybe that is the novel issue to talk about. With increasing the number of clusters, proposed algorithms achieve better results, reffering both localization and synchronization error.

When no cluster is formed, energy consumption is very large. When cluster is formed then energy consumption reduces according to the number of cluster. With increasing the number of clusters, the energy consumption increases slightly, but anyway for energy efficient algorithms like proposed ones it is very important result to be implemented, with an optimal number of clusters in the network, according to the optimizing methods. Implementing these solutions, at particular point energy consumption is minimum for a particular number of cluster. This is an important point for improving the energy efficiency in time-space localization for WSNs and for prolonging the sensor networks lifetime.

6 Conclusion

Due to the close integration of sensor networks with the real world, the categories time and location are fundamental for many applications of sensor networks, to interpret sensing results or for coordination among sensor nodes. Existing solutions for time synchronization and node localization do not cover all parts of space and time in wireless sensor networks problem.

In this paper, we have proposed and evaluated an original, energy-efficient algorithm for the time-space localization, a mobile beacon approach, for different network topologies, by clustering sensor networks. Using a mobile beacon, all sensor nodes are able to localize themselves both in time and space. The beacon node sends packets and all regular nodes are able to synchronize and compute their positions with a zero communication cost algorithm. The synchronization component uses the packets required by the positioning component to improve its performance. The positioning component gets use from the communication, required by the synchronization component to decrease errors. Synchronization is improved, in contrast to the scenarios in which these problems are solved separately. In these cases, communication and processing resources can be reduced, thus saving energy and network resources.

In conclusion, the proposed algorithm (by using mobile beacon equipped with GPS) for both time and space localization, under specific WSN topologies and beacon trajectories gives more reliable results than the existing solutions known in the literature. In addition, it was proved that a clustered WSN is superior from the point of view of energy efficiency.

References

1. Shikoska, U.R., Davcev, D.: Localization in Wireless Sensor Networks. In: Web Proceedings ICT Innovations, Ohrid, Republic of Macedonia, pp. 281–298 (2010) ISSN 1857 – 7288
2. Shikoska, U.R., Davcev, D.: Space and Time in Wireless Sensor Networks. In: PCO 2010, Kucing Borneo, Sarawak, pp. 73–83 (December 2010)
3. Rechkoska, U., Davcev, S.D.: Time and space localization for wireless sensor network. In: The Seventh Advanced International Conference on Telecommunications, AICT 2011, St. Maarten, The Netherlands Antilles, pp. 12–19 (March 2011) ISBN: 978-1-61208-004-8
4. Bulusu, N., Heidemann, J., Estrin, D.: GPS-less Low Cost Outdoor Localization for Very Small Devices. IEEE Personal Communications Magazine 7(5), 28–34 (2000)
5. Boukerche, A., Oliveira, H., Nakamura, E.F., Loureiro, A.F.: Localization in Time and Space for Wireless Sensor Networks: A Mobile Beacon Approach. In: IEEE ISCC 2008, 04594838, pp. 1–8 (August 2008)
6. He, T., Huang, C., Blum, B.M., Stankovic, J.A., Abdelzaher, T.F.: Range-Free Localization Schemes in Large Scale Sensor Networks. In: Proc. of ACM MOBI-COM 2003, pp. 81–95 (2003)
7. Niculescu, D., Nath, B.: Ad Hoc Positioning System (APS). In: Proc. of IEEE GLOBECOM 2001, San Antonio, vol. 5, pp. 2926–2931 (2001)

8. Niculescu, D., Nath, B.: DV Based Positioning in Ad-hoc Networks. Telecommunication Systems 22(1-4), 267–280 (2003)
9. Savarese, C., Rabay, J., Langendoen, K.: Robust Positioning Algorithms for Distributed Ad-Hoc Wireless Sensor Networks. In: Proc. of USENIX Technical Annual Conference, Monterey, CA, pp. 317–327 (June 2002)
10. Capkun, S., Hamdi, M., Hubaux, J.-P.: GPS-Free Positioning in Mobile Ad-Hoc Networks. In: Proc. of 34th Hawaii International Conference on System Sciences, vol. 9, p. 9008 (2001)
11. Maroti, M., Kusy, B., Simon, G., Ledeczi, A.: The flooding time synchronization protocol. In: SenSys 2004, pp. 39–49. ACM Press, Baltimore (2004)
12. Ping, S.: Delay measurement time synchronization for wireless sensor networks. Technical Report IRB-TR-03-013, Intel Research (June 2003)
13. Niculescu, D., Nath, B.: Ad hoc positioning system (aps). In: GlobeCom 2001, San Antonio, TX, USA, pp. 2926–2931 (November 2001)
14. Oliveira, H.A., Nakamura, E.F., Loureiro, A.A., Boukerche, A.: Localization in time and space for sensor networks. In: AINA 2007, Niagara Falls, Canada (May 2007)
15. Boukerche, A., Oliveira, H.A., Nakamura, E.F., Loureiro, A.A.: A novel lightweight algorithm for time-space localization in wireless sensor networks. In: MSWiM 2007, Chania, Crete Island, Greece (October 2007)
16. Sichitiu, M.L., Ramadurai, V.: Localization of wireless sensor networks with a mobile beacon. In: MASS 2004, Florida, USA, October 2004, pp. 174–183 (2004)
17. Pathirana, P., Bulusu, N., Jha, S., Savkin, A.: Node localization using mobile robots in delay-tolerant sensor networks. IEEE Trans. on Mobile Comput. 4(4) (August 2005)
18. Romer, K.: Time Synchronization and Localization in Sensor Networks. PhD thesis, ETH Zurich, Zurich, Switzerland (June 2005)
19. Gavalas, D., Pantziou, G., Konstantopoulos, C., Mamalis, B.: Efficient Active Clustering of Mobile Ad-Hoc Networks. In: Bozanis, P., Houstis, E.N. (eds.) PCI 2005. LNCS, vol. 3746, pp. 820–827. Springer, Heidelberg (2005)
20. Langendoen, K., Reijers, N.: Distributed Localization in Wireless Sensor Networks: A Quantitative Comparison. Computer Networks 43(4), 499–518 (2003)

Performance of Amplify-and-Forward Relay Systems with On-off Relaying

Katina Kralevska and Zoran Hadzi-Velkov

Abstract. Amplify-and-forward method with fixed gain relays is one of the most attractive cooperative scheme due to its low complexity. In this paper, we propose a new relaying scheme, called the on-off relaying, where the relay is utilized only if the quality of the direct link between the source and the destination is lower then the relayed link. For this scheme, we present asymptotic performance analysis for the outage probability and average error probability for high SNRs. Monte Carlo simulations are used to verify these results.

Keywords: Amplify-and-forward system, fixed gain relay, outage probability, error probability.

1 Introduction

Recently, relaying over wireless fading channels has attracted a lot of interest among researchers due to its potential to offer significant capacity and coverage improvement in wireless cooperative communication systems at virtually no additional cost [1]. It is one of the underlying principles of *user (partner) cooperation*, which is seen as a fundamental novel concept for achievement of very high bandwidths and communication reliability of future wireless networks. There are two main relaying techniques – amplify-and-forward (AF) and decode-and-forward relaying, the former being simpler to implement in practice.

Dual-hop AF relaying can be either variable-gain or fixed-gain according to whether the relay utilizes or does not utilize the channel state information (CSI), i.e., CSI-assisted (variable-gain) or "blind" (fixed-gain) relay, respectively [2]-[5]. CSI-assisted relays use the instantaneous CSI of the previous hop to control the gain introduced by the relay, while fixed-gain relays do not use the instantaneous

Katina Kralevska · Zoran Hadzi-Velkov
Faculty of Electrical Engineering and Information Technologies, Ss. Cyril and Methodius University, Skopje
e-mail: kkralevska@yahoo.com
zoranhv@feit.ukim.edu.mk

L. Kocarev (Ed.): ICT Innovations 2011, AISC 150, pp. 119–127.
springerlink.com

CSI of the previous hop and the relay amplifies the incoming signal from the source by a fixed gain.

The outage probability (OP) and error probability (EP) of various relay systems employing either variable-gain and fixed-gain AF relays have been studied extensively in recent years [1]-[5]. To achieve this, various approaches have been utilized. One important approach is to conduct asymptotic analysis of OP and/or EP, valid for high signal-to-noise ratios (SNR), because such an analysis typically results in compact expressions that are suitable for devising insights into the behavior of the analyzed system.

In this paper, we propose a new relaying protocol, called on-off relaying, which has the potential to improve the capacity of cooperative diversity systems [1]. According to this scheme, the relay is utilized only if the quality of the direct communication channel between the source and the destination is worse then the relay communication channel. For the on-off relaying scheme, we also present an asymptotic analysis of OP and EP in Rayleigh fading channel, which yields simple yet very useful expressions that match the exact values at moderate and high SNRs.

The remainder of this paper is organized as follows. Section 2 introduces the system and the channel models under consideration. Section 3 presents expression for the outage probability in high SNR regime and section 4 presents expression for the error probability. Simulation results and conclusion are given in Section 5 and Section 6.

2 System Model

We consider an on-off relaying system, where the communication between the source S and the destination D in a given time slot is either direct or via the half-duplex fixed-gain AF relay R (Fig. 1). The destination D has the CSI knowledge of all the three links, the S-D link, the S-R link and the R-D link, which uses to decide whether to receive and process the signal coming from the source or the repeated signal coming from the relay via the max criterion.

The communication between S and D is realized in two phases: in the first phase (i.e., the first transmission slot), the source broadcasts its signal to the destination and the relay. Then, using its CSI knowledge, the relay decides whether to use the direct signal from the source (received in the first phase), or will use the amplified and forwarded signal by the relay over the R-D link in the second phase.

Assume that the source S is transmitting a signal $s(t)$ with an average power ε_s. The received signal at the relay via the S-R path during the first phase is

$$r_R(t) = \alpha_1 s(t) + n_1(t) \ , \tag{1}$$

where α_1 is the fading amplitude of the channel between the source and the relay and $n_1(t)$ is an additive white Gaussian noise (AWGN) signal with a power of N_{0_1} at the input of the relay. In the first phase, the received signal at the destination via the direct path is

$$r_D(t) = \alpha_0 s(t) + n_0(t), \tag{2}$$

where α_0 is the fading amplitude of the channel between the source and the destination and $n_0(t)$ is an AWGN signal with a power of N_{0_0}.

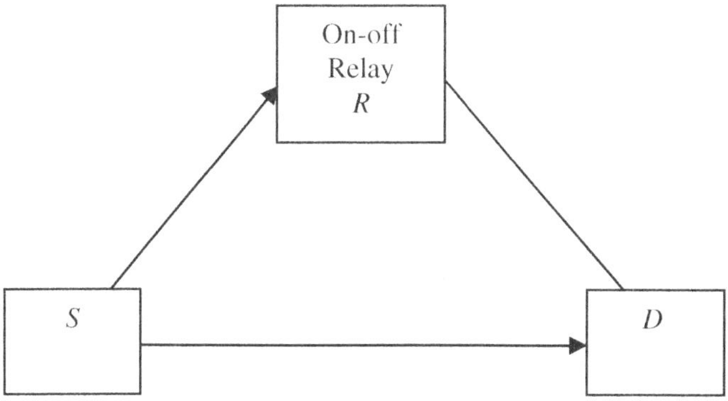

Fig. 1 Wireless communication system where the destination D is communicating with the source S via the relayed and the direct link

Depending on the decision of D whether to process the direct or the relayed signal, in the second phase, R amplifies the signal by the fixed gain G and forwards it to D. The received signal at the destination D is

$$r_D(t) = \alpha_2 G(\alpha_1 s(t) + n_1(t)) + n_2(t), \tag{3}$$

where α_2 is the fading amplitude of the channel between the relay and the destination and $n_2(t)$ is an AWGN signal with a power of N_{0_2} at the input of D. The end-to-end signal-to-noise ratio (SNR) of the signal received at D via R is written as [2, Eq. (3)]

$$\gamma_1 = \frac{\dfrac{\varepsilon_s \alpha_1^2}{N_{0_1}} \dfrac{\alpha_2^2}{N_{0_2}}}{\dfrac{\alpha_2^2}{N_{0_2}} + \dfrac{1}{G^2 N_{0_1}}}. \tag{4}$$

Let introduce the variables $X_1 = \alpha_1^2$ and $Y_1 = (N_{0_1}/N_{0_2})\alpha_2^2 G^2$ with average values $E[X_1] = \Omega_1$ and $E[Y_1] = \Omega_2 G^2$. Additionally, we assume without loss in generality $N_{0_0} = N_{0_1} = N_{0_2}$. Then (4) is rewritten as

$$\gamma_1 = \gamma_0 \frac{X_1 Y_1}{Y_1 + 1}, \quad (5)$$

where $\gamma_0 = \varepsilon_s / N_{0_1}$.

The end-to-end SNR of the received signal at the destination via the direct link is written as

$$\gamma_2 = \gamma_0 X_0, \quad (6)$$

where $X_0 = \alpha_0^2$ with $E[X_0] = \Omega_0$.

The overall end-to-end SNR is written as

$$\gamma_{eq} = \max_i \{\gamma_1, \gamma_2\}, \quad (7)$$

where γ_1 and γ_2 are given with (5) and (6).

The outage probability is defined as the probability that the instantaneous SNR, γ_{eq}, falls below a predetermined protection ratio, γ_{th}. The OP of the relayed link is calculated as

$$P_1 = P[\gamma_1 < \gamma_{th}] = P\left[\gamma_0 \frac{X_1 Y_1}{Y_1 + 1} < \gamma_{th}\right], \quad (8)$$

which is known in closed form according to [6, Eq. (3.324.1)], as

$$P_1 = 1 - 2\sqrt{\frac{\gamma_{th}}{G^2 \gamma_0 \Omega_1 \Omega_2}} e^{-\gamma_{th}/\gamma_0 \Omega_1} K_1\left(2\sqrt{\frac{\gamma_{th}}{G^2 \gamma_0 \Omega_1 \Omega_2}}\right), \quad (9)$$

where $K_1(\cdot)$ is the first-order modified Bessel function of the second kind defined in [7, eq. (9.6.22)].

The OP of the direct link is written as

$$P_2 = 1 - e^{-\gamma_{th}/\gamma_0 \Omega_0}. \quad (10)$$

3 Asymptotic Outage Probability

In this section, we present an asymptotic expression for the outage probability of our proposed communication scheme in the high SNR regime. Using [7, eq.(9.6.11)] and [7, eq.(9.6.7)], we rewrite (9) as

$$P_1 \approx 1-(e^{-\gamma_{th}/\gamma_0\Omega_1}+(2\delta-1)\frac{\gamma_{th}}{G^2\gamma_0\Omega_1\Omega_2}e^{-\gamma_{th}/\gamma_0\Omega_1}$$

$$+\frac{\gamma_{th}}{G^2\gamma_0\Omega_1\Omega_2}\ln\left(\frac{\gamma_{th}}{G^2\gamma_0\Omega_1\Omega_2}\right)e^{-\gamma_{th}/\gamma_0\Omega_1}), \tag{11}$$

where δ is the Euler constant. Note that γ_0 is the transmit SNR, thus in high SNR regime

$$\lim_{\gamma_0\to\infty} e^{-\gamma_{th}/\gamma_0\Omega_1}=1 \tag{12}$$

and

$$e^{-\gamma_{th}/\gamma_0\Omega_1}\approx 1-\frac{\gamma_{th}}{\gamma_0\Omega_1}\,. \tag{13}$$

Applying (13) over the second term of (11) and applying (12) on the third and the fourth terms of (11) yields to

$$P_1 \approx \frac{\gamma_{th}}{\gamma_0\Omega_1}+\frac{\gamma_{th}}{G^2\gamma_0\Omega_1\Omega_2}\left(1-2\delta-\ln\left(\frac{\gamma_{th}}{G^2\gamma_0\Omega_1\Omega_2}\right)\right). \tag{14}$$

This approximation is appropriate in high SNR regime.

Using (13), the outage probability for the direct link is

$$P_2 \approx \frac{\gamma_{th}}{\gamma_0\Omega_0}\,. \tag{15}$$

Now we concern on our case in which the destination receives the signal via the direct link or via the relayed link depending on which link contributes the most to the received SNR.

The outage probability for the system is calculated as [8, eq.(7.15)]

$$P_{out}\approx P\lfloor\gamma_{eq}<\gamma_{th}\rfloor=P\lfloor\max_i\{\gamma_1,\gamma_2\}<\gamma_{th}\rfloor=P_1P_2\,. \tag{16}$$

The outage probabilities for the direct and for the relayed link in high SNR regime are given with (14) and (15). According to (16) the asymptotic outage probability for the system is written as

$$P_{out}\approx\frac{\gamma_{th}}{\gamma_0\Omega_0}\left(\frac{\gamma_{th}}{\gamma_0\Omega_1}+\frac{\gamma_{th}}{G^2\gamma_0\Omega_1\Omega_2}\left(1-2\delta-\ln\left(\frac{\gamma_{th}}{G^2\gamma_0\Omega_1\Omega_2}\right)\right)\right). \tag{17}$$

4 Asymptotic Error Probability

The average error probability communication systems exposed to fading channels is expressed as

$$P_b(e) = \int_0^\infty p(e\,|\,\gamma) p_{\gamma_{eq}}(\gamma) d\gamma \ , \tag{18}$$

where $p(e\,|\,\gamma)$ is the conditional error probability of the considered system in absence of fading (for given SNR), while $p_{\gamma_{eq}}(\gamma)$ is the distribution of the instantaneous SNR.

For several Gray bit-mapped constellations with coherent demodulation, the error probability conditioned on the instantaneous received SNR is in form of $Q(\sqrt{\beta\gamma})$, where $Q(x)$ is the Gaussian Q-function defined as $Q(x) = \left(1/\sqrt{2\pi}\right)\int_x^\infty e^{-t^2/2} dt$ and β is a constant determined by the modulation. For example, $\beta = 2$ for binary phase shift keying (BPSK).

The integral in (18) can be evaluated using the technique described in [4] as $P_b(e) = \int_0^\infty Q(\sqrt{\beta\gamma}) p_{\gamma_{eq}}(\gamma) d\gamma$ as

$$P_b(e) = \frac{1}{\sqrt{2\pi}} \int_0^\infty F_{\gamma_{eq}}\left(\frac{t^2}{\beta}\right) e^{-t^2/2} dt \ , \tag{19}$$

where $F_{\gamma_{eq}}(\cdot)$ is the cumulative function (CDF) of the received SNR γ_{eq} .

Using the derived expression for the outage probability (17) in the previous section, the average error probability can be calculated. Substituting (17) into (19)

$$P_b(e) = \frac{1}{\sqrt{2\pi}} \int_0^\infty \frac{t^2}{\gamma_0 \Omega_0 \beta} \left(\frac{t^2}{\gamma_0 \Omega_1 \beta} + \frac{t^2}{G^2 \gamma_0 \Omega_1 \Omega_2 \beta} \left(1 - 2\delta - \ln\left(\frac{t^2}{G^2 \gamma_0 \Omega_1 \Omega_2 \beta} \right) \right) \right) e^{-t^2/2} dt \tag{20}$$

and using [6, eq.3.461.2] and [6, eq.4.352.1], we obtain the high SNR approximation of the EP for the considered on-off relaying system with a single fixed-gain AF relay as

$$P_b(e) \approx \frac{3}{2\beta^2 \gamma_0^2 \Omega_0 \Omega_1} - \frac{1}{2\beta^2 G^2 \gamma_0^2 \Omega_0 \Omega_1 \Omega_2} \left(5 + 3\delta + 3\ln\left(\frac{1}{2\beta G^2 \gamma_0 \Omega_1 \Omega_2} \right) \right) \tag{21}$$

5 Simulation Results

In this section, we verify the derived asymptotic expressions for OP and average EP through Monte Carlo simulation (Fig. 2 and Fig. 3).

Fig. 2 presents the OP vs. transmit SNR curves of the considered relay system under Rayleigh fading, which are obtained either from our asymptotic expressions and via simulation, for two different thresholds γ_{th} (0 dB and 5 dB).

The results for EP are illustrated in figure 3. It shows the average EP vs. transmit SNR of the considered relay system when the modulation is binary phase shift keying (β=2). The curves are obtained for different combinations of Ω_0, Ω_1 and Ω_2 and the threshold γ_{th} is not taken as a system parameter.

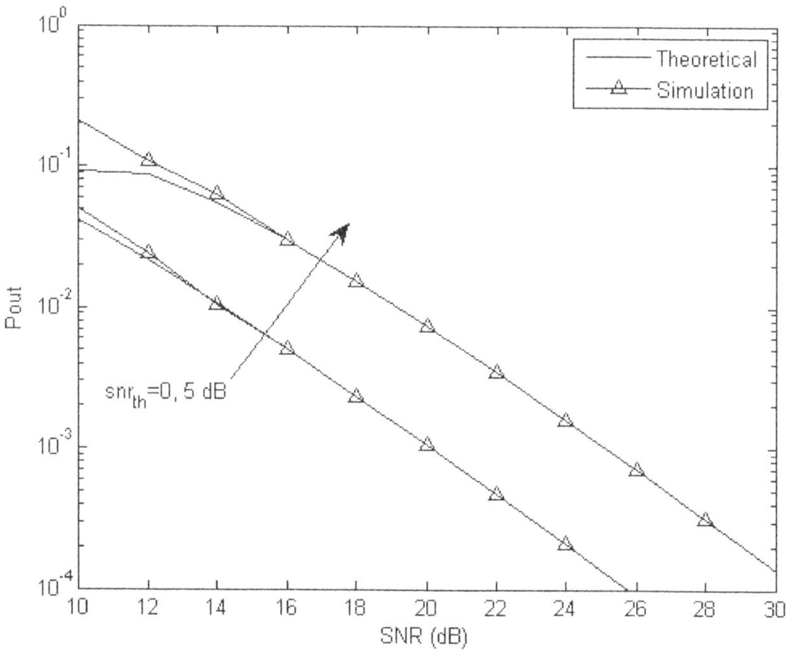

Fig. 2 OP for on-off relaying system with system parameters: thresholds are 0dB and 5dB and $\Omega_0 = \Omega_1 = \Omega_2$

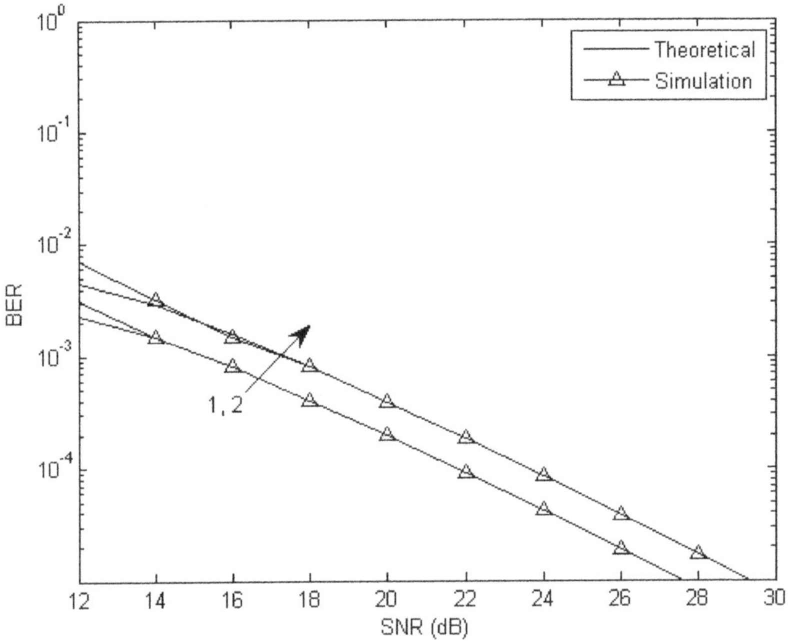

Fig. 3 Average EP for on-off relaying system with system parameters: (1) $\beta = 2$ and $\Omega_0 = 2\Omega_1 = 2\Omega_2$ and (2) $\beta = 2$ and $\Omega_0 = \Omega_1 = \Omega_2$

Fig. 2 and Fig. 3 show an excellent match between the exact and the approximate results for moderate and high SNR. This proves the validity of our derived expressions in the previous Section.

6 Conclusion

In this paper, we proposed a new relaying scheme employing fixed gain AF relay, where the relay is utilized only when the direct link has a lower quality then the relayed link. For this scheme, we presented an asymptotic performance analysis, asymptotic OP and asymptotic EP expressions, valid for high transmit SNRs. The simulation results validate the tightness of our approximations for moderate and high SNRs.

The proposed on-off relaying scheme has the potential to improve the channel efficiency, because the second transmission phase (slot) is necessary only when the relay is activated. We leave this analysis for future research.

References

1. Sendonaris, A., Erkip, E., Aazhang, B.: User cooperation diversity- Part I: System Description. IEEE Trans. on Commun. 51(11), 1927–1938 (2003)
2. Hasna, M., Alouini, M.S.: A performance study of dual-hop transmissions with fixed gain relays. IEEE Trans. Wireless Commun. 3(6), 1963–1968 (2004)
3. Zhao, Y., Adve, R., Lim, T.J.: Improving amplify-and-forward relay networks: optimal power allocation versus selection. In: Proc. IEEE International Symp. on Information Theory (July 2006)
4. Zhao, Y., Adve, R., Lim, T.J.: Symbol error rate of selection amplify-and-forward relay systems. IEEE Commun. Letters 10(11), 757–759 (2006)
5. Elkashlan, M., Yeoh, P.L., Louie, R.H.Y., Collings, I.B.: On exact and asymptotic SER of receive diversity with multiple amplify-and-forward relays. IEEE Trans. Vehicular Techn. 59(9), 4602–4608 (2010)
6. Gradshteyn, I.S., Ryzhik, I.M.: Table of Integrals, Series, and Products, 7th edn. Academic, San Diego (2007)
7. Abramowitz, M., Stegun, I.A.: Handbook of Mathematical Functions with Formulas, Graphs, and Mathematical Tables, 9th edn. Dover, New York (1970)
8. Papoulis, A.: Probability, Random Variables and Stochastic Processes, 2nd edn. McGraw-Hill (1965)

Secure Transport of Electronic Payment over GPRS

Ana Vuksanovic Vanevska and Vladimir Trajkovik

Abstract. This paper identified electronic security as a key component to the delivery of e-finance benefits in performing electronic transactions through GPRS type of communication. The fraud is known issue from the beginning of electronic payment and it is growing in line with the development of the technology. The paper describes the approach what GPRS POS should have in order to provide higher level of the security in GPRS network and to enhance the prevention in GPRS transactions in order to be more accepted by the customers.

1 Introduction

E-finance is the use of electronic means to exchange information, to transfer signs and representations of value, and to execute transactions in a commercial environment. E-finance comprises four primary channels: electronic funds transfers (EFTs), electronic data interchange (EDI), electronic benefits transfers (EBTs) and electronic trade confirmations (ETCs). EFT, which began in the early 1960s, is the oldest form of electronic money transmittal. The amount of money moving by EFT is $2 trillion per day and growing. The volume of EFT usage worldwide is 677,411,204 transactions. E-security can be described on the one hand as those policies, guidelines, processes and actions needed to enable electronic transactions to be carried out with a minimum risk of breach, intrusion, or theft. On the other hand, e-security is any tool, technique, or process used to protect a system's information assets. Information is a valuable strategic asset that must be managed and protected accordingly[1]. The degree of e-security used for any activity should be proportional to the activity's underlying value. Thus, security is a risk-management or risk-mitigation tool and appropriate security means mitigation of the risk for the underlying transaction is in proportion to its value. Traditional risks have thus been reshaped. In the physical environment,

Ana Vuksanovic Vanevska · Vladimir Trajkovik
Faculty of Computer Science and Engineering, Skopje, R.Macedonia
e-mail: Ana.Vuksanovic@gmail.com
trVlado@feit.ukim.edu.mk

L. Kocarev (Ed.): ICT Innovations 2011, AISC 150, pp. 129–146.
springerlink.com

frauds traditionally were paper-based or people -based, whereas the following are the means most often used to commit crimes online: message interception and alteration, unauthorized account access, identity theft15, manipulation of stocks and bonds, extortion, unauthorized system access (e.g., system damage, degradation, or denial of service), industrial espionage, manipulation of e-payment systems, credit card theft. Credit card theft is exploding via the Internet. Internet chat rooms facilitate the sale of stolen credit cards. Most hackers utilize online casinos to set up accounts, close them and then collect the illicit funds. GPRS offers the user an "always on" connection to Internet and Intranet. Services as financial transactions require high level of security. It is important to have strong focus on the security, so companies and persons that demand high level of security can take advantage of the services GPRS offered. What normally happens is that the services win against the security. This is in most cases adequate security for what a normal subscriber requires.

The GPRS has inherited most of the security threats that exists in the GSM system. In addition the GPRS encounter new and bigger challenges. This is since GPRS employs IP technology and it is connected to the Internet. Threats to the GPRS are not only from the insecure public network, Internet [2]. The attacks of the EFT/POS terminals that use GPRS for communication are usually on the data at the air interface.

The main objective of this paper is to provide approach of higher level of the security in GPRS network to enhance the prevention in GPRS transactions. In the second sector it gives the overview of the GPRS communication and single points of risks for fraud detection. In the third sector describes the characteristics that GPRS POS should have in order to be accepted by the customers. Next in the forth sector it provides the methods how to get higher level of the security in GPRS network, and at the end it provides actual methods that should be used to enhance the prevention in GPRS transactions.

2 GPRS

General Radio Packet System (GPRS) provides a network infrastructure to facilitate a range of data services that are provided by network operators worldwide. 3G is the common name for UMTS Universal Mobile Telephony Services), which uses GPRS to provide services at data higher than 57.6kbps (normally quoted at 384kbps or higher).

The main difference between a GPRS and 3G network infrastructure is the technology that handles the radio communication in a 3G infrastructure, the equipment is capable of operating at much higher bandwidths, however, the core infrastructure components are identical. Below is an overview of the purpose and connectivity of each of the primary components within the GPRS infrastructure as shown on Figure 1. *Mobile device* can be a phone or EFT/POS terminal or PDA or datacard; *Base Station System* consists of a Base Station Controller (BSC), and

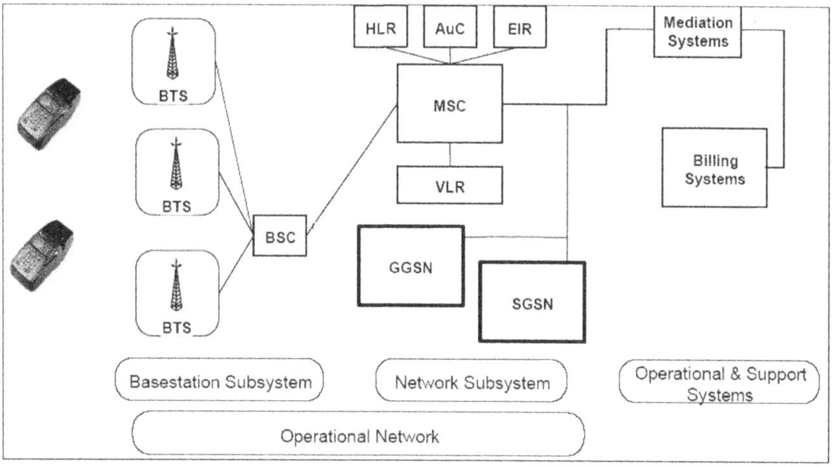

Fig. 1 GPRS infrastructure

Base Transceiver Station (BTS), which is shown in Figure 1. The BTS is the radio equipment that transmits and receives information over the air to provide communication with the Mobile Device. *Home Location Register and Visitor Location Register* are the databases that hold information for every individual who has a subscription with the GPRS operator. Information present in the HLR includes supplementary services, authentication credentials, APN (Access Point Names), the subscribers ISP (Internet Services Providers). For GPRS, subscriber information is exchanged between the HLR and SGSN. The VLR contains temporary subscriber information needed to provide services for visiting subscribers who are roaming overseas. *SGSN or Serving GPRS Support Node* forwards data to and from an Mobile Devices within the SGSN service area, and it also provides data routing and transfer to and from the SGSN services area. A GPRS subscriber may be served by any SGSN in the network all depending on location. The SGSN also provides: Ciphering and authentication, Session management and Mobility management. *GGSN or Gateway GPRS Support Node* is a primary component in the GSM network using GPRS [3]. The GGSN provides the interface towards the external IP packet networks. The GGSN therefore contains access functionality that interfaces external ISP functions like routers and RADIUS servers (Remote Access Dial-In User Service), which are used for security purposes. From the external IP network's point of view, the GGSN acts as a router for the IP addresses of all subscribers served by the GPRS network. The GGSN thus exchanges routing information with the external network. The GGSN also provides GPRS session management and communication setup towards external network, functionality for associating the subscribers to the right SGSN, output of billing data. The GGSN collects billing information for each MS, related to the external data network usage. Both the GGSN and the SGSN collect charging information on usage of the GPRS network resources. When SGSN and GGSN functionalities are combined in the same physical node (network element), structure made is called *CGSN (Co-located GPRS Support Node).* SGSN and GGSN contain GPRS backbone network protocol (IP) routing functionality, and they may be interconnected with IP routers. SGSN and GGSN can be located in different PLMNs. The two PLMNs will then be connected via Border Gateways (BG) for security and interoperability reasons. The Border Gateways are part of the GGSNs. The BG could consist of a firewall, security functions, and routing functions. BGs as well as their functionality are selected by the GPRS operators' mutual agreement to enable roaming. *EIR or Equipment Identity Register* is a database containing mobile equipment identity information, which helps to block calls from stolen, unauthorized, or defective MSs. *AuC or Authentication Center* includes information for identifying authorized users of the GPRS network and for preventing unauthorized use of the network. AuC is often a physical part of the HLR. AuC is a GSM-specified entity that provides triplets to the authentication and ciphering process used within GSM. The authentication for GPRS and for GSM subscribers is the same. The change in security for GPRS is related only to the new ciphering algorithm. This change does not require an update for AuC. *LIN or Lawful Interception Node* is used to collect information about some pre-defined subscriber or subscribers. The information could include, e.g., the data sent and received by the interception target, location information, and subscriber information. The lawful interception is an action based on the law, which is performed by the GPRS network. The GPRS network has to be able to deliver required user data and other network related information to the Law Enforcement Agency (LEA), whenever wanted. *GPRS backbone networks* can be either intra- or inter-operator network. The main function of the intra-operator backbone network is to connect the GSNs of a single operator. The inter-operator backbone network connects GPRS operators and provides international GPRS roaming. GPRS backbone networks are IP based.

3 GPRS POS

Globalisation has brought a new era to payments systems. Rapid innovations in finance have led to the introduction of new products and services, such as new payments systems, which can be defined as a

> "system of instruments and rules which permits agents to meet payment obligations and to receive payments owed to them" (Heffernan, 1996, p. 77) [5].

Payments are made by different methods and regulations differ between geographical areas. Although traditionally payment has been based on money transactions, there are now a number of other means of payment, including credit- and debit-cards. Technological development has allowed more efficient and secure payment systems through the Internet, sometimes known as e-payments using e-money. Technological development has also made it possible for institutions to provide payment services without actually being banks, or to separate payment services from other banking activities. At the same time, mobile communications have entered a mature period. Since its first launch (First Generation/1G), GSM has developed and extended features have been added until it has become 3G mobile communication. The GPRS is a GSM phase 2+ service and is also an essential first step towards third generation mobile network (UMTS). The standardisation of GPRS started in 1993 by the ETSI (European Telecommunication Standards Institute). Along with the maturing technology, the market has also progressed. To avoid market saturation, both mobile communications' manufacturers and operators have preserved their product and service innovations and this particularly applies to service innovations, since service has become the key to successful marketing rather than the technology itself. Technological innovation allows more sophisticated payment methods by combining the existing payment system with mobile technology and, at the same time, increases the efficiency of payments by reducing transaction costs. As a result, following the rapid growth in the mobile communication industry, mobile technology has extended to the banking and finance industry and particularly the payments industry. This is indicated by the migration of mobile technology to mobile commerce devices. The European Central Bank (ECB) refers to this as the electronification of payments, which means a migration towards the provision of payment services on a fully electronic and highly automated basis. A collaborative development of some technical frameworks, such as magnetic stripe and chip card, point-of-sale (POS) terminal and automated teller machine (ATMs) have become the base for even further development. Based on their technical capability, there are four main characteristics identified as being very important. Payment with POS terminals and especially with POS terminals that use GPRS connection must be *simple*. Consumers do not want to deal with complicated features when paying, especially for micro-payments. If consumers have to put a lot of effort in to making a payment, for instance, if they are required to login with complex PIN requirements before they proceed, there is a risk of technological failure and that consumers will refuse to use it. Therefore, it must be a user-friendly technology. Second, it must be *secure*. Providing security in electronic payments is not only a technological issue, but also must be a valid business model that is accepted by and is not too costly for consumers (ECB, 2003, p. 69). There are, actually, a number of standards for encryption systems to support the security, such as Data Encryption Standard (DES) and Rivest Shamir Adleman (RSA). In addition to digital signatures provided by RSA, the combination of these standards provides a secure digital envelope for sending encrypted messages to secure digital money. On the business side, the security issue is challenging for GPRS POS terminals developers, because they are dealing with consumers' funds. For players from the financial industry, since they are familiar with risk management, this security issue is relatively unproblematic although they are often faced with fraud. ECB lists five important security components, cited as follow:

1. *Availability*: the instrument provides an efficient and timely response, has adequate capacity to support acceptable performance and is able to recover quickly from disruptions.
2. *Authenticity and authorisation*: the instrument has appropriate means to authenticate the identity of and authorise customers using the service, and to make sure that all transactions are legitimate.

3. *Integrity*: the instrument has the appropriate means to protect the integrity of the data in e-payment transactions. This means that e-payment-related information in transit or in storage cannot be altered or deleted without authorisation.
4. *Non-repudiation*: the instrument uses transactions authentications methods that make repudiation difficult.
5. *Confidentiality*: the instrument has the appropriate means to preserve the confidentiality of relevant e-payment information. Key information should not be disclosed in such a way that it can be viewed or used by those unauthorised to do so. Third, GPRS POS terminal must be *convenient*. It has to be quick. It has to be easy-to-use so that consumers need to expend little effort and spend no time in the learning process. If it is easy to use, consumers will be encouraged to make more payments on that POS terminals. Convenience also means that GPRS POS Terminal must be accessible and operable without any geographical barriers.

Nowadays dial up EFTPOS terminals are more often undertaken with EFTPOS terminals with GPRS type of communication because of not needs of fix PSTN line, which makes serious problem in the merchant's side. Even more merchants refused to have dial up EFTPOS terminals. GPRS EFTPOS terminals or so called mobile POS terminals are suited to an organization whose business is on the move. It is also great as a Pay at Table terminal for restaurants, bars and cafés, giving the flexibility to take payments to customers tables. They have same capabilities as dial up terminals but because they use wireless way to transmit the transactions the completion of the transaction is faster but in the same way has to spend more time in development of secure transmitting.

4 Security

Security needs are growing fast. Business faces a large-scale and well-resourced criminal network. Additionally there is a growing threat of cyber espionage and a need to meet ever more stringent compliance benchmarks. At the same time, businesses are becoming more inherently vulnerable through the greater use of automated collaborative processes, and more flexible and mobile workforces [8].

E-security companies and vendors generally fall into three categories: access, use, and assessment. Today's industry includes companies that provide active content monitoring and data filtering, develop intrusion detection services, place firewalls, conduct penetration tests to expose hardware or software vulnerabilities, offer encryption software or services, and create authentication software or services that use passwords, tokens, keys, and biometrics to verify the identity of the parties or the integrity of the data. Data Encryption Standard (DES) and the newer version 3DES has been a worldwide standard for over 15 years and it has a fascinating history. It has help up to cryptanalysis by the most powerful of adversaries, and probably will continue to do so for a while yet. The security of DES/3DES is based not on the secrecy of its encryption algorithm, but on the secrecy of the key used to encrypt a given message. Encryption with DES and related algorithms is known as secret-key or symmetric-key cryptography, since the same secret must be used to encrypt and decrypt a message. RSA stands for Rivest, Shamir, Adleman, who are the inventors of

RSA. It was introduced in 1978. RSA is the first full-fledged public-key algorithm, meaning an algorithm that works for encryption as well as for digital signatures. Also, RSA is by far the easiest algorithm to understand and implement of all the public-key algorithms proposed over the years. RSA gains its security from the difficulty of factoring large prime numbers. RSA has become a part of many standards worldwide. For instance, ISO 9676 (from the International Organisation for Standardisation) cites RSA. So does the International Telecommunication Union (ITU) X.509 digital certificate standard [4]. RSA has become a de facto standard in the financial community. It is included France's ETEBAC 5 and Australia's AS2805.6.5.3 standards for digital signatures and electronic funds transfer. The existence of a de facto standard is important to the development of Internet commerce, regardless of what official standards may also exist. If one public-key system is available everywhere, then signed digital documents can be exchanged among users in many different nations, using different software on different platforms. If there is an accepted standard for digital signatures, it becomes possible to have, for instance, leases, wills, passports, college transcripts, checks, and voter registrations that exist only in electronic form. A paper version would be a copy of the digital electronic document rather than the other way around.

Confidentiality, Integrity and Authentication (CIA) are three different services that computer and network security should cover. All the three services have to be protected, an attack against one or some of them are possible. It is important to have strict control for who should have Access control and dispense with Denial-of-Service for the authorized users.

As mentioned above, confidentiality, the property of information that should not been disclosed to unauthorized parties, has traditionally been seen as the most formidable threat in the communications system. Encryption is used to provide confidentiality. In the other hand integrity is the property of information that should not been changed by unauthorized parties and is normally associated with error correction and retransmission techniques to ensure that data are not corrupted. Cryptographically checksum is a technique to ensure that data is not wilfully modified. Authentication, as provision of assurance of the claimed identity of an entity, is reference to the user identity verification. Challenge-Response is a common authentication mechanism that active challenge the user to claim that he is the right person, so the user has to give the right response. Other fundamental service is access control, which is prevention of unauthorized use of a resource, including the prevention of a resource in an unauthorized manner. Access control is to give access to services for authorized user and denying unauthorized user the same services. While access control is about denying the unauthorized user access to the services, Denial-of-Service can be seen as a security service to ensure that unauthorized users are denied access to the services [6].

The GPRS network offers various security related services on the radio path. This is to give the user a secure and reliable service. The functions offered can be divided in three groups:

- Confidentiality of the user identity
- Confidentiality of user data
- Confidentiality of signalling information elements and connectionless user data

As additional enhanced protection of the data transferred throw GPRS network can be used SSL security protocol. The Secure Sockets Layer (SSL) is a protocol designed by Netscape Communications to enables secure data transfer between two devices over a public network. SSL protects applications running over TCP, and is mostly utilized to protect HTTP transactions. It can be used for other applications like IMAP, POP etc. Nowadays SSL has been replaced by Transport Layer Security (TLS) which is more advanced version of it [7]. Overall SSL provides end-to-end encryption, source authentication, and integrity protection of application data over insecure public networks. The protocol requires a reliable bidirectional byte stream service. Typically, this service is provided by TCP, which guarantees that there is no duplication, loss, or reordering of bytes. After many years of public scrutiny and deployment experience SSL is the most widely trusted security protocol for all sorts of TCP/IP transactions. The addition of SSL capabilities to mobile devices would bring the same level of security to the wireless world. The protocol has two primary functions: privacy between client and server, and authenticaty the server to the client.

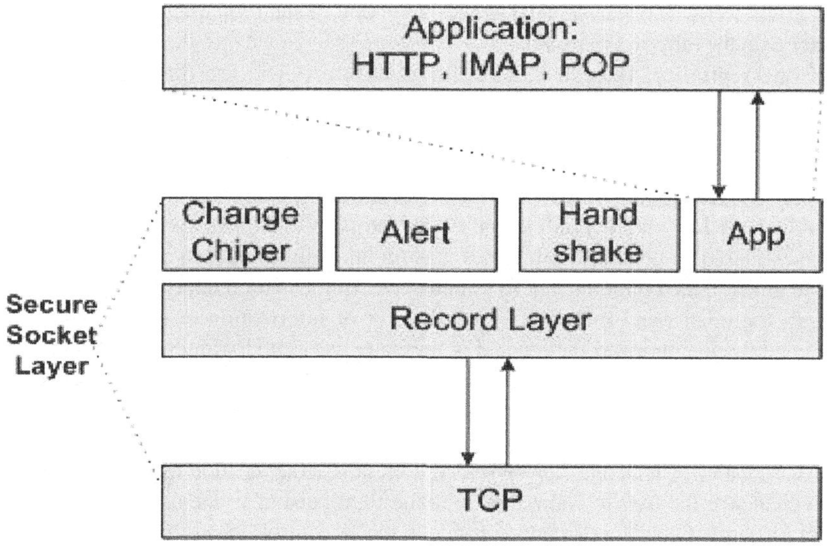

Fig. 2 SSL Architecture

As shown in the figure 2 above SSL is a layered protocol. The Record layer sits above the underlying transport and provides bulk encryption and authentication services using symmetric key algorithms. The keys for these algorithms are established by the Handshake protocol, which uses public-key algorithms to create a master secret between the SSL client and server. This master secret is further used to derive cipher keys, initialization vectors, and message authentication code (MAC) keys for use by the Record layer. Until these keys are installed, the Record layer acts as a simple bidirectional passthrough for all data. Conceptually, the Alert and Change Cipher Spec (CCS) protocols sit within the same layer as the Handshake protocol. The former is used for notification of any protocol failures. The latter is used to signal successful completion of the handshake, and the start of bulk encryption and authentication in an SSL stream. SSL is very flexible and can accommodate a variety of algorithms for key agreement (RSA, DH, etc.), encryption (RC4, 3DES etc.), and hashing (MD5, SHA, etc.). To guard against adverse interactions (from a security perspective) between arbitrary combinations of these algorithms, the standard specification explicitly lists combinations of these algorithms, called cipher-suites, with well-understood security properties. The Handshake protocol is the most complex part of SSL with many possible variations, Figure 3. In the following subsection, we focus on its most popular form, which uses RSA key exchange and does not involve client-side authentication. The SSL protocol allows both client- and server-side authentication. However, due to the unwieldy problem of maintaining client-side certificates, only the server is typically authenticated.

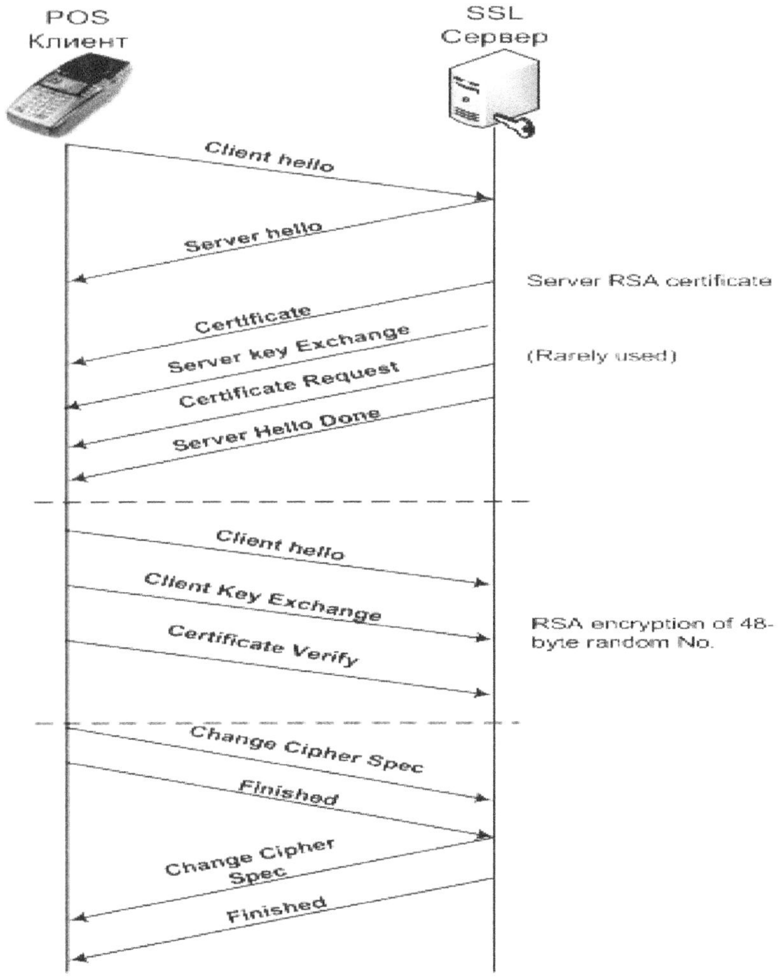

Fig. 3 Full SSL Handshake

Client authentication, in such cases, happens at the application layer above SSL, for example, through the use of passwords (one-time or otherwise) sent over an SSL-protected channel. The server's CertificateRequest message as well as the client's certificate and CertificateVerify messages, shown in Figure 3, are only needed for client-side authentication and rarely encountered in practice. Full SSL Handshake — The client initiates a new SSL session by sending a ClientHello message that contains a random number (used for replay protection), a session ID (set to zero), and a set of supported cipher-suites. If the server is unwilling to support any of the proposed cipher suites, it aborts the handshake and issues a failure notification. Otherwise, it generates a random number and a session ID and sends them in the *ServerHello* message along with the selected cipher suite. The *ServerHello* is followed by a *Certificate* message containing the server's RSA public key in an X.509 certificate. If this key is unsuitable for generating the *ClientKeyExchange* message (e.g., if the key is authorized for signing but not for encryption), the server includes another RSA public key in the *ServerKeyExchange* message and signs it with the private key corresponding to its certified public key. The client verifies the server's public key. It then generates a 48-byte random number, called the *pre-master secret*, and encrypts it with the server's public key. The result is sent to the server in the *ClientKeyExchange* message. The client also computes a *master secret* based on the pre-master secret, and the client and server random numbers exchanged previously. The master key is processed further to derive the symmetric keys used for bulk encryption and authentication. These keys are installed in the record layer and a *ChangeCipherSpec* message is sent to signal the end of in-the-clear communication. The *Finished* message is the first one from the client side to be secured by the negotiated cipher suite. It ensures that any tampering of prior handshake messages, sent in the clear, can be detected. Upon receiving the *ClientKeyExchange* message, the server decrypts the pre-master secret and follows the same steps as the client to derive the master secret and symmetric keys for the Record Layer. After installing these keys in the Record Layer, the server is able to validate the client's *Finished* message. The server also sends the *ChangeCipherSpec* and *Finished* messages to complete the handshake. From here on, application data can be exchanged and is protected by the encryption/ hashing algorithms negotiated in the handshake. Each direction of the traffic's flow uses distinct encryption and MAC keys.

Abbreviated SSL HandShake — The SSL specification also supports a feature called session reuse, which allows the client and server to reuse the master key derived in a previous session.

The abbreviated handshake protocol is shown in Figure 4. Here, the *ClientHello* message includes the nonzero ID of a previously negotiated session. If the server still has that session information cached and is willing to reuse the corresponding master secret, it echoes the session ID in the *ServerHello* message. Otherwise, it returns a new session ID, thereby signaling the client to engage in a full handshake. The derivation of symmetric keys from the master secret and the exchange of *ChangeCipherSpec* and *Finished* messages is identical to the full handshake scenario. The abbreviated handshake does not involve certificates or public key cryptographic operations so fewer (and shorter) messages are exchanged. Consequently, an abbreviated handshake is significantly faster than a full handshake.

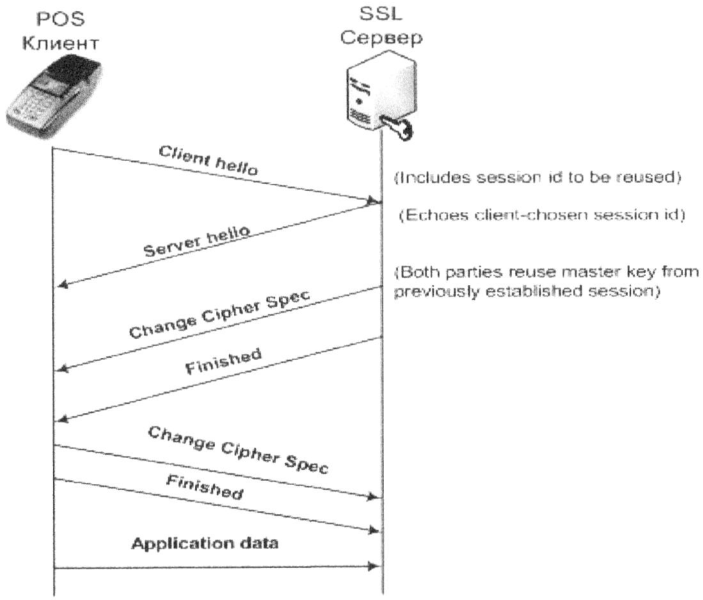

Fig. 4 Abbreviate SSL Handshake

SSL is commonly perceived as being too heavy weight for comparatively weaker CPUs and low bandwidth high-latency wireless networks. The need for RSA operations in the handshake, the verbosity of X.509 encoding, the chattiness (multiple round-trips) of the handshake protocol, and the large size of existing SSL implementations are all sources of concern and contribute to the perception that SSL is too "big" for small devices. However, we are not aware of any empirical studies evaluating SSL for small devices and a careful analysis of the protocol's most common usage reveals some interesting insights: Some constraints ease others. If the network is slow, the CPU doesn't need to be very fast to perform bulk encryption and authentication at network speeds;

A typical SSL client need only perform RSA public key, rather than private key, operations for signature verification and encryption. Their small exponents (typically no more than 65537, a 17-bit value) make public key operations much faster than private- key operations. It is worth pointing out that the performance of RSA public key operations is comparable to that of equivalent Elliptic Curve Cryptography (ECC [8]) operations [9] • There are several opportunities to amortize the cost of expensive operations across multiple user transactions. Most often, a client communicates with the same server multiple times over a short period of time; for example, such interaction is typical of a portal environment. In this scenario, SSL's session reuse feature greatly reduces the need to perform public key RSA operations. Although SSL can be used to secure any connection- oriented application protocol (SMTP, NNTP, IMAP), it is used most often for securing HTTP. The HTTP 1.1 specification encourages multiple HTTP transactions to reuse the same TCP connection. Since an SSL handshake is only needed immediately after TCP connection setup, this "persistent http" feature further decreases the frequency of SSL handshakes.

The EFT/POS terminals always use Full SSL handshake in the first transaction and then until the SSL session is existing, which is usually predefine parameter in minutes use abbreviated SSL Handshake. The SSL session should be long enough not use Full SSL Handshake protocol for every next transaction and prolong the complete performing transaction time, but short enough not to find the encryption key and abuse the massage and the communication. The banks usually calculate their needs of speeding the transactions, security exposed and merchant requirements and position the parameter for SSL session timeout. Most of the EFT/POS terminals supports fallowing list of features:

- *Keys*: Symmetric keys of different lengths and RSA and DSA public/private keys with modulus lengths up to and including 2048-bits.
- *Ciphers*: RSA (for key exchange) and RC4 (for bulk encryption). 3DES code is available.
- *Message digests*: MD5 and SHA.
- *Signatures*: RSA with both MD5 and SHA.
- *Certificates*: X.509 certificates containing RSA keys and signed using RSA with MD5 or SHA are supported.
- *SSL*: Client and server side authentication. All SSL versions are supported, including SSLv3.0 and TLS1.0.

5 Approach Gor Secure Transport over GPRS POS

Although GPRS is considered to be a GSM service, it has its own core network, with shared radio channel between these two services. GPRS core network is attached to GSM radio channel via open interface. Additionally, GSM may utilize GPRS to achieve better performance and GPRS users may use some of GSM supplementary services from the other side. There is a possibility to build a GPRS network separately from GSM. In such case GPRS network needs its own radio channel. There are five main areas shown on the next figure, where security of GPRS system is exposed:

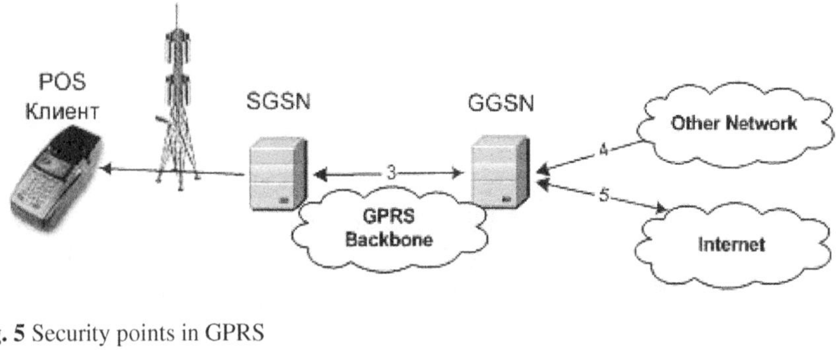

Fig. 5 Security points in GPRS

1. Mobile phone and SIM card security.
2. Authentication of subscriber identity in the network, data confidentiality over the air interface, and file access conditions are supported by security features of the SIM card.
3. Security mechanics between the MS and the SGSN. These includes air interface between MS and BSS.
4. GPRS backbone (traffic between SGSN and GGSN) network security.
5. Security between different operators.
6. Security between GGSN and the external connected networks, like Internet.

The user authentication procedures in GPRS are similar to procedure in GSM. The difference is that the procedures are executed from the SGSN instead of the MSC[9]. The primary function of the SIM in conjunction with a GPRS network is to authenticate an MS before it gets access to the network. The SIM contains the IMSI, Ki, the ciphering key generating algorithm (A8), the authentication algorithm (A3), as well as a Personal Identification Number (PIN). The GPRS A5 algorithm is implemented in the Mobile Equipment (ME) together with the International Mobile Equipment Identity (IMEI) that is physically secured in the ME. The subscriber authentication procedure is illustrated in Figure 6.

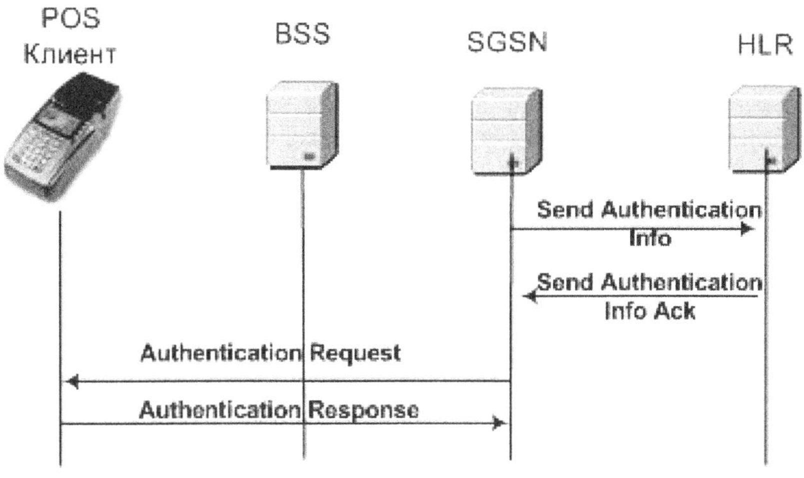

Fig. 6 GPRS authentication procedure

If the SGSN does not have previously stored authentication triplets, they are acquired from the HLR by sending message Send Authentication Info. The HLR responds with Send Authentication Info Ack message and now the SGSN has the authentication triplet. When the SGSN has the authentication triplet it sends message Authentication Request including RAND to the MS and the MS computes SRES from RAND and secret Subscriber authentication key Ki. The MS then sends Authentication Response including SRES to the SGSN and if the SRES computed by the HLR equals to the SRES computed by the MS, the MS is considered to be authenticated to use the network. During authentication procedure the SGSN informs to the MS whether ciphering is used or not. If ciphering is wanted to use the MS starts ciphering after sending Authentication Response message and the SGSN after receiving a valid Authentication Response. It is important to note that all security functions inside the GPRS network are based on the secrecy of the secret key Ki. Ki is Individual Subscriber Authentication Key. The length of Ki is 128 bits. It is stored into the SIM (Subscriber Identification Module) card and into the HLR at subscription time and it is not known by the subscriber. A3 is the authentication algorithm. Different operators have the choice to use the ETSI algorithm A3 or they can use an applicable A3 algorithm to their subscribers. The purpose of the algorithm A3 is to allow authentication of a mobile subscriber's identity. The algorithm A3 must compute an expected response SRES from a random challenge RAND sent by the network. For this calculation, algorithm A3 makes use of the secret authentication key Ki.

In GPRS data and signalling during data transfer are ciphered. The ciphering method is GPRS Encryption Algorithm (GEA) which is a secret algorithm. The scope of ciphering in GPRS is from the ciphering function at the SGSN to the ciphering function in the MS in contrast to the GSM ciphering which is a single logical channel between the BTS and MS. Mutual key setting is the procedure that allows the MS and the network to agree on the key Kc to use in the ciphering and deciphering algorithms. The Kc is handled by the SGSN independently from the MSC. If the MS is able to use both GSM and SGSN services then it have two different keys one in the MSC and one in the SGSN. Key setting is triggered by the authentication procedure, but the network may initiate key setting as often as the operator wishes. Key setting procedure is not encrypted and shall be performed as soon as the identity of the mobile subscriber is known by the network. The transmission of the Kc to the MS is indirect and uses the authentication RAND value. Kc is derived from RAND using algorithm A8 and Ki. The maximum length of Kc is only 64 bits. After computation, the key is stored by the MS until it is updated at the next key setting.

The A8 algorithm is using the Ki together with the 128 bits authentication RAND to generate the 64 bits Ciphering Key, GPRS- Kc. The GPRS-A5 algorithm is used for ciphering the data and signalling during data transfer [10]. The range of ciphering in GPRS is from the ciphering function at the SGSN to the ciphering function in the MS. The enciphering stream at one end, and the deciphering stream at the other end, must be synchronized for the enciphering bit stream and the deciphering bit streams to coincide. Synchronisation is guaranteed by driving GPRS-A5 algorithm's explicit variable input per established LLC (Logical Link Control) and direction. Input is the sequence number of the LLC packet and its initial value is selected by the network. Direction is either from the MS to the network or from the network to the MS allowing input to be identical in both directions. The output of the GPRS-A5 is exclusive with the clear text at the sending end and with the ciphered text at the receiving end.

The identity of the user is protected to avoid the possibility for intruder to identify which subscriber is using a given resource on the radio path by listening to the signalling exchange or the traffic. As a condition to accomplice this the IMSI (International Mobile Subscriber Identity) or any other information allowing a listener to derive the IMSI easily, should not normally be transmitted in clear text in any signalling message over the radio path. It is from a security point of view necessary that on the radio path a protected identifying method is used instead of the IMSI. The IMSI should not normally be used as addressing means. But when signalling procedures permit it, signalling information elements that can expose information about the mobile subscriber identity must be ciphered for transmission. Secure GPRS Interworking with Packet Data Network GPRS supports interworking with packet data networks (PDN) and more specifically with IP. These interworked IP networks may be either the Internet or intranets. GPRS is able to operate with IPv4 and IPv6.

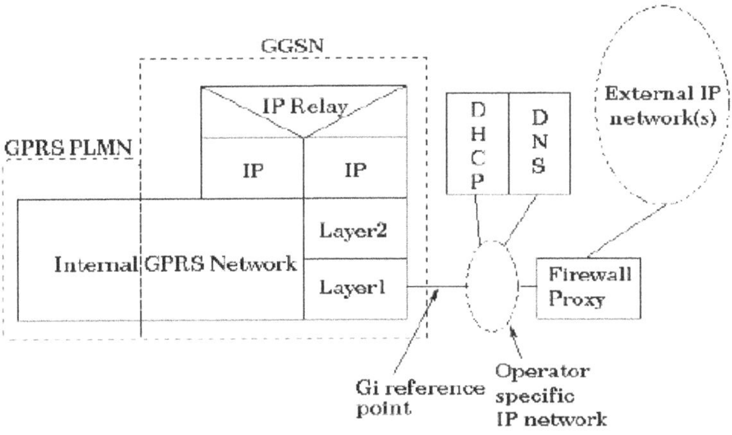

Fig. 7 IP network interworking

Figure 7 illustrates the Gi reference point and protocol stack needed for GPRS interworking with IP networks. The Gi reference point is located between the GGSN and the external IP network. From the external point of view, the GGSN is seen as a normal IP router and can been seen in Figure 7. The Layer1 and Layer2 protocols are operator specific and are negotiated between e.g. a GPRS operator and an external IP network operator. Between the GGSN and the external IP network the following assumptions are valid in generic case:

- A firewall is configured by the GPRS operator. Basically, all applications based on IPare supported but the GPRS operator may restrict their usage. Also, in most cases it is necessary to restrict access from the external IP networks to GPRS network.
- A domain name server is managed by GPRS operator or it can be managed by operator of external IP network operator.

The MS is given an address belonging to the operator's addressing space. The address is either a static address given at subscription time or a dynamic address given at PDP context activation. The received address is used as for packet forwarding between the Internet nodes and GGSN as to map packet within GGSN. The MS does not need to send any authentication request at PDP context activation or participate in user authentication or authorization processes. The user authentication and data encryption are completed within the "Intranet protocol" only if one of them is required.

The MS receives an IP address which belongs to the address space of the intranet or Internet service provider (ISP). In the same way as the transparent access, the address is either static given at subscription time or dynamic given at PDP context activation. This received address is used for packet forwarding within GGSN, Intranet or ISP. This requires a link between the GGSN and address allocation server belonging to the Intranet or ISP again. Information that is used for the authentication request from GGSN to Intranet or ISP comes from the user in the PDP context activation. The GGSN requests user authentication and configuration from a server. The connection between GPRS network and ISP can be arranged over any network, even through an insecure one, such as Internet. In case of an insecure connection a dedicated link or a special secured tunnel can be arranged using e.g. IPSec as a security protocol or above mentioned SSL. The security protocol is defined by mutual agreement between the GPRS PLMN operator and the Intranet or ISP administrator.

The security threats to GSM are quite limited, there are not too many hackers that can or attempt to crack the obscurity SS7 protocol. The GPRS systems are a much more exposed to intruders, because of it IP based backbone. There are a lot of people who have thoroughly knowledge about the TCP/IP in proportion to the SS7.

There are such threats as:

- Integrity of data which is main point in this paper
- Stolen terminal and SIM card
- Borrowed terminal and SIM card
- Eavesdrop, masquerade or manipulate
- Confidentiality of user data and authentication data
- Cloned SIM card
- Non-type approved terminals and defective equipment

Inside the GPRS network all information, such as subscriber information and routing tables, is in clear text format and not protected. Subscriber's information is confidential information. Also incorrect routing tables may cause huge economical losses for the GPRS operator. Therefore, it is very important to protect GPRS network from crackers making firewall (and GGSN) configuration very important. Another threat concerning configuration of the firewall is denial of service attacks. If cracker is able to deny service at GGSN (or any other network element) financial losses for the operator will be enormous. Inside GPRS network cracker would be able to send GPRS signaling messages also and thus affecting behavior of the GPRS network and connections.

IPSec is an „Intranet protocol" that can be used to offer the GPRS subscriber to communicate over insecure public networks securely. It consists of several open standards and the main purpose of it is to ensure secure private communication over IP networks, e.g. the Internet. IPSec offers encryption and authentication on network layer. In transport mode only the message (payload) of the IP packet is encrypted. It is fully routable since the IP header is sent as a plain text. Transport mode is used for host-to host communications. This mode has the advantage to add only a few bytes to each packet. In tunnel mode, the entire IP packet is encrypted. It must then be encapsulated into a new IP packet for routing to work. Tunnel mode is used for network-to network communications (secure tunnels between routers) or host-to network and host-to-host communications over the Internet. IPSec is an mandatory part of IPv6 and is optional for use with IPv4. While the standard is designed to be indifferent to IP versions, current deployment and experience concerns IPv4 implementations. Since the Internet Protocol does not basically provide any security capabilities, IPSec was introduced to provide security services such as:

- Encrypting traffic (so it cannot be read by non-intended for that parts)
- Integrity confirmation
- Authenticating the peers (ensuring that traffic is from a trusted part)
- Anti-replay (protection against replay of the secure session)

IPSec is implemented by a set of cryptographic protocols for securing packet flows and internet key exchange.

6 Conclusion

GPRS is „access networks" to other network which offer mobility. The system should offer authentication and security functions for possibility to use GPRS service for transmitting confidential or private data. GPRS offers ciphering function over the radio channel as well as authentication to the GPRS network. However, the operator must be very careful with its A3 algorithm, because if the authentication is not trustful a harmful system could be caused. It does not mean that authentication works against copying of SIM. If it can be copied, then unauthorized user can use the identity of the authorized user until the subscription invalidates. The security of the transmitted data can not be kept cryptographically excellent because the decryption takes place in the SGSN and not in the BTS. If new A5 ciphering algorithm will be implemented, security will be improved. It is also possible to get the identity of the subscribers using the false BTS. This problem can lead to eavesdropping the other user's traffic. Subscribers should not trust security of the GPRS networks when transferring confidential data more than they do using Internet for the same purpose. For that reason extra protection using SSL shows viable technology even for today's mobile devices and wireless networks. By carefully selecting and implementing a subset of the protocol's many features, it is possible to ensure acceptable performance and compatibility with a large installed base of secure Web servers while maintaining a small memory footprint. Our approach brings mainstream security mechanisms, trusted on the wired Internet, to wireless devices. The use of standard SSL/TLS ensures end-to-end security, an important feature missing from current wireless architectures. The mobile payment scene is a fast evolving. There are a number of market drivers that are forcing the rethinking of the current payment methods, changes in legislation, the growth of on-line commerce and the increase in on-line frauds. In the years to come many new attempts to offer a secure payment infrastructure will emerge. The security methods presented in this paper fulfills many of the generic criteria for secure transport of electronic payment.

References

1. Glaessner, T., Kellermann, T., McNevin, V.: Electronic Security: Risk Mitigatio in Financial Transactions, Public Policy Issues, The World Bank (2001)
2. Bjåen, G.S., Kaasin, E.: Security in GPRS. Agder University College (2001)
3. Halonen, T.: A System for Secure Mobile Payment Transactions, Department of Computer Science, Helsinki University of Technology (2002)
4. Rysavy, P.: Data Capabilities: GPRS to HSDPA (2004), `http://www.rysavy.com`
5. Hellmund, M.: Smart Personalization for Wireless Applications, Media and Computer Science, Department of Digital Media, University of Applied Sciences, Furtwangen, Germany (2003)
6. Hughes, L., Hay'awi, A.-K.: Analysis of Java-based end-to-end wireless internet security. In: IADIS International Conference WWW/Internet (2004)
7. Lim, A.S.: Power Battles in ICT Standards-Setting Process, Lessons from Mobile Payments, Eindhoven Centre for Innovation Studies, Technische Universiteit Eindhoven (2006)
8. Huovinen, L.: Authentication and Security in GPRS Environment: An Overview, Department of Computer Science and Engineering, Helsinki University of Technology (2007)
9. Financial Crime and Intelligence Division, Data Security in Financial Services (2008)
10. Orman, K.: GPRS Security (2008)

Collaborative Augmented Reality (AR) Mobile Phone Game for Co-located Players

Saso Koceski and Natasa Koceska

Abstract. Games have been an important activity among people for many years, but it has never been as interesting to explore as it is today. The main aim of our work is to investigate the influence of mobile phone games on collaboration and social interaction of physically co-located players, instead of focusing on the consequences of playing digital games that are antisocial in nature. Considering the performances of the contemporary converged mobile phones, we have developed and augmented reality (AR) TeamTwixT game for mobile phones. We have evaluated the system performances, user interface performances by adopting the Fitts' law and the usability of mobile phones for collaborative AR. The results have shown that, despite the small screen and limited input capabilities mobile phones and 3D AR technology can be used to naturally enhance collaboration of co-located players and increase their confidence in respect to other gaming conditions.

Keywords: Mobile Games, Augmented Reality, collaborative games, Human-Mobile Interaction.

1 Introduction

With the advances of new technologies playing and gaming nowadays has been transformed into a digital form, just as many other everyday peoples' activities. Moreover, the continuous research and development in Wireless Sensor Networks (WSN) and mobile technology has lead to increased development and usage of digital games on mobile and pervasive devices. When games got transformed into a digital form some people started to argue that these form of play were antisocial [1]. We will not discuss the validity of this argument but, we will try to analyze it from a different point of view. In this context, the main aim of our work is to

Saso Koceski · Natasa Koceska
Faculty of Computer Science, University, "Goce Delcev"-Stip,
bul. Krste Misirkov bb. 2000 Stip, Macedonia
e-mail: {saso.koceski,natasa.koceska}@ugd.edu.mk

L. Kocarev (Ed.): ICT Innovations 2011, AISC 150, pp. 147–157.
springerlink.com

design mobile phone game that make people collaborate and interact when they are co-located, instead of focusing on the consequences of playing digital games that are antisocial in nature.

Games are interactive processes which usually involve several players. Of course, there also exist successful games for single players, but the majority of games are played in a social setting among co-located people. Daft and Lengel [2] asserted that face-to-face is the richest form of information processing because it provides immediate feedback and with feedback, understanding can be checked and interpretations corrected. Several research initiatives have aimed to regain some of the social aspects of traditional game play. One example is the "Pirates!" multiplayer game [3] designed to be played among co-located people using handheld computers connected in a wireless local area network and proximity-sensor technology.

Augmented Reality (AR) is a technology that allows a user to see virtual imagery overlaid and registered with the real world. Traditionally the AR content was viewed through a head mounted display (HMD). Wearing an HMD leaves the users hands free to interact with the virtual content, either directly or using an input device such as a mouse or digital glove.

In the last decade AR applications have migrated to other platforms, including Tablet PCs [4], PDAs [5] and mobile phones [6]. Due to the rapid technological advances of mobile phones, the distinction between PDAs and mobile phones might soon be history. Today mobile phones are becoming platforms that have the potential to bring augmented reality to the general public. According to ITU statistics [7], mobile phone use has increased to over forty-five hundred million users worldwide, and the popularity rate of mobile phones has exceeded more than 90 percent in developed countries. The ratio of converged mobile phones in the worldwide mobile phone market has risen to 16 percent, and is estimated to reach 37 percent in 2014 [8]. Consequently, mobile services have become more and more important. Moreover, the current generations of mobile phones are developed in the same manner as personal computers. With integrated cameras, fast processors, full color displays, possibilities for wireless communication and even dedicated 3D graphics chips the mobile phones have been transformed into ideal mobile augmented reality devices.

Henrysson [9] and Moehring [6] have shown how mobile phones can be used for simple single user AR applications. In their work they create custom computer vision libraries that allows developers to build video see through AR applications that run on a mobile phone.

Recently, a variety of games were prototyped, both by academic researchers [10,11,12] and non-academics [13]. In many cases, these games have been created to demonstrate or test new technologies [14,15]. Many AR applications are created for supporting healthcare [16], learning and education activities [17,18]. Designing handheld AR mobile games and applications is challenging, both because the technology is difficult to work with, but also because we do not yet understand how to create effective AR play experiences. Unlike the previous work, in our study, we have been focused on exploring the affordances of AR for play, especially as a vehicle for social interaction and collaboration between co-located players.

The main aim of our work is to design mobile phone game that make people collaborate and interact in a face-to-face setting, instead of focusing on the consequences of playing digital games that are antisocial in nature. Considering the performances of the contemporary converged mobile phones, we have developed and augmented reality (AR) TeamTwixT mobile phone game. We have evaluated users' experience, system's performances, user interface performances by adopting the Fitts' law and the usability of mobile phones for collaborative AR. We report the qualitative and quantitative findings from a conducted user study.

2 Game Description

One of the main goals in our study is to find if and how AR board games can encourage collaboration and collaborative thinking and foster social interaction. Therefore, we have developed TeamTwixT an AR variant of the popular TwixT game [19]. The original TwixT is a two-player abstract board game, invented by Alex Randolph. It's one of the first "connection type" games. In this genre of games, "connection" usually refers to the object representing a path which connects something to something else. TeamTwixT is a game that tests communication and strategic skills between team members applying to the 9 principles of war [20]. It involves two teams of two players that have a shared AR view of a virtual 3D playing board and a common goal: building an uninterrupted chain of linked pegs that connect the teams' parallel borders. The board is a 24x24 square grid of holes, minus the corner holes (Fig. 1a). The holes along the edges are referred to as 'border rows'. Team members sit opposite one another and they are connected via a collaborative AR mobile phone tool (Fig. 1b).

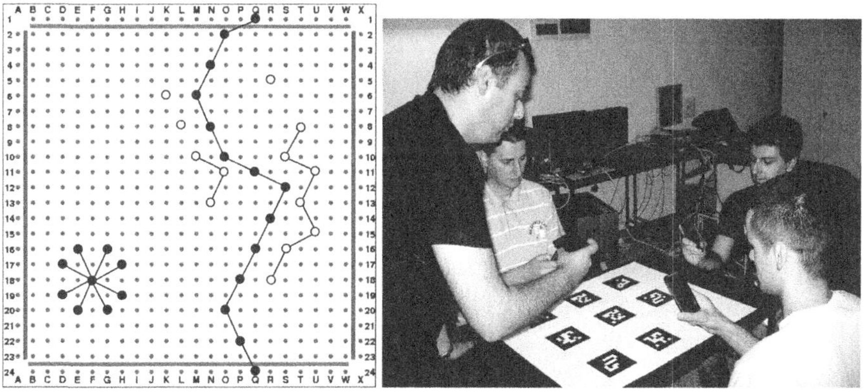

Fig. 1 a) 24x24 TeamTwixT board with 8 possible linking directions from a given central peg (bottom left corner of the board) and a winning situation (connected opposite border rows). b) Physical game setup with co-located players.

To encourage collaboration and social interaction, initially we have considered the possibility to allow one of the players to observe the game while the other plays it. But, we soon realized that such unequal division of labor is disruptive to the player experience and to the balance of the game. So, in our game scenario, play alternates clockwise around the board between teams and players. For example, one order of play would be: Red Team Player 1 followed by Blue Team Player 1 followed by Red Team Player 2 followed by Blue Team Player 2. During each move, a player places a peg on the board. No partner is limited to any one sector of the board. As the game proceeds, if pegs of the same color are in a position to be linked together, then they may be linked within the same move. Pegs may be linked only when the distance between them corresponds to the diagonal of a 6-holed rectangle (Fig. 1a). It is not necessary to place a peg in position to be linked in each move. Players may sometimes opt to place pegs in strategic positions so that they can be linked in a later move. The game ends when one team has a continuous uninterrupted chain connecting each of their borders (Fig. 1a). If neither team can complete such a chain, the game is drawn.

During the game, partners may communicate strategies to one another by any form of signaling. Once during each game one partner of either team may request a consultation of maximum 3 minutes. Once, and only once in each game, one partner may also call out, "Jump!" after making his own move. On claim of Jump, the opponent who would ordinarily move next must yield his turn to the other member of the team asked for Jump, giving that team two successive plays. The player to move next will be the partner of the opponent who yielded his turn. If the opponents have not been defeated by the Jump move, they may take their own Jump at that point or later in the game. Jump is invoked primarily for defensive reasons. If this option is carefully planned by a certain team it can easily lead to fast win and can make the game very challenging. This way the possibility of monopolizing the play of a team is minimized and at the same time the possibility to study collaborations still exists.

3 System Implementation

3.1 System Architecture

For the purpose of this study a multi user system was developed and its architecture is given on Fig. 2. We model this system as client-server architecture. The back end consists of a database server containing data sources, which can be queried over a network by the application or the web server. The application server, where most of the application specific functionality resides, links recognized targets and user actions to specific data queries, packaging the results as a virtual user interface. The mobile device then operates as a "thin client", rendering 3D GUI elements on top of its captured image and leaving the core application processing to the server.

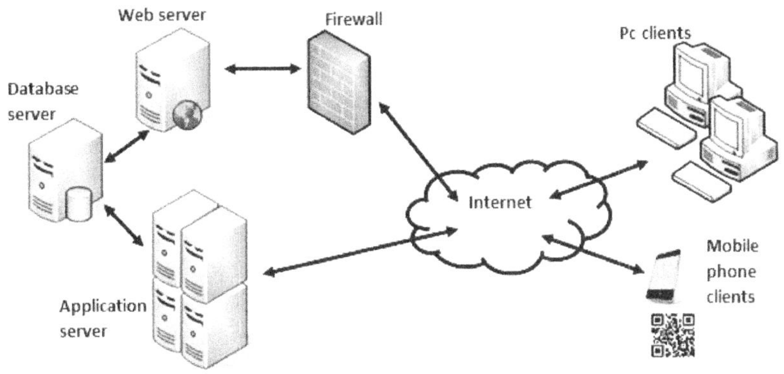

Fig. 2 System architecture.

The idea behind this system architecture is to enable playing multiple TeamTwixT games contemporary and to keep current state, as well as the history of each game on the server side, but also to enable spectators (users which are not part of any team) to watch and analyze the game and statistical data. Therefore, any user who wants to participate as a team member or as a spectator must register to the system through a special web application hosted on the web server. New TeamTwixT game can be created in two ways: by one of the users (called captain) who sends join invitation to other 3 users (one as his partner and two for the opposite team members) or by the system administrator who assigns the team membership and sends invitation to four game players.

The client mobile application was developed on an Android platform. The mobile phone is rendering 3D scenes using the OpenGL ES library. In our system, the real world images are shown as background and virtual object models are combined with the scene using own AR framework. An additional Internet communication module was also developed. Probably the most trilling part in mobile client implementation was the board game visualization using multiple fiducial markers. Namely, analyzing the TeamTwixT game and the user moves, we have observed that during the gameplay, a user may need to focus on a specific board part (for e.g. corners) of the virtual board to observe the situation more clearly, or select a peg to move. This action may cause the camera viewfinder to fall out of the fiducial marker and fail to calculate the marker posed transformation matrix. To solve this problem we designed a nine different marker solution, with each marker having a fixed distance and covering one of the nine board regions, for better localizations of the board part (Fig. 1b). By adopting this approach, when a user aims at one of the nine board regions (including corners), the viewfinder can still capture the corresponding marker. Thus, the transformation matrix can be calculated and the virtual game board can consequently be displayed in an appropriate position.

3.2 Interacting with the Game

Three different user interfaces have been explored to interact with the game, using a touch screen as the input device. Besides the traditional 2D interface, two additional AR user interfaces (Pointing and DirectTouch) were developed. In the Pointing AR user interface, to emulate mouse operations (mostly pointing and clicking), a targeted cursor (cross-hair) is displayed in the center of the mobile screen. The user can "pick up" a peg by positioning the crosshair on it and pressing a button. The user can then use the crosshair to assign the moving destination on the virtual game board. This pointing-based technique is directly compatible with traditional mouse input, and therefore requires least learning effort for users. However, precise freehand pointing is a difficult task due to hand jitter, as reported by Myers et al. [21]. Another way to interact with the AR game is by direct touching on the mobile screen. In the DirectTouch AR user interface, users can directly manipulate a peg by touching it on the touchscreen. Direct touch input can achieve a much higher precision than what is possible with Pointing UI. It also offers a tighter connection between the interface and the physical world.

In both AR user interfaces, to determine which peg is affected by a touch on the screen we have to transform a 2D screen position into a 3D space. To achieve this we use Ray-Picking method (Fig. 2): The object that is closest to the screen and intersects with a ray shot orthogonal to the screen through the touched point is considered picked. This ray is defined by the two points obtained through unprojecting the touched screen point to the far as well as the near plane.

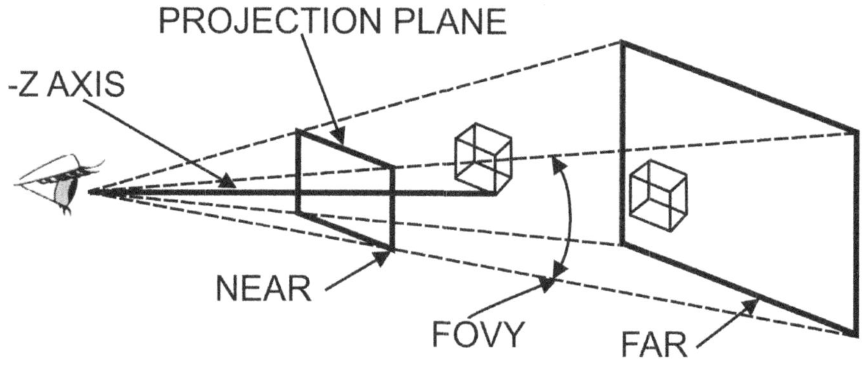

Fig. 3 Ray – picking method for touch screen mobile phone.

The code to determine the view ray from the touch point on the mobile screen is given in following:

```
Private float[] getViewRay(float[] tap)
{
    int[] viewport =
{0,0,nScreenWidth,nScreenHeight};
    float[] eye = new float[4];
  GLU.gluUnProject(tap[0],nScreenHeight-tap[1],0.9f,
mg.mModelView,0,mg.mProjection,0,viewport,0,eye,0);
    if(eye[3]!=0)
    {
        eye[0] = eye[0] / eye[3];
        eye[1] = eye[1] / eye[3];
        eye[2] = eye[2] / eye[3];
    }
    float[] ray = {eye[0] - posCamera[0], eye[1] -
posCamera[1], eye[2] - posCamera[2], 0.0f};
    return ray;
}
```

3 Evaluation Methodology

3.1 Participants

We recruited 50 participants, 40 were male and 10 female; 38 were students and 12 non-students. All of the participants were aged 21-38. We recruited each person separately and ensured that each pair of participants were not friends before the study. All participants were beginners to TeamTwiX game. All but five were unfamiliar with augmented reality technologies before the user study. All but four had the experience of playing with strangers in online games or sports.

Before the evaluation, users are asked to fill in a questionnaire regarding their experience of using mobile phones and 3D digital games. The opinions of users are rated through a 5-point Likert scale (5 for strongly agree; 1 for strongly disagree) for evaluation.

As shown in Table 1, most of the participants play mobile games an average less than 30 minutes a day. Users may feel impatient with mobile applications when they spend more than half an hour in one round. This result suggests a 30-minute time limit for one lesson of mobile edutainment, or one round of mobile games.

Table 2 shows that the selected participants prefer 3D games over 2D games (M=mean, SD=standard deviation.).

Table 1. Number of participants and their time per day spent for playing mobile games.

Average time per day spent for playing mobile games	Number of participants
Less than 30 minutes	32
Between 30 min. and one hour	10
More than 1 hour	8

Table 2. Opinions about 3D technologies.

Survey question	Mean value	Standard deviation
3D games are more interesting than 2D game	3.14	1.088
3D scenes grab my attention more than 2D	3.88	0,895

3.2 *User Study*

To evaluate how AR technology and physical co-location of players affects playing experience and social interaction between people, the user study included three stages. During the first stage, the players learned the game rules and application interface. The participants were given a phone, markers and demo AR application. An instructor gave short introduction to the game and game controls by following a pre-determined scenario and showing a pre-recorded instructional video presentation. Then the participants were introduced the concept of augmentation of the real world, as well as paradigms for human-mobile interaction. After this introductory session they have the opportunity to test various actions using a demo AR application.

During the second stage, the participants played the game together. Each team was given at most three times to play. This game sessions and player behavior was observed and notes were taken while the game was played.

In the final stage, the participants filled out the questionnaire independently after playing the game. Afterwards, a semi-structured interview was conducted to understand more about the incentives and experience.

3.3 *UI and Interaction Evaluation*

We have evaluated the performances of the developed user interfaces, through examination of the users' capacity for AR objects selection. For this purpose, a special program which permits the users to select a peg and move it arbitrary to another location was designed. The program records the selection time and the number of times the destination selection failed. The selection test for each user interface is performed 40 times with random movements in distance, by each of the twenty randomly selected users. To investigate the input performance of each user interface we adopted Fitts' Law:

$$MT = a + b * ID, \text{ where } ID = \log_2(A/W + 1) \quad (1)$$

Where MT is movement time, and ID is the index of difficulty, A is the target distance and W is the target width, a and b are regression coefficients. The results for the coefficients a and b obtained with our experiments are given in the Table 3. In average movement time was shorter in the traditional 2D interface than in both AR user interfaces.

Table 3. The coefficients in Fitts' law equation of the developed user interfaces.

Interface type	a	b
Traditional 2D UI	0.824	0.088
Direct touch UI	0.012	2.195
Pointing UI	0.123	3.123

4 Results and Discussion

The conducted semi-structured questionnaire with the participants after the game, revealed some interesting initial findings about collaborative computer games where the players are physically co-located. In the questionnaire, the users reported that the physical co-location fosters the collaboration between team members (4.82/5 and SD=0.498); they enjoyed the verbal and body sign communication during the game (4.52/5 and SD=0.782); they feel comfortable playing with an unknown teammate (4.64/5 and SD=0.683); and they have a neutral opinion about whether they think the play experience will be better with a friend (3.12/5 and SD=0.844). It is very interesting that the participants believed that their teammate was also willing to communicate (4.65/5 and SD=0.668).

We also observed that as users played longer, their conversation got longer. Almost all players started to talk more after the first round. The conversations that happened between rounds were typically reflections on the previous game, the "theories" about which strategy may work better, and the plan for the next round. The conversation during the game was concentrated on passing knowledge about the next move and possible game direction. After the first round we have observed that the players in three of the teams started to use body language and hand gestures to communicate strategies bout the next move.

Regarding their experience with the AR game interface, in the questionnaire users reported that they are impressed by the virtual objects in AR (4.20/5 and SD=0.638); they prefer AR UI more than traditional 2D interface (4.15/5 and AD=0.852) and they think that AR games are more fun than 2D games (3.94/5 and SD=0.832), finally they think that AR UI are rising the users' interest for gaming (4.18/5 and SD=0.784). So, although the evaluation of the developed user interfaces showed that the information transmission rate 1/b of two user interfaces, DirectTouch and Pointing AR, are quite low (with experimental values 0.456 bits/sec and 0.320 bits/sec, respectively) and that traditional 2D UI is the fastest for selection, users still prefer the three-dimensional AR display.

5 Conclusion

In this work we have investigated the potential of using contemporary mobile technologies and AR for improving the collaboration and social interaction between people. Our study shows that the mobile technologies and AR games are particularly suited and can be used as ice-breaking media to bootstrap communication between people. We also argue that by moving computational elements into the physical world eliminates the restrictions on the traditional games demanding specific physical conditions and devices. Therefore, we believe that, with full mobility and immersion in the combined physical/ virtual world, AR games have the potential to transform the way people interact and collaborate with each other.

References

1. Setzer, V.W., Duckett, G.E.: The Risks to Children Using Electronic Games. In: Asia Pacific Information Technology in Training and Education Conference and Exhibition, Brisbane, Australia (1994), http://www.ime.usp.br/~vwsetzer/video-g-risks.html (Online visited, June 10, 2011)
2. Daft, R.L., Lengel, R.H.: Organizational Information Requirements, Media Richness and Structural Design. Management Science 32(5), 554–571 (1986)
3. Bjork, S., Falk, J., Hansson, R., Ljungstrand, P.: Pirates! Using the Physical World as a Game Board. In: Proceedings of Interact 2001 (2001)
4. Träskbäck, M., Haller, M.: Mixed reality training application for an oil refinery: user requirements. In: ACM SIGGRAPH International Conference on Virtual Reality Continuum and its Applications in Industry, VRCAI 2004, Singapore, pp. 324–327 (2004)
5. Wagner, D., Schmalstieg, D.: First steps towards handheld augmented reality. In: Proc. of the 7th International Symposium on Wearable Computers (ISWC 2003), pp. 127–137. IEEE Computer Society, White Plains (2003)
6. Moehring, M., Lessig, C., Bimber, O.: Video See-Through AR on Consumer Cell Phones. In: Proc. of International Symposium on Augmented and Mixed Reality (ISMAR 2004), pp. 252–253 (2004)
7. ITU Corporate Annual Report (2009), http://www.itu.int/dms_pub/itu-s/opb/conf/S-CONF-AREP-2008-E06-PDF-E.pdf (Online, visited June 10, 2009)
8. Smartphone forecast by Pyramid Research, http://www.pyr.com/pr_prlist/PR121009_SMART.htm (Online, visited June 10, 2009)
9. Henrysson, A., Ollila, M.: UMAR - Ubiquitous Mobile Augmented Reality. In: Proc. Third International Conference on Mobile and Ubiquitous Multimedia (MUM 2004), College Park, Maryland, U.S.A. October 27-29, pp. 41–45 (2004)
10. Henrysson, A., Billinghurst, M., Ollila, M.: Artennis. In: SIGGRAPH 2006: ACM SIGGRAPH 2006 Sketches, vol. 13, ACM, New York (2006)
11. Peitz, J., Bjork, S., Jäppinen, A.: Wizard's apprentice gameplay-oriented design of a computer-augmented board game. In: ACE 2006: Proceedings of the 2006 ACM SIGCHI International Conference on Advances in Computer Entertainment Technology, vol. 79. ACM, New York (2006)
12. Mulloni, A., Wagner, D., Schmalstieg, D.: Mobility and social interaction as core gameplay elements in multiplayer augmented reality. In: DIMEA 2008: Proceedings of the 3rd International Conference on Digital Interactive Media in Entertainment and Arts, pp. 472–478. ACM, New York (2008)
13. Szentgyorgyi, C., Terry, M., Lank, E.: Renegade gaming: practices surrounding social use of the Nintendo ds handheld gaming system, pp. 1463–1472 (2008)

14. Löchtefeld, M., Str, W., Rohs, M., Krüger, A.: LittleProjectedPlanet: An Augmented Reality Game for Camera Projector Phones. Artificial Intelligence, 15–27 (2009)
15. Wagner, D., Reitmayr, G., Mulloni, A., Drummond, T., Schmalstieg, D.: Real-time detection and tracking for augmented reality on mobile phones. IEEE Transactions on Visualization and Computer Graphics(2010)
16. Botella, C., Breton-López, J., Quero, S., Baños, R.M., García-Palacios, A., Zaragoza, I., Alcaniz, M.: Treating cockroach phobia using a serious game on a mobile phone and augmented reality exposure: A single case study. Computers in Human Behavior 27(1), 217–227 (2011)
17. Liu, T.-Y., Tan, T.-H., Chu, Y.-L.: QR Code and Augmented Reality-Supported Mobile English Learning System. In: Jiang, X., Ma, M.Y., Chen, C.W. (eds.) WMMP 2008. LNCS, vol. 5960, pp. 37–52. Springer, Heidelberg (2010)
18. Wang, K.-H., Chen, L.-C., Chu, P.-Y., Cheng, Y.-M.: A Study on the Interaction Styles of an Augmented Reality Game for Mobile Learning in a Heritage Temple. Learning. USA Publishing (2010)
19. Bush, D.: TwixT Tactics Part 2. Abstract Games Magazine (8), 14–16 (2001)
20. Army Field Manual FM-3 Military Operations, `http://www.globalsecurity.org/military/library/policy/army/fm/3-0/` (Online, visited June 10, 2011)
21. Myers, B.A., Bhatnagar, R., Nichols, J., Peck, C., Kong, D., Miller, R., Long, A.C.: Interacting at a distance: measuring the performance oflaser pointers and other devices. In: ACM CHI Conference on Human Factors in Computing Systems (2002)

Framework for Prototyping and Evaluation of Recommendation Algorithms in Mobile Applications

Vasil Vangelovski and Sonja Gievska

Abstract. This paper describes a framework for prototyping and evaluation of techniques by which, machine learning can be brought into the realm of mobile augmented reality and location aware applications in general. Two realistic scenarios will be presented and considerations for the inclusion of particular recommendation techniques will be discussed. The results of our exploratory evaluation have pointed out that the framework could be used in evaluating similar adaptable augmented reality systems and location aware mobile applications using it as a building block for future exploration.

Keywords: Augmented reality, context-aware computing, collaborative filtering.

1 Introduction

Advances in accurate positional sensors, cameras, high-resolution displays, coupled with the widespread adoption of unified and flexible mobile software platforms have brought a wave of context aware mobile applications. Augmented reality (AR) interfaces are no longer exclusive to military or entertainment system; smart phone technology makes it affordable for everyday consumers. At the age, when persistent connection to the Internet is a customary, not an exceptional commodity, there is an enormous opportunity for adding new qualities to the traditional mobile interfaces.

World wide web users have become accustomed to personalized content, useful recommendations and adaptive responses from applications that exploit the knowledge about individual and collective web usage, users' behavior and preferences. Online recommendation engines are based on a variety of algorithms, especially

Vasil Vangelovski · Sonja Gievska
Faculty of Electrical Engineering and Information Technologies,
Rudger Boskovic bb, 1000 Skopje, Macedonia
e-mail: vvangelovski@gmail.com
sonjag@gwu.edu

L. Kocarev (Ed.): ICT Innovations 2011, AISC 150, pp. 159–169.
springerlink.com

from the category known as collective intelligence. We explore the possibilities of expanding such recommendation engines to augmented reality (AR) and location aware interfaces. While recommendation engines employed by commercial location aware mobile applications are just starting to rise on the market [1], [2], the advances in interface design are yet to reach their potentials. Significant contributions are expected from the relevant research in the areas, such as, intelligent technology, distributed computing, and novel interaction and presentation techniques.

Recent research trends directed toward integration of context and user preference querying capabilities within a geospatial and relational database system [3] have made the creation of generic data storage for personalized location aware services more realistic. The wide scalability of such tightly integrated systems is difficult to achieve, while keeping the complexity of the system within manageable constraints. The need for distributed data processing and databases will pose challenges that require appropriate solutions for the implementation of an efficient recommendation engine.

The most challenging part in implementing recommendation engines is selecting the most suitable algorithm or combination of algorithms for a given scenario, as every technique will perform differently based on the characteristics and size of the available data. Additional requirements, such as hardware constraints and availability, real-time data processing and the level of user's needs and expectations may further narrow the choice of applicable techniques. Hence, continuous evaluation and step-wise refinement of the proposed algorithms is a crucial design issue in implementing successful recommendation engines.

This paper presents a framework, which can be used as a basis for integrating and evaluating recommendation engines within handheld AR interfaces. The architecture is modular in nature, incorporates technologies related to geospatial databases and intelligent techniques that are feasible to implement. In addition, a versatile interface for mobile devices has been developed to support our current and future evaluation studies.

Section 2 describes the key components of the proposed modular architecture, while Section 3 discusses the categories of machine learning algorithms most often employed by recommendation engines. Section 4 sets the results of the exploratory evaluation of different recommendation techniques for two realistic scenarios. The proposed framework has the potential to be easily extended into a more feature-rich system that can be used in large-scale usability evaluations.

2 System Overview

The conceptual view of our prototyped system architecture is shown in Figure 1. The handheld application, denoted as *AR browser*, is based upon an open source Augmented Reality browser Mixare [4]. The AR browser was developed for Android mobile device, which needs to be equipped with the location sensors, such as, GPS, compass, gyroscope and camera. In order to provide personalized data and security, it was extended with authentication support. A user has the option of selecting a radius of interest within which the suggested places of interest will be displayed.

The location of the user is sensed by the mobile device and delivered to a *web Service*, which responds with points of interests (POIs) located in the surrounding area. The recommended POIs are then serialized as a JSON HTTP response and displayed to the user. The original version of Mixare supports several web services with public data such as Wikipedia and Twitter. To minimize data transfer from and to our web *Service*, a more complex data format was implemented for the purpose of displaying different types of POIs in different ways from aggregated data in one response. Different types of POIs are represented by different icons within the AR browser. As the mobile device is moved around, POIs are presented as superimposed icons over the current camera image. Each point of interest is accompanied by its name and distance from the user in accordance to its relative position in space. In addition, POIs can also be displayed as an overlay on a map. The web-search information for each place is only a touch away. A notification service, which periodically provides audible notifications to the user about the near-by POIs that might be of interest, was integrated to provide a more engaging user experience.

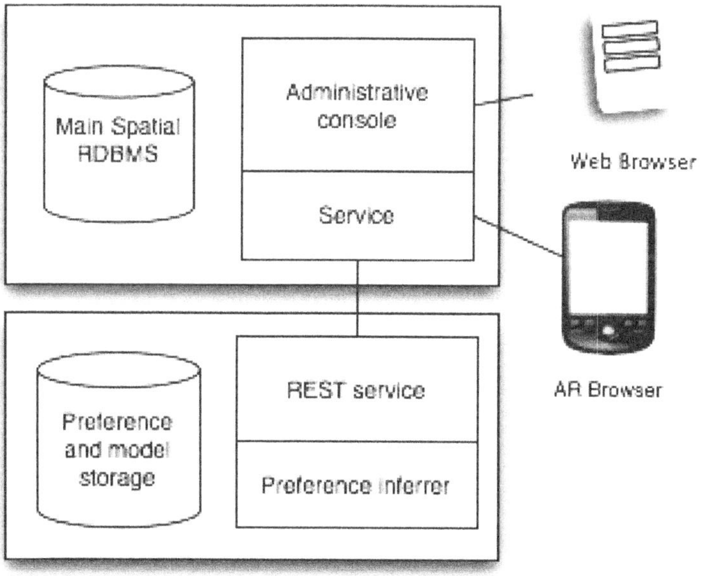

Fig. 1 Conceptual framework of the prototyping system.

A special web-based interface, *Administrative Console*, was implemented to offer the administrator a convenient interface for adding new simulation data for testing and calibration purposes. The web browser and the frontend web service were written in Python, and have easily-extendable architecture. New types of POIs and related data can be added on request, to accommodate additional evaluation scenarios and future exploratory studies.

One important consideration in location-aware applications is the ability to easily extract points of interest within a given circular or rectangular area. In a realistic scenario where there might be millions of possible POIs scattered around the world there is an unavoidable need for a database with spatial indexing abilities. PostgreSQL was used as the main relational data storage engine coupled with the PostGIS spatial extensions to support utilization of various filtering techniques.

The key component of the system is the *preference inferrer*, which helps the system decide on the set of POIs that might be of interest to a particular user. Efficient implementation of recommendation engines relies on distributed data processing solutions for many reasons. Highly optimized data structures are used for storing and processing data models required by the underlying recommendation algorithms. Real-time recommendations require a vast number of data to be stored in the working memory. In addition, most of the real-world scenarios would require numerous adaption and step-wise refinement of the implementation parameters that would generate an amount of data exceeding the working memory of one computer.

To satisfy the requirement for distributed data processing, Apache Mahout library [5] was selected for the implementation of the preference inferrer. The library provides a sound foundation of data structures for efficient implementation of a wide category of machine learning algorithms. It includes a set of basic implementations of collaborative filtering, clustering and classification algorithms, as well as tools for their evaluation and investigation of new domain-specific solutions.

The core of the described system is a wrapper around Mahout based on a representational state transfer (*REST*) service. The service provides convenient means for storing preferences and content data in appropriate formats for the machine learning algorithms. Modularity and data encapsulation in our design can easily accommodate future changes to inference mechanisms.

3 Personalizing Content

Recommendation engines provide personalized content creation and useful recommendations to a user by employing different statistical and machine learning techniques based on the history of previously recorded ratings and user behavior. The most exploited methods for generating recommendations fall under four general categories.

Collaborative filtering (CF) techniques [6] help the system recommend a list of items to a user based upon the ratings assigned by many people for many items. Preferences are traditionally expressed as ratings on a 5-point Likert scale. Two types of collaborative filtering techniques [7] are widely in use, both of which are included in our initial exploratory evaluation. **User-based similarity CF** technique provides recommendations for a list of items to a user, based on the fact that "similar" users have previously rated them higher on the preference scale. The underlying assumption is that people with similar preferences would most likely rate items in a similar way. **Item-based similarity CF** approach finds similar items to the ones a particular user prefers, which implies that a certain model of relationships between items has been established and used. **Slope One** algorithm as initially proposed in [9], have been implemented and included in our set of CF techniques for evaluation. A special challenge for CF techniques is the so-called "cold start" problem, namely, (1) no recommendation can be made for a new user before he assigns any preferences; and (2) a new item will be excluded from recommendations, when no preferences are associated with it. Suitable techniques could be employed to overcome the issue.

Content-based recommender would suggest items to a particular user that are similar to what they have previously liked. Similarity is determined exclusively on the basis of comparison of a selected set of features characteristics of the items. The choice of a suitable content-based algorithm (i.e., vector similarity, clustering, classifiers) is determined by the nature of the items (i.e. text, image) and the type of information that is available for them.

In a **rule-based** system, recommendations are based on a set of rules determined to create a model about a stereotype or a demographic group on the basis of users' past usage. The rules are often modeled using decision trees. The potential drawback for rule-based techniques are the privacy concerns, related to collecting and keeping personal and demographic information about users.

Co-occurenece-based recommender discovers patterns in historical usage data and use them to predicting user preferences for a particular item. Methods such as sequential patterns [8] and association rules [7] fall into this category.

4 Evaluation Metrics

Given larger predefined datasets, recommendation engines can be evaluated automatically in terms of performance, accuracy and the relative relevance of their recommendations. On the other hand, usability studies may be useful for evaluating user acceptance and trust, although conducting large-scale evaluations may require considerable resources; time, money, people, design effort being some of them. In our research, we have advanced the view that both qualitative and qualitative evaluations are needed to determine ways in which such systems may be integrated in the contexts of interest. Two types of metrics have been considered in our investigations: accuracy metrics and information retrieval metrics as a decision-support measure (i.e. relevance).

Accuracy is defined as the statistical average difference between the values predicted by the system and the actual user ratings. For accuracy estimates, the available dataset is usually divided into two partitions: training and test set. Root means square (RMS) could also be used as an alternative measure of accuracy.

Precision is the number of relevant results provided by the system divided by the total number of results in a given response. *Recall* is the number of relevant results retrieved in a given response divided by the total number of relevant results known to the system. A derived measure, *fallout,* which presents the probability that a system will give any number of non-relevant results in a response, was also considered in our performance evaluations. In order to apply precision and recall as measures of a CF system, a relevance threshold has been introduced to measure the relevance; users' ratings above this threshold would be considered relevant.

5 Evaluation Scenarios

We discuss two scenarios used in evaluating the appropriateness of the proposed framework and consider ways in which personalized content can be provided in each. In a location-aware system, only a limited number of items that are available in the immediate surrounding of the user can be taken into consideration, as opposed to a web-based e-commerce system where any product known to the system can be recommended to any user. This fact poses a significant challenge in finding a suitable technique for giving recommendations, especially when relying on collaborative filtering techniques.

5.1 Recommending Movies

The first scenario under consideration uses the AR-based application to make recommendations for movies shown in the nearby theaters. A public dataset of movie ratings, MovieLens, has been used to evaluate different CF-based recommendation techniques for their performance and appropriateness. For testing purposes, additional information for new movies was used.

When evaluating accuracy, only 10% of the ratings for a particular user is selected and used for comparison. Precision, recall and fallout are evaluated for 5% of users limiting the number of recommendations to 10.

Our performance benchmarks were conducted using a 64b x 86 CPU with 2 cores and 4 threads coupled with 8GB of RAM. As most default implementations of recommenders in Mahout automatically scale with the number of CPU cores/threads, the load in our evaluation runs was evenly balanced across 4 threads. One should note that the speed of every recommender improves by orders of magnitude when the dataset is reduced in size. However, the performance results can serve to signify the relative difference in performance between each set of algorithms.

Table 1 shows the evaluation results for user-based CF techniques with different similarity metrics. Rather than taking a fixed predefined number of the most similar users, the implementation relies on taking into account all users within a certain similarity threshold. The relatively fast response times in the range of 200ms as presented in Table 4 could be attributed to the fact that only a limited number of items (ones rated by the most similar users) were taken into account.

Item-based CF gives slightly less accurate results as shown in Table 2. The performance times were orders of magnitude inferior in the range of 1600 – 2000 ms (Table 4), which could be attributed to the fact that the entire dataset was taken into consideration.

User-based CF techniques proved as inappropriate for this scenario, since they only take into consideration a limited number of items from the set of items rated by the most similar users. They were unable to give recommendations in most cases. Item-based CF proved is a better candidate in this regard, since more items are taken into consideration when making recommendations, although they prove to be less accurate and have poor real-time performance (Table 4).

In terms of accuracy and speed, Slope One proves to strike a nice balance, satisfying the requirements of the problem at hand. It was able to give recommendations in most situations where enough ratings for a new movie were provided (Table 3), though it still suffers from the cold start problem. The problem was especially noticeable, since recommendations are mostly provided for newly introduced movies in this scenario. The internal model variables for Slope One (i.e., diffs) are recalculated periodically and stored in memory, though incremental updates are also supported. Once the internal variables are calculated, the time needed to generate a recommendation is considerably reduced. This responsiveness is at the expense of high memory usage for storing internal variables, which can be decreased by decreasing the amount of data stored in memory.

Precision and recall computations performed with a different similarity threshold in each run require considerable time, because of their high algorithmic complexity. For our evaluation runs, the Mahaout implementation automatically selects the relevance threshold. Considering the size and the sparsity of the datasets, it was evident that the chosen threshold was not appropriate; hence the calculated precision and recall were bellow 0.1 (rounded to 0.0).

Table 1. Evaluation of user-based CF (similarity threshold is 0.7 ; dataset contains 1 million MovieLens ratings).

Algorithm	Avg. difference	Precision	Recall	Fallout
Pearson	0.59	2.3	1.96	0.19
Euclidian Distance	0.7871	4.8	6.8	0.2
Uncentered cosine	0.78	1.6	1.7	0.2

Table 2. Evaluation of item-based CF (dataset contain 1 million MovieLens ratings).

Algorithm	Avg. difference	Precision	Recall	Fallout
Pearson	0.71	0.0	0.0	0.27
Euclidian Distance	0.8258	0.0	0.0	0.27
Uncentered cosine	0.824	0.0	0.0	0.27

Table 3. Evaluation of Slope One (dataset contain 1 million MovieLens ratings).

Algorithm	Avg. difference	Precision	Recall	Fallout
Slope One	0.70	0.0	0.0	0.27

Table 4. Average time per recommendation (TPR) and memory usage (dataset contains 1 million MovieLens ratings. IB denotes item-based similarity algorithm).

Algorithm	TPR (ms)	Memory (MB)
User-based	200	100
Pearson (IB)	2000	100
Euclidian distance (IB)	1700	50
Uncentered cosine (IB)	1600	60
Slope One	300	700

5.2 Recommending Restaurants

A second scenario, considers the prototype system being used for recommending restaurants in the vicinity of a user location. The dataset used in this evaluation has been scraped from Yelp.com over a period of more than two weeks. The scraping process started with a group of around 5000 restaurants located within three areas in the USA: parts of California, Florida and areas around Washington DC. Data collection was conducted by incrementally scraping the user profiles containing rating for the initial restaurants and other restaurants they have rated. This process was repeated several times. Ratings span from 1 to 5 on a five-point scale. Information for restaurants such as location and category (cuisine type) has also been collected. Collecting personal user information has been kept to a minimum. The final dataset amounts to around 24000 restaurants, 37000 users and 6 million ratings. Compared to the MovieLens dataset, the user/item/rating ratios is much sparser, so it reflects a more realistic situation. Same reference hardware was used for the evaluation tests, executed in manner similar to the previously discussed scenario.

The results presented in Table 5 show that user-based similarity using Pearson correlation gives the most accurate results. Other user-based algorithms were frequently unable to produce recommendations because of the sparsity of the dataset so their accuracy was impossible to determine.

Item-based techniques still give accurate results as presented in Table 6, although they exhibited higher response times (Table 7), because of the size of the dataset.

Table 5. Evaluation of user-based CF algorithms (similarity threshold is 0.7 ; dataset contains 6 million Yelp ratings).

Algorithm	Avg. difference	Precision	Recall	Fallout
Pearson	0.589	3.8	1.61	0.0

Table 6. Evaluation of item-based CF algorithms (dataset contains 6 million Yelp ratings).

Algorithm	Avg. difference	Precision	Recall	Fallout
Pearson	1.147	1.52	2.35	0.0
Euclidian Distance	0.75	0.0	0.0	0.27
Uncentered cosine	0.9	0.0	0.0	0.27

When a user is found in an area where he/she has rated some restaurants, both item-based and user-based CF techniques were able to provide recommendations with satsifying accuracies. As the recommender provides recommendations for only a relatively small set of restaurants in user's vicinity, response times increased to a few hundred milliseconds.

Table 7. Average time per recommendation (TPR) and memory usage (dataset contains 6 million Yelps ratings).

Algorithm	TPR (ms)	Memory (MB)
User-based	2700	540
Pearson (IB)	8200	560
Euclidian distance (IB)	9100	550
Uncentered cosine (IB)	9000	550

When the application was used in a new location from a user point of view (i.e., a geographical area for which the user had not given any ratings), none of purely CF- based techniques were able to provide recommendations most of the time. The explanation could be found in the fact that users who are considered similar by the algorithms would most likely have rated restaurants in one geographic area. Restaurants, which are considered "similar" according to the item-based CF algorithms, are geographically close too.

We propose a mixed-technique approach to be used to overcome this problem. A solution was implemented that combines item-based CF and content-based similarity of restaurants based on their categories. The Tanimoto coefficient similarity of restaurants (according to their categories) was added to the calculated item-based similarity. This proved to enable the preference inferrer to give recommendations in the simulated problematic scenario. However, the accuracy, compared to purely item-based CF was decreased by 0.8.

Another solution, which chooses an appropriate location-based recommendation technique was implemented to alleviate the problem. When a user was found in the area for which he did not yet provided any ratings, a purely content-based recommender was used; Pearson correlation user-based recommendations were given otherwise. We deferred the exploration of the appropriateness and the effectiveness of the proposed methods for future analyses.

The quality of the current implementation of the AR browser is heavily tight with the limitations posed by the embedded sensors of the handheld device. The most limiting factor in current handheld devices proved to be the compass as it is prone to magnetic interference and often takes some time to be recalibrated. Uncalibrated compass may have significant impact on system accuracy. The power requirements of the sensors pose additional limitations on the availability of the mobile application. However, when all the sensors function properly, the AR browser shows high accuracy up to several meters. Further advances in relevant technologies are expected to solve some of the problems in near future.

In terms of performance, the prototyped system has exhibited satisfactory results for the scenarios of interest. The response time was in range of hundreds of milliseconds, namely, 300 to 500 ms. By caching the response of the preference inferrer or each recommender separately, response times can be significantly reduced to approx. 200ms.

6 Conclusions and Future Work

This paper described a framework for rapidly prototyping a recommendation engine and its integration within a mobile location-aware application. A study, evaluating the system performance in two different scenarios was presented to show, in detail, how one might use the proposed framework. The focus in our research study was given to the recommendation component of the system, though the prototype system may also be used as a basis for more complex systems.

Even though the overall performance of the system may be considered satisfactory for the presented cases, improvement and utilization of advanced recommendation technique such as ones based on singular value decomposition [10] or stacking based on supervised learning invite further empirical explorations.

The overall system performance was evaluated with relatively small test datasets. The challenges in providing real-time recommendations in similar systems for larger datasets may require consideration of additional techniques and their combinations. The proposed framework provides enough flexibility to accommodate support for parallel processing in a distributed manner. However, this would introduce high latencies in the response of the real-time recommenders. The possibility of pre-calculation and off-line processing invites further investigation and experimentation.

From a usability perspective, augmented reality interfaces for modern handheld devices have received a positive user feedback in terms of improved quality of experience. Our research is currently focused on providing more insights into user's quality of experience with the handheld AR application. A design of an extensive usability study is in place to help us refine the system design and recommendation engine.

References

1. Google Hotpot, `http://www.google.com/hotpot/`
2. Qype, `http://www.qype.co.uk/`
3. Levandoski, J.J., Mokbel, M.F., Khalefa, M.E.: CareDB: A Context and Preference-aware Location-based Database System. In: VLDB Endow. (2010)

4. Mixare, http://www.mixare.org/
5. Apache Mahout, http://mahout.apache.org/
6. Su, X., Khoshgoftaar, T.M.: A Survey of Collaborative Filtering Techniques. Advances in Artificial Intelligence (2009)
7. Agrawal, R., Imielinski, T., Swami. A. N.: Mining association rules between sets of items in large databases. In: SIGMOD (1993)
8. Agrawal, R., Srikant, R.: Mining sequential patterns. In: Proc. ICDE (1995)
9. Lemire, D., Maclachlan, A.: Slope One Predictors for Online Rating-Based Collaborative Filtering. In: SIAM Data Mining (2005)
10. Brand, M.: Fast Online Svd Revisions for Lightweight Recommender Systems. In: 3rd SIAM International Conference on Data Mining (2003)

A Protocol for Secure SMS Communication for Android OS

Smile Markovski, Aleksandra Kuzmanovska, and Milivoj Simeonovski

Abstract. By its nature, the SMS communication is insecure and the message information can be viewed of many interested parties. Here we propose a protocol for secure SMS end-to-end communication between two mobile devices. Our protocol is based on a symmetric encryption of the message context and on secure key exchange. The security of our protocol is based on the usage of already secure cryptographic algorithms: AES, SHA-2, DH-EC, and others. We have considered several possible attacks on our secure protocol and we could conclude that it is resistant of them.

Keywords: End-to-end Security, Encryption, Mobile Communication, Secure SMS.

1 Introduction

Short Messaging Service (SMS) is a communication service, originally developed as part of the Global System for Mobile Communications (GSM). Today it is one of the most widely used mobile services, with million messages exchanged on a daily basis.

Their present-day uses are far different from the initial idea. SMS has now become a popular means of communication by individuals and businesses. Banks worldwide are using SMS to conduct some of their banking services. People sometimes exchange confidential information such as passwords or sensitive data amongst each other. The mobile commerce is everyday growing.

SMS provides many conveniences in our everyday lives, but is it really secure?

Smile Markovski · Aleksandra Kuzmanovska
Faculty of Computer Science and Engeeniering, Ss. Cyril and Methodius University, Skopje
e-mail: {smile.markovski,kuzmanovska.aleksandra}@gmail.com

Milivoj Simeonovski
NLB Tutunska Banka
e-mail: milivojs@gmail.com

L. Kocarev (Ed.): ICT Innovations 2011, AISC 150, pp. 171–178.
springerlink.com

SMS messages are sent via a store-and-forward mechanism to a Short Message Service Centre (SMSC), which will attempt to send the message to the recipient and possibly retry if the user is not reachable at a given moment. Transmission of the short messages between SMSC and phone is via the Signalling System Number 7 (SS7) within the unencrypted protocol allowing employees within the cellular providers network, which has access to SS7 network, to eavesdrop or modify SMS messages.

To protect our privacy, as well as to protect our confidential data, we propose a protocol that SMS messages from the source to the destination will be sent encrypted and, accordingly, only the final (end user) recipient can read it. If someone sniff the channel or wants to read the contents of the SMS messages directly from the base, he can only see the text with meaningless incomprehensible characters.

Android is a mobile phone platform developed by the Google-led Open Handset Alliance (OHA). The platform is popular amongst the community for its open source nature and adoption by telecommunication providers world-wide. This OS is based on Linux with middleware presented to developers and core applications. The platform itself focuses on applications, and much of the core phone functionality is implemented as applications in the same fashion used by third-party developers. SMS program is one of the core applications, but it does not support any security for SMS communication. SMS does not use TCP as the transport protocol, so it cannot rely on HTTPS to secure the transmission.

In our protocol, a secure channel is established between two Android devices, using SMS as the transmission medium. To ensure that the SMS remains confidential, a symmetric-based cipher is used to encrypt the message's content. The encryption can be made by any secure symmetric cipher, like AES. We emphasise that our paper is an extension of the ideas given in Hassinen and Markovski's paper [5], where the encryption was defined by using quasigroup transformations of messages. Their solution provides end-to-end confidentiality but they did not propose any method for exchanging the secret key and assumed that users can handle those themselves before communicating. Herein we developed complete secure protocol that work autonomously, including the secret key exchange and allows arbitrary many users. The communication is establishing online, provided that the users have installed the needed components of our protocol.

Nowadays, exists a lot of proposed solutions to secure the sms communication by using public key cryptography such as [6], [8],[9],[7],[3]. All the solutions are based on the server architecture and someone, the mobile operator or service provider, controls the servers. The servers in such architecture are controlling the cryptographic key generations such as key distributions and authenticate the users as well. If the servers are compromised, the whole security system will fail. However, public key cryptography also requires a lot of computational and storage resources to calculate and store public and private keys.

Our solution is based on non-server architecture due to its independency from third party (the mobile phone network operator or service provider). The user does not need to make any aditional agreement with the mobile operator or other third party and all the cryptographic operations are achieved on the users mobile phone.

2 Protocol for Secure SMS Communication

We propose a novel protocol for secure SMS communication based on AES (Advanced Encryption Standard)[4] [2] for encryption/decryption and Diffie-Hellman Elliptic Curve algorithm for key exchange [1][10]. For hashing purposes we use the SHA-2 algorithm.

The protocol is composed of three basic steps:

1. Authentication
2. Key agreement
3. Encryption/Decryption

3 PIN Authentication

By its nature, mobile devices are easy target for thieves. Therefore, authentication with personal identification number(PIN) has very important role in our protocol. Our protocol design includes PIN for authentication and it is required every time when protocol is in use. The PIN is a private key for encryption algorithm and encrypts all the data stored in the application database. If the user enters wrong PIN, application will be opened, but data will be displayed in unreadable format. The PIN has no relation with other elements in the protocol and it is used only for encryption and appropriate decryption of the content data.

The PIN is stored in the phones private memory, that is associated and managed by the application and it is stored for limited amount of time. After expiration, application went to inactive state and the PIN has to be typed again. A user can choose how many minutes the application will be active after the last activity.

The authentication process is independent and it consists of two phases:

1. The first phase is hashing the PIN using SHA-2 algorithm and the resulting string represents the users private key;
2. The second phase is encryption of the database content.

Using this private key and the encryption algorithm AES (or some other), the protocol encrypts all content data.

4 Key Agreement

For establishing secure, i.e. encrypted communication, parties in a communication process must exchange secret keys. Those keys are used by the symmetric algorithm.

The protocol needs the following two types of keys to be used:

- There are two types of Initial Session Keys used for establishing initial secure communication:
  - Initial Session Key for sending SMS messages,
  - Initial Session Key for receiving SMS messages.

- There are two types of Working Session Keys used for sending and receiving secure SMS messages:
  - Working Session Key for sending SMS messages,
  - Working Session Key for receiving SMS messages

We are introducing two session keys (for sending and receiving messages) to avoid possible collision, in a scenario when both parties sends messages at the same time.

4.1 Initial Session Key Exchange

The Initial Session Key is in fact a secret Initial Vector (IV) used for starting the secure communication between two users. From this IV we derive two initial session keys, IV_0 and IV_1. We take $IV_0 = IV$ and IV_1 is the reverse order of IV. The length of IV is at least $n = 16$ characters, in order to be used for AES-128. (It can be set to any value $n \geq 10$, in order to obtain at least 80 bits of security). In the remaining text of the paper, when we mention IV we mean of both IV_0 and IV_1.

$$IV = x_1\ x_2\ x_3\ \ x_n,\ \mathrm{n} \geq 10,$$
$$IV_0 = x_1\ x_2\ x_3\ \ x_n,\ \mathrm{n} \geq 10,$$
$$IV_1 = x_n\ x_{n-1}\ x_{n-2}\ \ x_1,\ \mathrm{n} \geq 10.$$

For Initial Session Key exchange, the DH-EC algorithm is used. Elliptic curve algorithm is chosen because the protocol is primary designed for SMS messages and their limitation is 160 characters, i.e. 1280 bits.

The initial session key IV is defined to be the secret that is shared between two parties. The initiator of the communication sets $IV = IV_0$ as his primary key for sending encrypted messages and IV_1 as a key for decryption of received messages. Using the same IV, the other party in a communication scenario sets IV_0 as a session key for decryption and, appropriately, IV_1 for encryption.

In the proposed protocol, elliptic curve point is defined and it is taken to be a 256 bit number for the X coordinate and a 256 bit number for the Y coordinate. This is unvarying value recommended by NIST and it is used for multiplication with the private key [1]. The private key, i.e. the IV, is defined using random number generator.

The multiplication process and the derivation of the shared key is not explained in this paper, because we use the standard DH-EC algorithm [10]. After establishing the IV, for any two users, the protocol creates pairs of their phone numbers and the secret initial session keys. In such a way a fixed user can send/receive secure messages to/from several users. From one user, the initial session key is used only once for encryption and only once for decryption purposes. After the first application of IV_0 and/or IV_1, it will be changed by the corresponding working session key.

5 Working Session Key

The protocol is working with messages whose length is 160 characters. If a user wants to send a message M larger than 160 bits, it will be separated to several messages $M_1, M_2, \dots, M_n$ ($M = M_1||M_2||\dots||M_n$), each one of length ≤ 160 characters, and for every message M_i a corresponding working key will be generated. The original SMS message M will be padded to a message M_p of 160 characters, in the case when its length is shorter than that. The padding is done so that the original message with length of $l < 160$ characters is concatenated with the character of the ASCII code 199, and after that with $159 - l$ randomly generated characters. Hence, the communication between two users A and B will be by SMS of length 160 characters. When the user B will get an encrypted message from the user A, the user B will not see the random characters after decryption, because the application will not display them. We choose the ASCII code 199 because it cannot be typed from the mobile phone keypad, so the user A cannot type it by accident.

5.1 Establishment of New Session Key

After each successfully exchanged message M', the new session key, called a working session key, will be generated from M' (Figure 1). Let's suppose that user A sends the message M to the user B. The user A already has its working session key $W_{AB,0}$ for sending encrypted messages to the user B and $W_{AB,1}$ for decrypting the messages obtained from B. $W_{AB,1}$ is the reverse order of $W_{AB,0}$. The user B also has its working session key $W_{BA,0}$ (=$W_{AB,1}$) for sending encrypted message to the user A and $W_{BA,1}$(= $W_{AB,0}$) for decrypting the message obtained from A.

After sending the encrypted message M', new session key will be produced for the user A as follows. By using the hash function SHA-2, a 256 bits hash value $h(M')$ will be produced. Then, the last 128 bits of $h(M')$ will define the new session key $W'_{AB,0}$ for encryption purposes of the user A, and the reverse order, $W'_{AB,1}$, of $W'_{AB,0}$ will be used for decryption purposes. (Here we suppose that AES-128 will be used, so we take 128 bits from the hash value.) The user B will generate the key for the new session after decryption of the obtained message, i.e. after getting the message M'. He will produce a 256 bits hash value $h(M')$ and the last 128 bits will define the new key $W'_{BA,1}$ for decryption of the next message obtained from A, while the reverse order, $W'_{BA,0}$, of $W'_{BA,1}$ will define the new session key for encryption of the next message to A.

After completion of this phase, both users have a new temporary working session key. This working session key will be used for encryption/decryption purposes of the next message only. In such a way, the working session key will be changed when the next message will be exchanged between them.

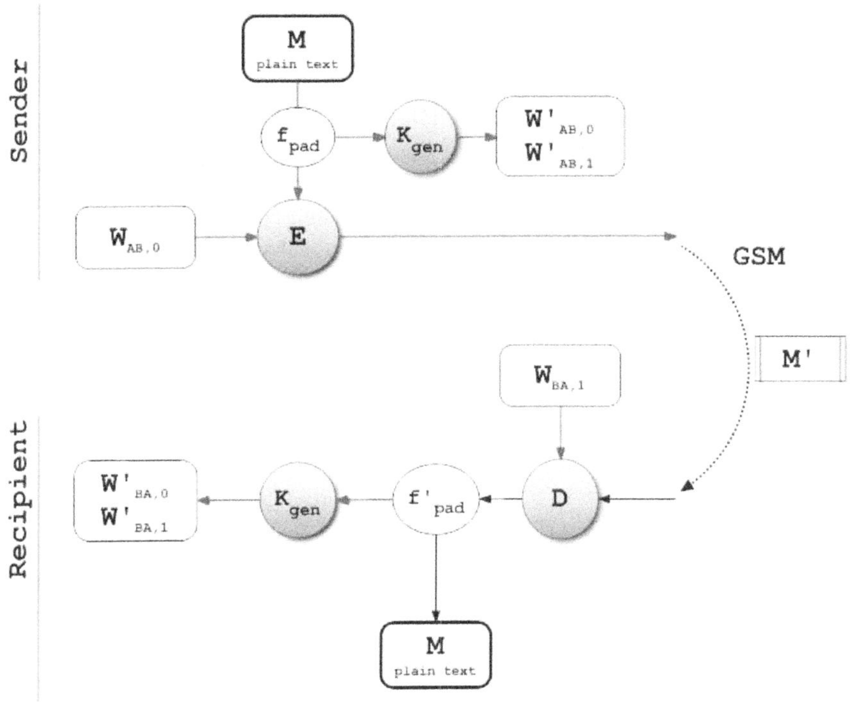

Fig. 1 Secure transmission and key generation

6 Attacks on Our Protocol

Our crypto system for secure SMS messages contains three types of secret keys: An IV, working session key and a PIN. The IV is used only ones, the PIN is permanent, while the working session key is temporary and it is changed after each exchanged message.

The initial session key, IV, is most vulnerable and special attention for its security is given. It should be exchanged between two parties by public channel and that is why the Diffie-Helman with Elliptic Curve Algorithm is used. Hence, the security of IV exchange is based on the security of DH-ECC. So, it is vulnerable of man in the middle attacks. To avoid the man in the middle attacks, a digital signature should be applied. This will restrict the usability of our protocol, but have to be applied in the case when higher security is needed, like bank transactions. The working session keys are generated from encrypted messages that are also sent through insecure public channel. To generate the working session key, the attacker has to know the previously generated working session key, since at first he/she should decrypt the encrypted message. By a recursive procedure, this means that the attacker should know the IV.

For the encryption and the decryption functions we use the standard 128 bits AES, and the security is based on the security of AES. We find that in such a way protection against linear, differential and other types of attacks, is achieved.

All secret data of our protocol are stored in the application database in encrypted form. If an attacker somehow can get the database from the phone memory of a user A, then the security of messages is based on the security of encryption algorithm. That will not affect the security of the other users of our secure protocol.

7 Implementation of the Protocol for Android OS

We have developed an application CyptoSMS that is a fully functional messaging application, and it provides us with high level of security when sending and receiving SMS text messages. CryptoSMS has been developed for the Android platform and it works on all phones and PDA's that have Android version 2.1 and higher. A list of such devices grows constantly with new models appearing. In order to be able to use the application, both the sending and receiving devices must have the application installed. The messages are sent as binary messages and the device make decision whether an incoming message is for CryptoSMS or not.

SMS receiver is part of the application that is working as a background process with task to listen for the incoming messages. The receiver catches all incoming messages and checks if the message is for CryptoSMS application. If this check fails, receiver leave the message to be received by the default phone application, otherwise receiving is interrupted and the message is processed by the application.

In our implementation for android, we use SQLite database to store all of the data. The database content is encrypted by a symmetric algorithm for encryption and SHA-2 hashing algorithm for the PIN authentication.

8 Conclusion

We have proposed a new protocol for secure SMS communications between two parties with dynamical working key generation. It is based on standard secure algorithms and hence it has the needed security. In fact, depending on the needs of the users, there are three possible levels of security, and they depend on the Initial Session Key exchange. For easy establishment of a fast secure communication between two users by public channel, the DH-EC algorithm can be used. For more secure communication by public channel, a digital signature should be applied. Finally, by using a private channel for Initial Session Key exchange, the best security can be obtained.

We proposed above usage of standard cryptographic algorithms, AES, SHA-2, DH-EC, but the protocol can be implemented by using other cryptographic algorithms as well.

References

1. Recommended elliptic curves for federal government use. National Institute of Standards and Technology (1999), http://csrc.nist.gov/encryption
2. Specification for the advanced encryption standard (aes). Federal Information Processing Standards Publication 197 (2001), http://csrc.nist.gov/publications/fips/fips197/fips-197.pdf
3. Albuja, J.P., Carrera, E.V.: Trusted sms communication on mobile devices. Proceedings of the IEEE, 165–170 (2002), http://profesores.usfq.edu.ec/vinicioc/papers/wtr09.pdf
4. Daemen, J., Rijmen, V.: The design of Rijndael:AES — the Advanced Encryption Standard. Springer (2002)
5. Hassinen, M., Markovski, S.: Secure sms messaging using quasigroup encryption and java sms api. In: SPLST, p. 187 (2003)
6. Herlin, H.: Method for secure communication in a telecommunications system (2000)
7. Lisonek, D., Drahanský, M.: Sms encryption for mobile communication. In: SecTech 2008, pp. 198–201. IEEE Computer Society, Washington, DC (2008), http://dx.doi.org/10.1109/SecTech.2008.48
8. Luo, T.: Method for sending a secure message in a telecommunications system (1999)
9. Ratshinanga, H., Lo, J., Bishop, J.: A security mechanism for secure sms communication. In: Conference of the South African Institute of Computer Scientists and Information Technologists
10. Research, Cetricom. Standards for efficient cryptography, SEC 1: Elliptic curve cryptography (2000)

On the Feasibility of Automatic Topic Detection in IM Chats

Katerina Packova and Sonja Gievska

Abstract. This paper presents an initial exploration into the feasibility of automatic topic detection in instant messaging applications. We have developed a prototype system that employs a set of preprocessing techniques, weighted scheme for word combination and suitable word classification and extraction of non-relevant words. The topic detection is based on the descriptive metadata for the selected keywords provided by an existing public knowledge base. In addition, the selected keywords have been linked with popular tweets to provide supplemental information. An exploratory user study has been conducted to gather some insights into the performance and usability metrics related to the proposed approach.

Keywords: Real-time topic detection, preprocessing techniques, IM chat message characteristics.

1 Introduction

Social networks, web-based and pervasive collaborative and communication tools have become commodities we heavily rely upon. Software artifacts used in domains such as: on-line discussion groups and forums, learning, collaboration, personal communications, could greatly benefit from a user interface that adds relevant context information to provide easy navigation and search in the vast amount of content information. Intelligent technologies are expected to play a prominent role in bringing these applications to a new level of usability.

We could envision more effective context annotations, topic detection, visualization and presentation modalities that go beyond the currently offered chronological

Katerina Packova
InterWorks, Karpos bb, 7000 Bitola, Macedonia
e-mail: kpackova@yahoo.com

Sonja Gievska
Faculty of Computer Sciences and Engineering, Rudger Boskovic 16, 1000 Skopje, Macedonia
e-mail: sonjag@gwu.edu

L. Kocarev (Ed.): ICT Innovations 2011, AISC 150, pp. 179–188.
springerlink.com

ordering and user tagging. Our main research goal is to enhance the on-line conversational space by providing context information, such as automatic topic detection in on-going chats. While our current attention has been given to instant-messaging (IM) chats and on-line forums, the proposed design approach is applicable to the general category of systems that support on-line communication.

The problem of topic detection has received a considerable research attention in the intelligence community, especially in areas such as, natural language processing and text mining [1, 2, 3]. There is an extensive set of machine learning techniques that have been proposed and used for extracting topics in conversations, such as, TFIDF classifier [4], naive Bayes classifier [3, 5], Associative Classification [5], Support Vector Machine [5, 6], and K-Nearest neighbor [3, 6]. In the absence of unified guidelines and recommendations for implementing effective topic detection, we recognize the importance of targeted evaluation of various intelligent techniques and their combinations.

Two closely related research works, [3] and [5], have adopted the supervised learning approach for determining the conversation topic. While both studies include substantial analyses of chat message characteristics, their evaluation studies have been conducted for off-line detection of five preset topics (e.g., sport, travel, love), each of them described by a predetermined set of indicative (feature) words. A selection of suitable machine learning algorithms acting as classifiers for detecting a topic are proposed to support the required social and semantic inference of a conversation.

The motivation for our research differs, and it is directed toward augmentation of the chat conversational space with real-time topic detection to provide immediate context awareness for the current and new participants, as well as to support users when searching through chat history. We have advanced the view that semantic and contextual topic detection could be realized by utilizing the existing context maps, ontologies and semantic knowledge bases offered by some public on-line resources. The collaborative and global initiative for constructing public databases containing contextual meanings of words accompanied by meta attributes have produced useful and effective online resources, such as Freebase [7] and WordNet [8].

Our explorations and experimental studies are designed differently from the related works by giving equal importance to performance metrics as well to users' acceptance and satisfaction. Our research is comprised of series of empirical evaluations that serve as a guidance in the step-wise refinement of the proposed design model. An exploratory user study was conducted to validate the effectiveness of the proposed framework by measuring its quantitative and qualitative gains in a particular experimental scenario.

In this paper, we examine the feasibility of automatic real-time topic detection in online chats. The paper is organized as follows. Section 2 presents the proposed design approach, with a particular focus on the employed processing techniques. The results of our initial exploratory study of the performance and potential usability advantages of the proposed method will be presented in Section 3. Section 4 discusses several improvements and future evaluation plans for strengthening our research. We conclude this paper with few remarks given in Section 5.

2 Approach

2.1 Message Characteristics

Instant messaging systems were originally designed for text-based communications only, though insertion of emoticons and hyperlinks are very common in messages. Our proposed prototype system has been designed and tested for chat messages written in English. However, due to the real-time and informal conversational environment of IM systems, chat messages are written in a very different way from conventional English. The use of acronyms, synonyms, short forms, and misspelled words poses additional requirements for preprocessing of the message content. Average message length spans in the range of 40 to 100 characters, though the statistics show that 78% of the messages contain less than 50 words. The interactive and dynamic nature of the IM conversations is characterized by rapid and frequent change of the conversation topic. One should consider that a particular chat session might contain multiple topics interleaving during the duration of one conversation [9].

2.2 Topic Detection

The process of topic detection goes through several stages as presented in Figure 1, each one requires a selection of suitable technique or algorithm that needs to be thoroughly validated and adjusted with real data. The processing starts with a set of *Preprocessing techniques* to extract the words that are candidates for later consideration in the topic detection. The next stage, *Meaningful word combinations* stage is dedicated to discovering meaningful combinations of successive words to provide more contextual clues for establishing the topic of a particular conversation. A weighted scheme has been established to give more relevance to such combinations. The initial set of words is further filtered by *Extraction of irrelevant words*, such as verbs and adverbs. The result of the preprocessing is a set of keyword candidates used to determine the topic.

Our system searches for the contextual meaning and descriptive features of the keywords by consulting the public resources of Freebase [7]. The topic is presented to the user as a set of attributes that appear in the chat window title bar. One should note that the topic clarification might require several messages to be processed; each additional detail bringing new depth to the process of revealing the topic. We have already pointed out that the dynamic changes of topics during a conversation would result in multiple changes of the topic over the course of one chat. Our current version of the prototype application records the last detected topic as the final topic of the processed chat. One should postulate that keeping multiple topics would provide additional assistance to the users when searching for a particular excerpts in chat history. We are currently improving our implementation in that regard and designing a more scientifically controlled environment for future evaluation purposes.

Preprocessing Techniques. The incorporated techniques for preprocessing have already been proposed by related research work [6]. They include suitable techniques for session detection, tokenization, exclusion of acronyms, short forms, icons and hyperlinks. Session of a chat is comprised of the entire chat activity generated before a pause, defined as a 20-minutes-inactivity period. Tokenization is used to extract words from messages that includes elimination of punctuation symbols, blank spaces, and special symbols. The list of words is then further filltered from acronyms, short forms, icons and hyperlinks. We use dedicated files for storing the lists of these particular types of symbols. The relevance of emoticons and hyperlinks are not to be neglected in the process of topic detection. We postpone their inclusion in our approach for future explorations.

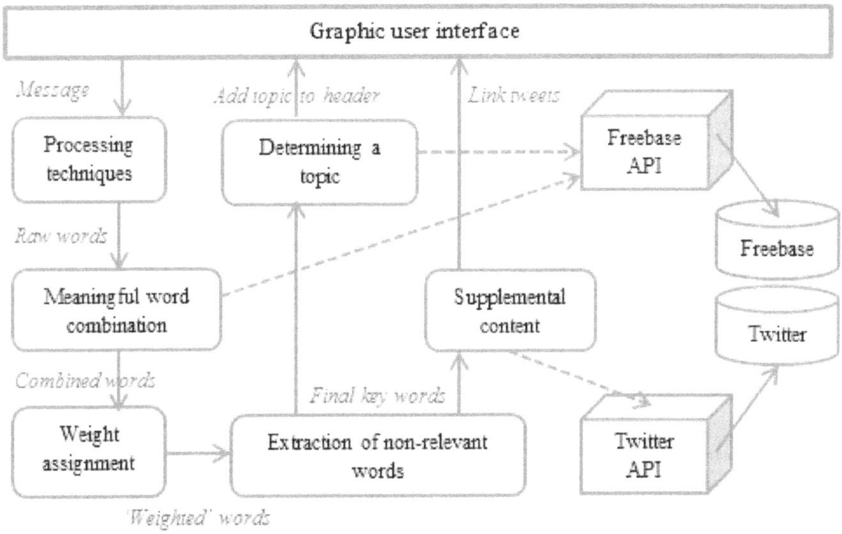

Fig. 1 Conceptual framework for topic detection.

Meaningful Word Combinations. The list of words generated by the preprocessing phase, retains the order of their appearances in the original message. We combine successive words, starting from the first word in a sentence, to discover meaningful combinations that are meant to be interpreted as groups not as separate words. This process has a special importance in distinguishing the names of people, places and phrases that contain words that would have completely different interpretation in another context.

Weight Assignment. The process of finding the descriptive categories according to Freebase is performed for every token in the list of combined words, making it a keyword-candidate for topic determination. We have introduced a simple weight assignment to distinguish between groups that could have more impact when determining the topic, namely, longer combinations are assigned higher weights, when found as an entry in Freebase.

Extraction of Non-relevant Words. In our initial investigation, only the nouns that appear in the chat content are included in the keyword processing and the determination of a chat topic. All verbs and adverbs, currently denoted as non-relevant words, are removed from the resulting list of words, though their presence in meaningful word combinations have been taken into account. Levenstain distance algorithm [10] was applied to find similarity between different forms of these types of words (e.g., play and playing are considered represent the same word and removed appropriately).

Determining a Topic. In our current implementation and evaluation, the contextual meaning of the relevant words distilled in the previous stages is determined using an online collaborative knowledge base, known as Freebase [7]. It provides an easy interface to collaboratively-created structured database that contains approximately 11.5 million topics as of April 2010. Each topic description is represented as a set of ontology-based categories (i.e., *types*), which show the scope of a word (e.g., music, science, author). For example, the freebase entry for US president, Barack Obama, would detect categories such as: a person, politician, award and celebrity.

We distill the initially gathered categories by selecting the ones with most occurrences. There is a considerable probability that each of these categories would participate in the topic formation, based on the fact that Freebase is frequently updated resource and it reflects the common usage and contextual meaning of everyday words. Context maps represented as decision trees containing sets and categories, which were revealed during the Freebase search, were stored and used in the final topic determination, as proposed in [11]. Our approach based on topic categories, frequency of occurrences and context maps represents a simplified implementation of text categorization [13].

Supplemental Content. In our initial explorations we have included another feature to extend and enrich the chat content, namely each detected keyword is linked to the most popular tweet, if such exists. Tweets are messages of 140 characters or less posted on Twitter [12], a micro blogging service. The content presentation in the chat window has been adjusted to accommodate this supplemental feature. All keywords are underlined, while a user can see the associated tweet information by positioning the mouse over the particular keyword.

2.3 Prototype Application

A prototype application that incorporates our approach and proposed techniques for topic detection has been developed in C# and ASP.NET. The system represents a chat client offering the possibility of peer-to-peer as well as group chats. After pressing the submit button at the end of the message, the back end application process it in the way discussed in previous subsections.

The process of topic detection includes communication with Tweeter and Freebase web services, using their corresponding APIs (Figure 2). Freebase is queried using MQL (Metaweb Query Language); it uses JSON objects as queries via standard HTTP requests and responses.

TweetSharp [14] is a client library for communicating with Twitter and works as a simple service, interface, or LINQ provider. For the current implementation only the daily popular tweets were taken into account. Daily tweets are returned as JSON object, to provide easy manipulation and conversion of types. Then, an appropriate comparison of the tweet object name field and the marked keywords in our application is performed to determine their correlation and relevance for the chat conversation.

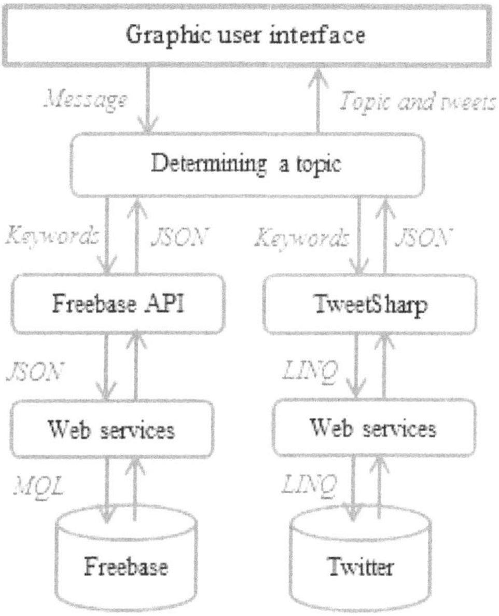

Fig. 2 Architecture of the prototype application.

3 Exploratory User Study

An exploratory study has been conducted to gather some evidence on user acceptance and the effectiveness of the prototype application. The effectiveness of the proposed approach was measured in terms of improved user performance when searching through the chat history to locate a specific topic. Users' subjective opinions and ratings regarding the impact of the inclusion of topic detection and supplemental tweet information were also gathered and statistically analyzed.

The statistical analyses have included a pool of approximately 21,000 chat messages exchanged during a period of one week. The general research question of the evaluation experiment was "Are there any differences in user quality of experience when automatic real-time topic detection is employed?" It was anticipated that the users would be more effective in searching through the chat history, while the novel context-related features, topic header and supplemental tweet information, would promote more engaging conversations.

3.1 Participants and Equipment

A total of 31 volunteers have participated in the study. The participants were a mix of young IT professionals employed by a local IT company and a group of people with various professional background. The age of the participants span between the ages of 15 to 70. A background questionnaire has been administered to each participant before his/her engagement with the prototype application. The demo application was deployed on users' personal or work computers.

3.2 Treatments and Procedure

The experiment involved comparing objective and subjective measures during two experimental treatments, with and without topic detection. Each participant was introduced and trained with the prototype application and the experimental scenario. The experiment lasted for 1-2 hours per day, for 3 days, not including the time to complete the questionnaire. The qualitative measures were solicited after each chat by an exit questionnaire. After the experiment was finished, the participants were asked to answer several questions, measuring their overall impressions of the prototype system.

3.3 Method

A 12-item exit questionnaire was designed to solicit self-reported assessment of subjective perceptions and feelings experienced overall and during two different treatment conditions. The questions concerning individual treatments were administered after each experimental treatment to avoid potential confusion. A five-point Likert scale was used for a cluster of questions, which were collecting feedback on the quality of experience. Another cluster of questions were answered by Yes or No.

3.4 Discussion of Results

Subjective judgments showed a positive user experience with the extensions made to a traditional chat client application. A number of statistical differences were found when comparing treatments, indicating that by providing more context-related features, the user performance and satisfaction were improved with respect to many subjective metrics. Based on the extensive experimenter's observations and performance results few weaknesses of the current prototype have been revealed and possibilities for improvements have been suggested. The results of the subjective measures analyses and observations are summarized and discussed in the following statements:

- The performance time for searching a particular topic in the chat history have been reduced by 80% compared to the treatment where topic header was not marked (approx. 3 minutes per search).
- Participants were not distracted by the inclusion of the two novel features; the theme and the content of chats were not affected by them as reported.
- In 80% of the cases, the participants inspected the selected topic at the end of a chat session. The topic changes were monitored more at the beginning of the chat than later. The perceived usefulness of the associated topic header was mostly correlated with chat history search.
- The duration of the chat conversation were considerably longer for younger users due to the time needed for data transfers with the corresponding web services. The time needed for topic detection, tens of seconds for each message, was unobservable in more mature age groups who, nevertheless, seem to take longer time in writing a message.
- The inclusion of supplemental content in the form of related tweets was accepted better by younger people and found more appropriate for informal themes, which was something that we have intuitively expected. The option of implementing the tweet links as a gadget that can be turned on and off is suggested and considered in our future adjustments.

It was observed that there was a difference in users' attitudes and behavior correlated to their age, as shown in Table 1. We are considering several design adjustments and adaptive features to target stereotypical users' groups.

Table 1. Impact of topic header and tweet links across different age groups.

Age group	Number of contacts	Average chat duration (minutes)	Augmented Chat duration (minutes)	Attention to new features	Persistence of interest in new features
15-18	60	10-30	15-35	On the first appearance	Throughout a session
18-26	30	10-15	15-25	On the first appearance	During the first half of a session
26-50	30	5-10	10-20	After several appearances	Lost interest in the first quarter of a session
50-70	10	20-30	20-30	After it was pointed out by the experimenter	Topic header checked at the end of a session Not interested in tweets

4 Future Work

For the problem under investigation, the semantic approach to keyword selection and subsequently topic creation implies an existence of a suitable contextual and semantic knowledge base. The design and creation of such a database was out of the scope of our research work, hence the selection of an existing online resource such as Freebase. The results of our initial exploration, although shows satisfying results, it also points out to few drawbacks regarding precision and exactness of topic creation.

We are investigating the feasibility of improving the system and its inferring capabilities by using the WordNet lexical database [8]. Its structure, which includes synonym words accompanied by description for every *synset*, has already been proven to accommodate the requirements of our application in a small trial we have conducted. Additional capabilities such as: sentence tagging to help determine the types of the words used in a message, and algorithms for semantic similarity [15] are expected to strengthen our prototype application.

The time needed for data transfer from and to the two public web services could significantly undermine the usefulness of the proposed approach. Further improvements are considered using approaches such as periodic, off-line and on-demand topic detection as opposed to the real-time option. Designing and using additional locally-stored context maps and ontologies invite some further research efforts. We are also looking into alternative techniques to optimize the preprocessing and processing phase in terms of time and effectiveness.

Considering daily tweets have often resulted in linking keywords with meaningless or irrelevant information. The prototype application may require additional consideration to properly address the issue of providing relevant and useful supplemental information.

5 Conclusions

In this research, we have explored the feasibility of extending instant messaging applications with context-related features, namely automatic topic detection and linking keywords to related tweets. We have proposed a set of preprocessing techniques, weighted word combination, and semantic topic formation based on the contextual data provided by an online collection known as Freebase. The prototype version of a client chat application was implemented to explore the design space, and identify the limitations and potential adjustments to the proposed topic detection framework.

The performance and the usability of the proposed approach have been evaluated in a targeted exploratory study. The experimental data analysis have shown that our prototype implementation have improved the time for searching a particular topic by 80% compared to the traditional chats with no topic inclusion. The results of our initial investigation are encouraging evidence that by enhancing the conversational space of IM chats with automatically-detected topic headers and related daily tweets, the quality of user experience could be improved.

References

1. Bengel, J., Gauch, S., Mittur, E., Vijayaraghavan, R.: Chattrack - Chat room topic detection using classification. In: 2nd Symposium on Intelligence and Security Informatics, Tucson, Arizona (2004)
2. Hansen, L.K., Larsen, J., Kolenda, T.: Blind Detection of Independent Dynamic Components. In: IEEE International Conference on Acoustics, Speech, and Signal Processing, vol. 5, pp. 3197–3200 (2001)
3. Ozyurt, O., Kose, K.: Chat Mining: Automatically Determination of Chat Conversations Topic in text based chat mediums. Journal on Selected Areas in Expert Systems with Applications ESWA 4843 (2010)
4. Seymore, K., Rosenfeld, R.: Using Story Topics for Language Model Adaptation. In: Eurospeech 1997 (1997)
5. Dong, H., Hui, S.C., He, Y.: Structural analysis of chat messages for topic detection. Online Information Review 30(5), 496–516 (2006)
6. Zhang, J., Yang, Y., Carbonell, J.: New Event Detection with Nearest Neighbor, Support Vector Machines, and Kernel Regression. CMU Technical Report CMU-CS-04-118, CMU-LTI-04-180 (2004)
7. Freebase, repository of structured data (accessed March 20, 2011), `http://www.freebase.com`
8. WordNet®, a lexical database for English, `http://wordnet.princeton.edu.` (accessed August 10, 2011)
9. Masaitene, D.: Analyzing Casual Conversations. Respectus Philogicus 1(16) (2001)
10. Navarro, G.: A guided tour to approximate string matching. ACM Computing Surveys 33(1), 31–88 (2001)
11. Apte, C., Damerau, F., Weiss, S.: Automated learning of decision rules for text categorization. ACM Transactions on Information Systems (1994)
12. Twitter, an online social networking and microblogging service, `http://twitter.com/` (accessed August 15, 2011)
13. Apte, C., Damerau, F., Weiss, S.: Text mining with decision rules and decision trees. In: Conference on Automated Learning and Discovery, Workshop 6: Learning from Text and the Web (1998)
14. TweetSharp,NET library for micro-blogging platform, `http://tweetsharp.codeplex.com` (accessed March 20, 2011)
15. Jiang, J., Conrath, D.: Semantic similarity based on corpus statistics and lexical taxonomy. In: International Conference on Research in Computational Linguistics, pp. 19–33 (1997)

Modality Classification Using Texture Features

Ivan Kitanovski, Katarina Trojacanec, Ivica Dimitrovski, and Suzana Loskovska

Abstract. Medical image classification based on the image modality is one of the most important and crucial tasks in the medical image analysis. Due to its importance, the aim of the paper is to investigate modality medical image classification problem by using a combination of several classification techniques and feature extraction algorithms over a set of medical images. Four feature extraction methods were used in this paper: LBP, GLDM, GLRLM, Haralick texture features. Additionally we concatenated all four features in one single feature to assess their joint performance. The feature extraction algorithms are tested over three classifiers: SVM extended for multiclass classification based on one against all strategy, k-nearest neighbor and C4.5 algorithm. This examination was conducted over a set of medical images provided by ImageCLEF. The dataset contains 18 classes of images, with a total of 988 images. The distribution of number of images per class is not uniform, so this additionally burdens the task. The best results were provided when the images were described with concatenated descriptors and classified with SVM classifier, with a classification accuracy of 73%.

Keywords: Medical images, modality based classification, feature extraction, LBP, GLDM, GLRLM, Haralick, SVM, k-nearest neighbor, C4.5.

1 Introduction

The amount of medical images produced daily in healthcare centers, hospitals and research institutes has rapidly increased. Consequently, the effective image retrieval systems have received more attention in recent years and have become one of the most challenging research topics. In such systems, images need to be annotated and organized efficiently.

Ivan Kitanovski · Katarina Trojacanec · Ivica Dimitrovski · Suzana Loskovska
Faculty of Electrical Engineering and Information Technologies, Skopje, Macedonia
e-mail: {ikitanovski,katarina,trojacanec,
ivicad,suze}@feit.ukim.edu.mk

L. Kocarev (Ed.): ICT Innovations 2011, AISC 150, pp. 189–198.
springerlink.com

At the beginning, the task of image annotation has been performed manually. However, according to the enormous amount of images nowadays, the manual image annotation has several serious disadvantages. In fact, it is tedious, impractical, highly subjective and, as a result, prone to errors. Therefore, the necessity for automating the annotation process has arisen. Several machine learning techniques are commonly used to automate the process of annotation. A collection of previously annotated images is used to train a classifier that afterwards would be able to annotate previously unseen images. Wide research is continuously performed towards retrieval systems that incorporate automated image classification. Because these systems are based on the features directly extracted from the visual image content itself, they are referred to as content based image retrieval (CBIR) systems [1].

Medical image databases used to perform medical image retrieval often consists of images of various modalities. Image modality is an important and fundamental image characteristic. Thus, the modality detection is one of the basic tasks involved in the medical image retrieval systems. It is intended to detect the acquisition modality of the images in a large medical image databases and can be used to improve the medical image retrieval process [2].

Medical image classification based on the image modality has recently been investigated. The efficiency of two different systems (MedIC and MedGIFT) in classifying the medical image modality has been compared in [3]. In [4], the authors compared their modality classifier and a NaiveBayes Nearest Neighbor technique for the modality classification using visual features. Authors in [2] used a neural network-based hierarchical scheme for the modality classification problem.

However, taking into account the specific nature of medical images, not all visual features can be used to efficiently and correctly describe the visual information in the image itself. Several feature extraction methods are investigated for solving modality classification problem. An improved version [5], [6] of the Fisher vector [7] is used for feature extraction from the medical images in [8]. In [9] the authors use three global texture features for feature extraction: LBP, Tamura and Gabor texture features. Combination of texture features and color intensity histograms was used in [10].

The paper is organized as follows. Section 2 provides a brief explanation of the classification techniques for modality classification. Section 3 describes the importance feature extraction techniques used for analysis. In section, 4 the dataset used for evaluation is presented, while section 5 provides the experimental results. The concluding remarks are given in section 6.

2 Classification Techniques

Various classification techniques are subject of continuous research such as support vector machines (SVM) [11], the k-nearest neighbor algorithm [12], C4.5 algorithm [13], [14], neural networks [12], Bayesian classifier [12] etc. For the purpose of our research we used the most popular classification algorithms: SVM, K-NN and C4.5. We use these classifiers, because they are mostly used for this task [2], [3], [8].

2.1 Support Vector Machines

Support Vector Machines (SVM) are proposed by Vapnik and his colleagues [15], [16], [17]. SVMs are amongst the most efficient classifiers that exist nowadays. Although, they are designed basically for binary classification, they can be extended to solve multiclass classification problems as well. There are many strategies to combine SVMs for multiclass classification problem. One possible strategy, that we implemented for our paper, is the one-against-all (OvA) scheme. It includes decomposition of the M-class problem (M>2) into series of two-class problems. The basic concept is to construct M SVMs where the *i*-th classifier is trained to separate the class *i* from all other (M-1) classes. This strategy has a few advantages such as its precision, the possibility for easy implementation and the speed in the training phase and the recognition process.

2.2 K Nearest Neighbors Classifier

The k-nearest neighbor algorithm (k-nn) is one of the most simplest and efficient classification techniques [18]. The classification process behind k-nn algorithm is performed by finding the closest training samples to the query one. Then the class of the unlabeled query sample is decided by a voting procedure. K is a parameter which can be adjusted and is usually an integer number. It specifies the number of nearest neighbors, which are taken into consideration for the voting process. The main issue at this point is the way how the distance is calculated between samples.

The main disadvantage of k-nn method is that the classes with far greater number of examples than other classes, tend to dominate the prediction process i.e. objects have a greater probability to be labeled as members of the dominant classes.

2.3 C4.5 Algorithm

C4.5 algorithm is a widely used classification algorithm in pattern recognition. The algorithm builds decision trees for a given training set by using of the concept of information entropy [19], [20]. It is similar to ID3 algorithm [21], but with few improvements.

C4.5 is made to handle both continuous and discrete attributes. In the case of continuous values C4.5 creates a threshold and then splits the list into the values of the attributes which are above the threshold and the values which are below or equal to it. The algorithm also handles nominal attributes. Once the tree is constructed the algorithm scans the whole tree again, removes the braches which do not influence the decision making process, and replaces them with leaf nodes.

The decision tree is fast to train and built, but on the other hand, it requires a large amount of data to develop relevant decision making abilities.

3 Feature Extraction

In the field of image processing, there are many ways to extract visual content from images. Features can generally be extracted by using color, texture or shape of the image. The research has shown [3] that visual content of medical images is best described using texture and shape features. Our research is focused on analyzing the performance of texture-based descriptors. We are using four well known statistical texture-based feature descriptors: LBP (Local Binary Patterns), Haralick texture features, GLRLM (Gray Level Run Length Method) and GLDM (Gray Level Difference Method). LBP and Haralick features are used in [9] and [10] for the same task as ours. We decided to compare them along two other statistical texture features: GLDM and GLRLM. Concatenation was made with the intent to incorporate the benefits of all descriptors.

3.1 LBP

Local Binary Patterns is a method that basically uses gray-scale invariant texture statistics. The method associates a binary number (binary pattern) to every pixel. The binary number associated with the pixel is based on the result of the thresholding between the center pixel and his neighbor pixels. The basic version of the LBP is considering only the eight neighbor pixels of the pixel [22], but there are also implementations where all circular neighborhoods are considered [23]. Our LBP descriptor is generated using (8,1) neighborhood using uniform patterns, thus, generating 59 values for each image.

3.2 Haralick Texture Descriptor

Haralick texture descriptor is based on the co-occurrence matrix calculated from the image. The co-occurrence matrix is used for displaying the gray level spatial dependency along different angular relationships, vertical and horizontal directions in the image. From the calculated co-occurrence matrix many different texture features can be produced. Haralick [24] proposed a combination of 13 features derived from the matrix for defining the image texture information: angular second moment, contrast, correlation, sum of squares, inverse difference moment, sum average, sum variance, sum entropy, entropy, difference variance, difference entropy, information measures of correlation, maximal correlation coefficient. The Haralick descriptor is calculated in four directions: 0, 45, 90, 135 and additional average matrix of all four is calculated. It means that we get 5 matrices. For each of the matrices, we calculate the features and we concatenate them all in one feature vector.

3.3 GLDM

To describe the gray level difference method (GLDM) [25], [26], let $g(n,m)$ be the digital picture function. For any given displacement $\delta = (\Delta n, \Delta m)$, where Δn and

Δm are integers, let $g_\delta(n, m) = |g(n, m) - g(n+ \Delta n, m+ \Delta m)|$. Finally, let $f(\cdot \mid \delta)$ be the estimated probability density function associated with the possible values of g_δ, i.e., $f(i \mid \delta) = P(g_\delta(n, m) = i)$. For our experiment four possible forms of the vector δ will be considered, $(0, d)$, $(-d, d)$, $(d, 0)$, $(-d, -d)$, where d is the inter sample spacing distance. We will refer to the functions $f(\cdot \mid \delta)$ as the gray level difference density functions. From each of these density functions, five texture features are defined: contrast, angular second moment, entropy, mean and inverse difference moment. Our GLDM feature descriptor is calculated for five displacements (1, 2, 3, 4 and 5) and, thus, our implementation has 25 features.

3.4 GLRLM

Gray Level Run Length Method is based on calculating the number of gray level runs of various lengths [25], [27]. The gray level runs are characterized by the direction of the run, the length of the run, and the grey level of the run. For the direction of the runs, we have considered four directions: 0, 45, 90, 135 degrees. The length of the run is defined as the number of consecutive pixels having the same gray level value. For every direction of the runs, gray level run length matrices are computed for each image. Based on these four matrices, eleven features are computed for each matrix: short run emphasis, long run emphasis, gray level distribution, run length distribution, run percentages, low gray-level run emphasis, high gray-level run emphasis, short run low gray-level emphasis, short run high gray-level emphasis, long run low gray-level emphasis, long run high gray-level emphasis. Our GLRLM feature descriptor computes 44 features, because it is calculated in four directions, so each feature is calculated for every matrix.

4 Dataset Description

For the purposes of our research, we used the CLEF2011 dataset [28]. The original dataset consists of 238000 medical images. The dataset contains 18 image modalities: 3D reconstruction (3D), angiography (AN), computed tomography (CT), magnetic resonance imaging (MR), x-ray (XR), ultrasound (US), retinography (RN), general photos (PX), histopathology (HX), dermatology (DM), electron-microscopy (EM), endoscopic imaging (EN), gross pathology (GR), gel (GL), graphs (GX), fluorescence (FL), drawing (DR) and compound figures (containing more than one type of image - CM). But, for the purposes of modality classification, a subset of 988 images was selected from original dataset. For, each of these images, a textual annotation was provided, which described the modality of the image. The images were organized in the scheme shown on Fig. 1.

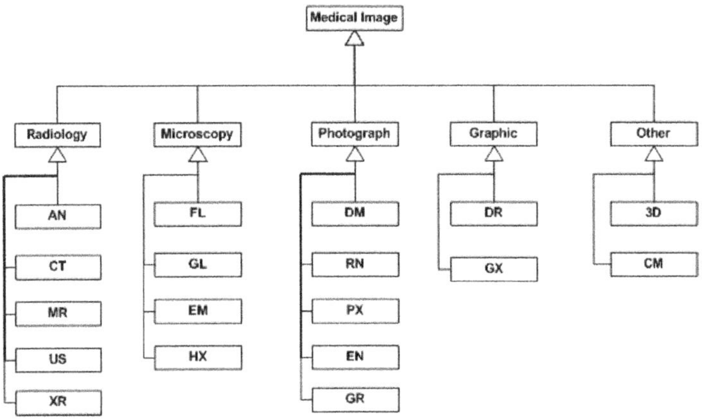

Fig. 1 Hierarchical organization of the dataset.

Table 1. Distribution of the images through the classes

Class	Train set	Test set	Total
3D	24	8	32
AN	10	1	11
CM	10	7	17
CT	58	12	70
DM	5	2	7
DR	40	3	43
EM	13	3	16
EN	8	2	10
FL	33	11	44
GL	45	5	50
GX	129	32	161
GR	40	3	43
HX	163	45	208
MR	13	4	17
PX	131	34	165
RN	3	2	5
US	24	6	30
XR	52	7	59
TOTAL	801	187	988

The first level is the general medical image class, which includes all images in the dataset. In the second level, the images are divided in logically similar origins. But, according to the hierarchy on Fig, 1, we can conclude that not all second level image classes are visually coherent, i.e. not all image subclasses of a second level class have similar visual characteristics. For example, subclasses of the Other class, can be visually different since they can present a variety of visual content.

The distribution of images within the classes is another key factor, which must be taken into consideration. Table 1 depicts the distribution of the number of images through the classes. From, the table it can be noted that the dataset has been split into the train and test subset. The dataset was split randomly, with 80% of its images belonging to the train subset and the other 20% were used as test images.

5 Experimental Results

The aim of the paper is to compare the performance of four texture-based feature extraction algorithms. Their performance is measured by classifying the extracted feature vectors using three classification algorithms. This should provide information which combination of feature extraction algorithm and classifier will yield the best performance.

We used four texture-based descriptors for feature extraction from the visual content of the image: Local Binary Patterns (LBP), Haralick texture features, Gray Level Difference method (GLDM) and Gray Level Run Length Method (GLRLM). The result of the feature extraction process is separate feature vector obtained for each of the images belongs to both, the train and the test set, for each descriptor. The results are then normalized using min-max normalization technique, for every descriptor separately.

The classification process was performed using three classification techniques, previously described: Support vector machines based on one-against-all (OvA) scheme, K - nearest neighbor classifier and C4.5 algorithm.

We implemented the SVMs for multiclass classification in Matlab R2010a using the LibSVM library [29]. The SVMs use 1-vs-All strategy for multiclass classification. For the k nearest neighbor classifier and C4.5 algorithm we used Weka [30] implementation. The optimization of k-nn and C4.5 was performed using CVParameter selection option in Weka. In the case of the SVMs, the parameters σ and c were provided using grid search.

The classification accuracy calculated for each classifier is depicted in Table 2.

Table 2. Classification Accuracy for Feature Extraction Algorithm per Classifier

Classification Accuracy for Feature Extraction Method per Classfication Algorithm	Feature Extraction Methods					
	LBP	GLRLM	GLDM	Haralick	Combined	
Classification Algorithms	SVM	64,1711	68,4492	56,1467	67,3797	73,262
	K-NN	56,1497	57,754	49,7326	63,6364	65,7754
	C4.5	51,3369	51,8717	42,7807	49,7326	56,1497

Evidently, the best classification accuracy was achieved in the case of SVM classifier when the concatenated descriptor was used for feature extraction of the images. The classification accuracy in this case is 73.262%, which is satisfactory taking into consideration the non-uniform distribution of the number of images per classes. Furthermore, in the case of other classifiers, K-NN classifier and C4.5 algorithm, the concatenated descriptor also gives the best performance compared to the other feature extraction algorithms.

6 Conclusion

In this paper, we investigated the performance of four texture-based feature extraction techniques, by using three classification methods. We have added the fifth, aggregated descriptor, obtained by the concatenation of the four descriptors. This aggregated descriptor provides the best performance in the case of all classifiers. The best classification accuracy (73.262%) in general was achieved when the combination of the aggregated descriptor and SVM classifier with one-vs-all strategy was used. Among the other descriptors we should emphasize that the Haralick descriptor and GLRLM yielded significant performance of 67.37% and 68.44%, respectively, when the SVM classifier was used for classification.

Modality classification of medical images is a very complex and sensitive problem. However, even the slightest improvement in this field is of great clinical importance.

References

1. Müller, H., Michoux, N., Bandon, D., Geissbuhler, A.: A review of content-based image retrieval systems in medical applications – clinical benefits and future directions. International Journal of Medical Informatics 73, 1–23 (2004)
2. Kalpathy-Cramer, J., Hersh, W.R.: Automatic Image Modality Based Classification and Annotation to Improve Medical Image Retrieval. In: MedInfo, pp. 1334–1338 (2007)

3. Florea, F., Müller, H., Rogozan, A., Geissbühler, A., Darmoni, S.: Medical image categorization with MedIC andMedGIFT. In: Medical Informatics Europe (2006)
4. Kalpathy-Cramer, J., Hersh, W.: Multimodal medical image retrieval: image categorization to improve search precision. In: MIR 2010: Proceedings of the International Conference on Multimedia Information Retrieval, pp. 165–174. ACM, New York (2010)
5. Perronnin, F., Sanchez, J., Liu, Y.: Large-scale image categorization with explicit data embedding. In: CVPR (2010)
6. Perronnin, F., Liu, Y., Sanchez, J., Poirier, H.: Large-scale image retrieval with compressed fisher vectors. In: CVPR (2010)
7. Perronnin, F., Dance, C.: Fisher kernels on visual vocabularies for image categorization. In: CVPR (2007)
8. Muller, H., Kalpathy-Cramer, J., Eggel, I., Bedrick, S.: Overview of the clef 2010 medical image retrieval track. In: Working Notes of CLEF 2010, Padova, Italy (2010)
9. Wu, H., Hu, C., Chen, S.: UESTC at Image. CLEF 2010 Medical Retrieval Task (2010)
10. Kalpathy-Cramer, J., Hersh, W.: Automatic image modality based classification and annotation to improve medical image retrieval. Studies in Health Technology and Informatics 129(2), 1334 (2007)
11. Christopher Burges, J.C.: A Tutorial on Support Vector Machines for Pattern Recognition. Data Mining and Knowledge Discovery 2, 121–167 (1998)
12. Wareld, S.K., Kaus, M., Jolesz, F.A., Kikinis, R.: Adaptive, template moderated, spatially varying statistical classification. Med. Image Anal. 4(1), 43–55 (2000)
13. Boser, B.E., Guyon, I.M., Vapnik, V.N.: A training algorithm for optimal margin classifiers. In: Fifth Annual Workshop on Computational Learning Theory, pp. 144–152. ACM, Pittsburgh (1992)
14. Kotsiantis, S.B.: Supervised Machine Learning: A Review of Classification Techniques. Informatica 31, 249–268 (2007)
15. Vapnik, V.: The Nature of Statistical Learning Theory, 2nd edn. Springer, New York (1999)
16. Burges, C.J.C.: A tutorial on support vector machine for pattern recognition. Data Min. Knowl. Disc. 2, 121 (1998)
17. Joachims, T.: Making large-scale SVM learning practical. In: Schölkopf, B., Burges, C.J.C., Smola, A.J. (eds.) Advances in Kernel Methods—Support Vector Learning, pp. 169–184. MIT Press, Cambridge (1999)
18. Wu, X., Kumar, V., Quinlan, J.R., Ghosh, J., Yang, Q., Motoda, H., McLachlan, G., Liu, B., Yu, P., Zhou, Z., Steinbach, M., Hand, D., Steinberg, D.: Top 10 algorithms in data mining. Knowledge and Information Systems 14(1), 1–37 (2008)
19. Quinlan, J.R.: C4.5: Programs for Machine Learning. Morgan Kaufmann Publishers (1993)
20. Quinlan, J.R.: Improved use of continuous attributes in c4.5. Journal of Artificial Intelligence Research 4, 77–90 (1996)
21. Quinlan, J.R.: Induction of decision trees. Machine learning 1(1), 81–106 (1986)
22. Ojala, T., Pietikainen, M., Harwood, D.: A Comparative Study of Texture Measures with Classification Based on Feature Distributions. Pattern Recognition 29(1), 51–59 (1996)
23. Ojala, T., Pietikainen, M., Maenpaa, T.: Multiresolution Gray-Scale and Rotation Invariant Texture Classification with Local Binary Patterns. IEEE Trans. Pattern Analysis and Machine Intelligence 24(7), 971–987 (2002)

24. Haralick, R.M., Shanmugam, K., Dinstein, I.: Textural features for image classification. IEEE Transactions on Systems, Man, and Cybernetics, 610–621 (1973)
25. Weszaka, J.S., Dyer, C.R., Rosenfeld, A.: A comperative study of texture measures for terrain classification. IEEE Trans. on Syst., Man, Cyber., 269–285 (1976)
26. Conners, R.W., Harlow, C.A.: A theoretical comparison of texture algorithms (1980)
27. Galloway, M.M.: Texture analysis using gray level run lengths, Comput. Graphics Image Processing 4, 172–179 (1975)
28. Muller, H., Kalpathy-Cramer, J., Eggel, I., Bedrick, S., Kahn Jr., C. E., Hersh, W.: Overview of the CLEF 2010 medical image retrieval track. In: The Working Notes of CLEF 2010 (2010)
29. Joachims, T.: Making large-scale SVM learning practical. In: Advances in Kernel Methods, pp. 169–184. MIT Press, Cambridge (1999)
30. Witten, I., Frank, E.: Data Mining: Practical machine learning tools and techniques, San Francisco (2005)

Estimation of Minimal Path Vectors of Multi-state Systems from Failure Data

Marija Mihova and Zaneta Popeska

Abstract. Consider a multi-state system with n-components multi-state components with graduate failure. We give an algorithm for determining the minimal path set to level j for given minimal cut set of level j and it's estimate for a given subset of failure vectors. We get an upper bound of the reliability function.

Keywords: Multi-state system, minimal path set, minimal cut set, reliability function.

1 Introduction

A multi state system with multi-state components, MSMC, is a system such that it and its components can operate in more than one level of performance. The first attempts to formally define multi-state monotone systems and multi- state coherent systems are made by Natvig, [[2]], [[3]] and [[4]], where he gives some initial mathematical models for them. Some statistical approach is given in [[5]]. In this paper we regard a multi-state system with n components, and the set state S. Suppose that the i-th component can be in one of the $M_i + 1$ levels, where M_i is the level of perfect work of the component and 0 is the level of its total failure. We define a **state vector** $\mathbf{x} = (x_1, x_2, \ldots, x_n)$ where x_i is the state of the i-th component of the system, for $0 \leq x_i \leq M_i$ and $1 \leq i \leq n$. Let $E = \{\mathbf{x} | 0 \leq x_i \leq M_i,\ 1 \leq i \leq n\}$ be the set of all state vectors of the system, the **state set** of the system. The function $\phi : E \to S$ is called the structure function. By (E, ϕ) we will denote the system with state set E and structure function ϕ. The ordering on the set E is defined as $\mathbf{x} \leq \mathbf{y}$ iff $\forall i\ 1 \leq i \leq n,\ x_i \leq y_i$. The system is coherent if $\phi(\mathbf{x})$ is monotone and every component is essential to the system. The vector $\mathbf{x} \in E$ is called **path** (**cut**) vectors to level

Marija Mihova · Zaneta Popeska
"Ss Cyril and Methodius" University, Faculty of Informatics Sciences and Computing Engineering, Ruger Boshkovikj 393, 1000 Skopje, Makedonija
e-mail: {marija.mihova, zhaneta.popeska}@finki.ukim.mk

L. Kocarev (Ed.): ICT Innovations 2011, AISC 150, pp. 199–206.
springerlink.com

j if $\phi(\mathbf{x}) \geq j$ ($\phi(\mathbf{x}) < j$). A vector $\mathbf{x}$ is a **minimal path vector to level** j iff it is a path vector to level j and all vectors smaller than $\mathbf{x}$ are cut vectors to level j. A vector $\mathbf{x}$ is a **minimal cut vector to level** j iff it is a cut vector to level j and all vectors greater than $\mathbf{x}$ are path vectors to level j. The set of all minimal path vectors to level j is called **minimal path set to level** j, $\mathscr{P}_j$, and the set of all minimal cut vectors to level j is called **minimal cut set to level** j, $\mathscr{C}_j$. From the pervious discussion it is clear that the structure function for an system (E,ϕ), may be specified by minimal path (or cut) vectors of all levels j, as $(E,\{\mathscr{P}_j|, j=\overline{1,M}\})$ (or shortly $(E,\{\mathscr{P}_j\})$) in the case of minimal path vectors, and as $(E,\{\mathscr{C}_j|, j=\overline{1,M}\})$ (or shortly $(E,\{\mathscr{C}_j\})$) in the case of minimal cut vectors. When the interest is focused only on one level j, we will use notation $(E,\mathscr{P}_j)$ (or $(E,\mathscr{C}_j)$).

In this paper we analyze the minimal path and cut vectors for systems with coherent structure. Moreover we assume that the system is consists of components with **graduate failure**, i.e. the process of failure is moving one "step" at a time, by changing one component by one level down at each move. Let $p_{\mathbf{x}}$ be the probability that the system during its work is found in the state $\mathbf{x}$, for $\mathbf{x} \in E$, then because we regard systems with graduate failure, $p_{\mathbf{x}} > 0$, for all $\mathbf{x} \in \mathscr{P}_0 \cup C_{E0}$.

Let $\mathbf{x}$ be the state vector of the system at the moment of failure under some level j. This vector is a cut vector to level j, but it does not mean that it is also a minimal cut vector to level j. On the other side, there are cut vectors to level j in which the system can not be at the moment of failure under level j. For that reason we define a fatal vector.

Definition 1. The vector $\mathbf{y}$ is a **fatal vector** to level j iff it is a cut vector to level j and there is a path vector $\mathbf{x}$ to level j such that $\mathbf{x}-\mathbf{y}=\mathbf{e}_i$, where $\mathbf{e}_i$ is the unit n-vector with 1 as a i-th component. The set of all fatal vectors to level j is called **fatal set to level** j, $\mathscr{C}_{Ej}$.

In the Section 2, we give an algorithm for obtaining the minimal path set of the system from the minimal cut set. This algorithm can be used to estimate system reliability, when the system structure is not known, but we know the state of the system in the moment of failure, i.e. failure data [[1]]. Similar consideration about binary systems is given in [[1]].

2 Obtaining the Minimal Path Set from the Minimal Cut Set

Let (E,ϕ) be a MSMC. For $\mathbf{x},\mathbf{y} \in E$ we define an operation $\otimes$ by

$$\mathbf{x} \otimes \mathbf{y} = (\max(x_1,y_1), \max(x_2,y_2), \ldots, \max(x_n,y_n)). \tag{1}$$

Similarly, for two sets $A,B \subseteq E$ we define product $A \otimes' B$ by:

$$A \otimes' B = \{\mathbf{x} \otimes \mathbf{y} | \, \mathbf{x} \in A, \mathbf{y} \in B\}. \tag{2}$$

The set $A \otimes B$ is obtained from $A \otimes' B$ by removing all non-minimal vectors (and keeping the other elements only once). It is clear that $\otimes$ on state vectors is commutative and associative operation, so the operation $\otimes$ on sets of state vectors is commutative and associative also.

For each vector $\mathbf{x} \in E$ we define the set

$$T(\mathbf{x}) = \{(0,0,\ldots,0,x_i+1,0,\ldots,0)|\ x_i \neq M_i\ 1 \leq i \leq n\}.$$

The following Lema determines the set of all minimal path vectors to level j for the system $(E,\mathscr{C}_j^{(s)})$, where $\mathscr{C}_j^{(s)}$ is some s-element subset from $\mathscr{C}_j$.

Lemma 1. *Let $(E,\{\mathscr{C}_r\})$ be a multi-state coherent system with graduate failure and $\mathscr{C}_j^{(s)}$ is a s-element subset from $\mathscr{C}_j$. Then the minimal path set to level j for $(E,\mathscr{C}_j^{(s)})$ is*

$$\mathscr{P}_j^{(s)} = \prod_{\mathbf{x}\in\mathscr{C}_j^{(s)}} T(\mathbf{x}). \tag{3}$$

Proof. Without lose of generality we may suppose that $\mathscr{C}_j^{(s)}$ is the set of the first s elements of $\mathscr{C}_j$.

For $s=1$, $\mathscr{C}_j^{(1)} = \{\mathbf{x}^{(1)}\}$. We need to proof that the minimal path set to level j for this system is $\mathscr{P}_j^{(1)} = T(\mathbf{x}^{(1)}) = \{(0,\ldots,0,x_i^{(1)}+1,0,\ldots,0)|x_i < M_i, 1 \leq i \leq n\}$. It is clear that the vectors in $\mathscr{P}_j^{(1)}$ are not smaller than $\mathbf{x}^{(1)}$. Moreover, all vectors smaller then some of the vectors in $\mathscr{P}_j^{(1)}$, are smaller then $\mathbf{x}^{(1)}$ too. So, each vector from $\mathscr{P}_j^{(1)}$ is a minimal path vectors to level j. Remain to proof that all minimal path vectors to level j are in $\mathscr{P}_j^{(1)}$. Let $\mathbf{a}$ be a minimal path vector to level j. Because $\mathbf{a} \leq \mathbf{x}^{(1)}$, there is a coordinate $i=\overline{1,n}$, such that $a_i > x_i^{(1)}$. Now $\mathbf{a} \geq (0,\ldots,0,a_i,0,\ldots,0) \geq (0,\ldots,0,x_i^{(1)},0,\ldots,0)$. But, $(0,\ldots,0,x_i^{(1)},0,\ldots,0)$ is a minimal path vector to level j, so it must be true $\mathbf{a} = (0,\ldots,0,x_i^{(1)},0,\ldots,0) \in \mathscr{P}^{(1)}$.

Suppose that for all $s<k$ the set $\mathscr{P}_j^{(s)}$, defined as in (3), is a minimal path set to level j for MSMC $(E,\mathscr{C}_j^{(s)})$, $\mathscr{C}_j^{(s)} = \{\mathbf{x}^{(1)},\mathbf{x}^{(2)},\ldots,\mathbf{x}^{(s)}\}$. Let $\mathscr{P}_j^{(s)} = \{\mathbf{a}^{(1)},\mathbf{a}^{(2)},\ldots,\mathbf{a}^{(t)}\}$. Then, for each $l \in \{1,2,\ldots,t\}$ and each $1 \leq i \leq n$, such that $x_i < M_i$,

$$\mathbf{a}^{(l)} \otimes (0,...,0,x_i^{(s+1)}+1,0,...,0) = (a_1^{(l)},...,a_{i-1}^{(l)},\max(a_i^{(l)},x_i^{(s+1)}+1),a_{i+1}^{(l)},...,a_n^{(l)}).$$

We have that

$$\mathbf{a}^{(l)} \otimes (0,\ldots,0,x_i^{(s+1)}+1,0,\ldots,0) \geq \mathbf{a}^{(l)}$$
$$\mathbf{a}^{(l)} \otimes (0,\ldots,0,x_i^{(s+1)}+1,0,\ldots,0) \geq (0,\ldots,0,x_i^{(s+1)}+1,0,\ldots,0).$$

This proves that all vectors obtained by (3) are paths to level j for the system $(E,\mathscr{C}_j^{(s+1)})$. Next we are going to proof that all minimal path vectors to level j for

$(E,\mathscr{C}_j^{(s+1)})$ are in $\mathscr{P}_j^{(s+1)}$. Keeping in mind that by our procedure all non-minimal vectors are removed, that is will prove that $\mathscr{P}_j^{(s+1)}$ is the minimal path set to level j.

Suppose that the vector $\mathbf{y}$ is a minimal path vector to level j for $(E,\mathscr{C}_j^{(s+1)})$, which is not in $\mathscr{P}_j^{(s+1)}$. This means that $\mathbf{y}$ may not be obtained by (3). It is clear thet $\mathbf{y}$ is a path vector for $(E,\mathscr{C}_j^{(s)})$.

Since $\mathbf{y}$ is minimal path vector to level j, we have that $\forall i=\overline{1,n}$ and $y_i \neq 0$, the vectors $(y_1,\dots,y_i-1,y_{i+1},\dots,y_n)$ are cut vectors, i.e.

$$\exists l_i \in \{1,2,\dots,s+1\}, (y_1,\dots,y_i-1,y_{i+1},\dots,y_n) \leq \mathbf{x}^{(l_i)}. \tag{4}$$

Two cases are possible:

1. When $\forall i=\overline{1,n}$, $\exists l_i \in \{1,2,\dots,s\}$ such that $(y_1,\dots,y_i-1,y_{i+1},\dots,y_n) \leq \mathbf{x}^{(l_i)}$, then $\mathbf{y}$ is a minimal path vector to level j for the system $(E,\mathscr{C}_j^{(s)})$, i.e. $\mathbf{y} \in \mathscr{P}^{(s)} = \{\mathbf{a}^{(1)},\mathbf{a}^{(2)},\dots,\mathbf{a}^{(k)}\}$. We claim that $\exists i=\overline{1,n}$, such that

$$\mathbf{y} \otimes (0,\dots,0,x_i^{(s+1)}+1,0,\dots,0) = \mathbf{y}.$$

Namely, if that is not true, i.e. when $\forall i=\overline{1,n}$ and $x_i < M_i$, $\mathbf{y} \otimes (0,\dots,0,x_i^{(s+1)}+1,0,\dots,0) \neq \mathbf{y}$, then

$$\mathbf{y} \otimes (0,\dots,0,x_i^{(s+1)}+1,0,\dots,0) = (y_1,\dots,y_{i-1},x_i^{(s+1)}+1,y_{i+1},\dots,y_n).$$

In fact, $\forall i=\overline{1,n}$, $y_i < x_i^{(s+1)}+1$, i.e. $y_i \leq x_i^{(s+1)} \Rightarrow \mathbf{y} \leq \mathbf{x}^{(s+1)} \Rightarrow \mathbf{y}$ is a cut vector to level j for $(E,\mathscr{C}_j^{(s)})$. This is in contradiction with the assumption that $\mathbf{y}$ is path vector to level j for $(E,\mathscr{C}_j^{(s+1)})$.

2. When $\exists i=\overline{1,n}$, $\forall l_i \in \{1,\dots,s\}$, $(y_1,\dots,y_i-1,y_{i+1},\dots,y_n) > \mathbf{x}^{(l_i)}$, than (4) must be true for $\mathbf{x}^{(l_i)} = \mathbf{x}^{(s+1)}$ i.e. $(y_1,\dots,y_i-1,y_{i+1},\dots,y_n) \leq \mathbf{x}^{(s+1)}$. So, $y_j \leq x_j^{(s+1)}, j \neq i$ and $y_i \leq x_i^{(s+1)}+1$.

   If $y_i < x_i^{(s+1)}+1$, then we have that $y_i \leq x_i^{(s+1)} \Rightarrow \mathbf{y} \leq \mathbf{x}^{(s+1)}$, i.e. $\mathbf{y}$ is a cut vector to level j, which is in contradiction with our assumption for $\mathbf{y}$. So, it must be true $y_i = x_i^{(s+1)}+1 \Rightarrow \mathbf{y} = (y_1,\dots,x_i^{(s+1)}+1,\dots,y_n)$. Because we was assumed that such a vector may not be obtained by (3), $\mathscr{P}_j^{(s)}$ does not contains a vector of the form $(y_1,\dots,\widehat{y}_i,\dots,y_n)$, $\widehat{y}_i \leq x_i^{(s+1)}+1$.

   Now we have that all vectors of the form $(y_1,\dots,\widehat{y}_i,\dots,y_n), \widehat{y}_i = \overline{0,M_i}$ are either cut vectors or non-minimal path vectors to level j for the system $(E,\mathscr{C}_j^{(s)})$. If all of them are cut vectors to level j, then $\mathbf{y} = (y_1,\dots,x_i^{(s+1)}+1,\dots,y_n)$ is also a cut vector to level j, which is a contradiction. So, there is a path vector to level j for $(E,\mathscr{C}_j^{(s)})$ of this form and from the previous discussion there is not minimal path vector to level j. This means that there is a path vector $\mathbf{z}$, such that $\mathbf{z} < \mathbf{y}$. Now.

$$\mathbf{z}' = \mathbf{z} \otimes (0,\dots,0,x_i^{(s+1)}+1,0,\dots,0) < \mathbf{y}$$

and $\mathbf{z}'$ is a path vector to level j for the system $(E,\mathscr{C}_j^{(s+1)})$, which is in contradiction with our assumption that $\mathbf{y}$ is a minimal path vector to level j for $(E,\mathscr{C}_j^{(s+1)})$.

This proves that all minimal path vectors are members in (3), and the proof is completed.

The following Theorem is a direct consequence of the preceding Lemma:

Theorem 1. *The minimal path set to level j for the multi-state coherent system with graduate failure $(E,\{\mathscr{C}_j\})$ is*

$$\mathscr{P}_j = \sqcap_{\mathbf{x}\in\mathscr{C}_r} T(\mathbf{x}). \tag{5}$$

Proposition 1. *If* $\mathbf{x} < \mathbf{y}$ *then* $T(\mathbf{x}) \otimes T(\mathbf{y}) = T(\mathbf{y})$.

Proof. From $\mathbf{x} < \mathbf{y}$, follows that $\forall i = \overline{1,n}, x_i \le y_i$. This implies that $\forall i = \overline{1,n}$, $(0,\dots,x_i,\dots,0) \otimes (0,\dots,y_i,\dots,0) = (0,\dots,y_i,\dots,0)$.

On the other side, if $i \neq j$, $(0,\dots,x_i,\dots,0) \otimes (0,\dots,y_j,\dots,0) \geq (0,\dots,y_j,\dots,0)$. This elements will be removed from $T(\mathbf{x}) \otimes T(\mathbf{y})$.

Next Theorem shows that (3) will remain true if instead minimal cut vectors we use fatal vectors.

Theorem 2. *Let $\mathscr{C}_{Ej}$ be the fatal set to level j, for the multi-state coherent system with graduate failure $(E,\{\mathscr{C}_r\})$. Then,*

$$\prod_{\mathbf{x}\in\mathscr{C}_j} T(\mathbf{x}) = \prod_{\mathbf{x}\in\mathscr{C}_{Ej}} T(\mathbf{x}).$$

Proof.

$$\prod_{\mathbf{x}\in\mathscr{C}_j} T(\mathbf{x}) = \prod_{\mathbf{x}\in\mathscr{C}_{Ej}} T(\mathbf{x}) \prod_{\mathbf{x}\in\mathscr{C}_j\setminus\mathscr{C}_{Ej}} T(\mathbf{x}).$$

Each element from $\mathscr{C}_j \setminus \mathscr{C}_{Ej}$ is a smaller then at least one element from $\mathscr{C}_{Ej}$. Using Proposition 1 we have that

$$\prod_{\mathbf{x}\in\mathscr{C}_j\setminus\mathscr{C}_{Ej}} T(\mathbf{x}) \subseteq \prod_{\mathbf{x}\in\mathscr{C}_{Ej}} T(\mathbf{x}).$$

Example 1. Consider a system with $n=3$ components, each has 4 possible states, 0,1,2 and 3, with minimal cut set to level 2, $\mathscr{C}_2 = \{(1,3,1),(0,2,3),(1,1,2),(2,0,2),(3,0,1),(1,0,3)\}$. The procedure for obtaining its minimal path set is:

Step 1. $\mathscr{C}_2^{(1)} = \{(1,3,1)\}$, $\mathscr{P}_2^{(1)} = \{(2,0,0),(0,0,2)\}$
Step 2. $\mathscr{C}_2^{(2)} = \mathscr{C}_2^{(1)} \cup \{(0,2,3)\}$ $\mathscr{P}_2^{(2)} = \{(2,0,0),(1,0,2),(0,3,2)\}$
Step 3. $\mathscr{C}_2^{(3)} = \mathscr{C}_2^{(2)} \cup \{(1,1,2)\}$ $\mathscr{P}_2^{(3)} = \{(2,0,0),(1,2,2),(1,0,3),(0,3,2)\}$
Step 4. $\mathscr{C}_2^{(4)} = \mathscr{C}_2^{(3)} \cup \{(2,0,2)\}$ $\mathscr{P}_2^{(4)} = \{(3,0,0),(2,1,0),(1,2,2),(1,0,3),(0,3,2)\}$
Step 5. $\mathscr{C}_2^{(5)} = \mathscr{C}_2^{(4)} \cup \{(3,0,1)\}$ $\mathscr{P}_2^{(5)} = \{(2,1,0),(1,2,2),(1,0,3),(0,3,2)\}$
Step 6. $\mathscr{C}_2^{(6)} = \mathscr{C}_2^{(5)} \cup \{(1,0,3)\}$ $\mathscr{P}_2^{(6)} = \{(2,1,0),(1,2,2),(2,0,3),(1,1,3),(0,3,2)\}$

The obtained minimal path set is $\mathscr{P}_2 = \{(2,1,0), (1,2,2), (2,0,3), (1,1,3), (0,3,2)\}$.

3 Upper Boundary for the Reliability Function

When some real system is under consideration, we usually do not have a full control over the working states of the system. But it is always possible to detect the state at the moment of total failure of the system. Sometimes it is possible to know the state of the system in the moment when it fails one level down. At that moment, the state vector of the system is one of the vectors in the fatal set to level j.

Suppose that k independent systems with the same structure function operating in the field, generate raw data. Whenever a system fails under some level j, the autopsy of this failed system is performed and the state of the system at that moment is determined. The obtained state vector is a fatal vector to level j. Let $S_j^{(k)}$ be the set of all obtained state vectors, and $D_j^{(k)}$ the set $\prod_{\mathbf{x}\in S_j^{(k)}} T(\mathbf{x})$.

Next theorem shows that sufficiently large number of experiments leads to almost all elements in $\mathscr{C}_{Ej}$:

Theorem 3. *Let k multi-state systems with graduate failure with coherent structure $(E,\{\mathscr{C}_r\})$ and set of fatal vectors to level j, $\mathscr{C}_{Ej}$, be on test until failure under level j, and $S_j^{(k)}$ be the observed data set. Then,*

$$\lim_{k\to\infty} P\left(\mathscr{C}_{Ej} = S_j^{(k)}\right) = 1 \tag{6}$$

and

$$\lim_{k\to\infty} P\left(\mathscr{P}_{Ej} = D_j^{(k)}\right) = 1 \tag{7}$$

Proof. It is clear that $S_j^{(k)} \subseteq \mathscr{C}_{Ej}$.

Let $\mathbf{x} \in \mathscr{C}_{Ej}$. The probability to get this vector as a failure data in one experiment is $p_{\mathbf{x}} > 0$, so the probability that this vector is not obtaining as a failure data in the first k experiments is $(1-p_{\mathbf{x}})^k$. Now,

$$\lim_{k\to\infty} P(x \notin S^{(k)}) = \lim_{k\to\infty} (1-p_{\mathbf{x}})^k = 0, \Rightarrow \lim_{k\to\infty} P(x \notin S^{(k)}) = 0.$$

So, $\lim\limits_{k\to\infty} P(\mathscr{C}_E \subseteq S^{(k)}) = 1$, which proves (6). From this,

$$\lim_{k\to\infty} P\left(\mathscr{P}_{Ej} = D_j^{(k)}\right) = \lim_{k\to\infty} P\left(\prod_{\mathbf{x}\in S_j^{(k)}} T(\mathbf{x}) = \prod_{\mathbf{x}\in\mathscr{C}_{Ej}} T(\mathbf{x})\right) = 1.$$

The last Theorem can be used to approximate the minimal path set to level j:

1. Enter the first failure vector $\mathbf{x}$;
2. $\mathscr{P}_j = T(\mathbf{x})$;
3. Enter the next failure vector $\mathbf{x}$;
4. $\mathscr{P}_j = \mathscr{P}_j \otimes T(\mathbf{x})$;
5. Go to the step 3.

Es a illustration we give the following example:

Example 2. Refer to the Example 1. The set of failure data for level 2 is: {(1,1,0), (1,2,0), (1,1,1), (1,3,0), (1,2,1), (1,3,1), (1,1,2), (2,0,0), (2,0,1), (2,0,2), (3,0,0), (3,0,1), (0,2,2), (0,2,3), (1,0,3), (0,1,3), (1,0,2), (0,3,1) }, so the data we shell get during the experiments are from this set only.

By 7 independent experiments we were obtained:

Step 1. $S_2^{(1)} = \{(1,1,2)\}$ $\mathscr{P}_2 = \{(2,0,0),(0,2,0),(0,0,3)\}$
Step 2. $S_2^{(2)} = S_2^{(1)} \cup \{(1,3,1)\}$ $\mathscr{P}_2 = \{(2,0,0),(0,2,2),(0,0,3)\}$
Step 3. $S_2^{(3)} = S_2^{(2)} \cup \{(1,1,1)\}$ $\mathscr{P}_2 = \{(2,0,0),(0,2,2),(0,0,3)\}$
Step 4. $S_2^{(4)} = S_2^{(3)} \cup \{(3,0,1)\}$ $\mathscr{P}_2 = \{(2,1,0),(2,0,2),(0,2,2),(0,0,3)\}$
Step 5. $S_2^{(5)} = S_2^{(4)} \cup \{(0,1,3)\}$ $\mathscr{P}_2 = \{(2,1,0),(2,0,2),(0,2,2),(1,0,3)\}$
Step 6. $S_2^{(6)} = S_2^{(5)} \cup \{(1,1,1)\}$ $\mathscr{P}_2 = \{(2,1,0),(2,0,2),(0,2,2),(1,0,3)\}$
Step 7. $S_2^{(7)} = S_2^{(6)} \cup \{(2,0,2)\}$ $\mathscr{P}_2 = \{(2,1,0),(3,0,2),(2,0,3),(0,2,2),(1,1,3)\}$

The estimated minimal path set is: {(2,1,0), (3,0,2), (2,0,3), (0,2,2), (1,1,3)}.

As we expected, the estimated system has more path vectors then the real one; the vectors (0,2,2) and (3,0,2) are extra.

The result of Theorem 3 can be used to estimate the upper boundary of the reliability function. The reliability R_j to level j of the coherent multi-state system $(E,\{\mathscr{C}_r\}) = (E,\{\mathscr{P}_r\})$ is, [[2]], [[5]]

$$R_j = \left(1 - \sum_{\mathbf{x}\in\mathscr{C}_j} P(\mathbf{X} \le \mathbf{x})\right) = \sum_{\mathbf{x}\in\mathscr{P}_j} P(\mathbf{X} \ge \mathbf{x}).$$

Let $R_j^{(k)}$ be define as

$$R_j^{(k)} = \left(1 - \sum_{\mathbf{x}\in S_j^{(k)}} P(\mathbf{X} \le \mathbf{x})\right) = \sum_{\mathbf{x}\in D_j^{(k)}} P(\mathbf{X} \ge \mathbf{x}).$$

Since $S^{(k)} \subseteq \mathscr{C}$, we have that for all $k \in \mathbf{N}$, $R_j^{(k)} \ge R_j$.

Theorem 4. *Let k multi-state systems with coherent structure* $(E,\{\mathscr{C}_r\}) = (E,\{\mathscr{P}_r\})$ *and a set of fatal vectors* $\mathscr{C}_{Ej}$ *be on test until failure under level j, and* $S_j^{(k)}$ *be the observed data set. Then,*

$$\lim_{k\to\infty} P\left(R_j^{(k)} = R_j\right) = 1$$

Proof. From the Theorem 3,

$$\lim_{k\to\infty} P\left(R_j^{(k)} = R_j\right) = \lim_{k\to\infty} P\left(\sum_{\mathbf{x}\in\mathscr{P}_j^{(k)}} P(\mathbf{X}\geq\mathbf{x}) = \sum_{\mathbf{x}\in\mathscr{P}_j} P(\mathbf{X}\geq\mathbf{x})\right) = 1.$$

4 Conclusion

In this paper we analyze the relationship between minimal path and cut sets to a given level j for a coherent multi-state system with graduate failure of the components. In the first part we propose and prove an algorithm for obtaining minimal path set to level j from the minimal cut set to level j. Moreover we give a method for estimation of the path and cut sets to level j, that uses a failure data consisting of the system states in the moment of failure under level j. The estimated minimal path set is used to get an upper boundary for the reliability function. This upper boundary is an approximation for the real value of the reliability.

References

1. Dharmadhikari, A., Kulathinal, S.B., Mandrekar, V.: An algoritam for estimation of minimal cut and path sets from field failure data. Statistics & Probability Letters 58, 1–11 (2002)
2. Natvig, B.: Multi-state Reliability Theory, Statistical Research Report, Dept. of Math, vol. (1). University of Oslo (January 2007) ISSN 0806-3842
3. Natvig, B.: Two suggestions of how to define a multistate coherent system. Advances in Applied Probability 14, 434–455 (1982)
4. Natvig, B., Srmo, S., Holen, A., Hgsen, G.: Multistate Reliability Theory - ACase Study. Advances in Applied Probability 18, 921–932 (1986)
5. Crowder, M.J., Kimber, A.C., Smith, R.L., Sweeting, T.J.: Statistical analysis of reliability Data. Chapman and Hall, London (1991)

Novel Inverse Sigmoid Fuzzy Approach for Water Quality Diatom Classification

Andreja Naumoski, Svetislav Krstić, and Kosta Mitreski

Abstract. The prediction accuracy of the fuzzy diatom models depends on both the manner of defining the fuzzy sets used (their number, shape and the parameters of the membership function (MF)) and the kind of the similarity metric used. In this paper, we define new similarity metric, which takes into the account the maximum number of diatoms' abundance in specific environmental parameter range. The inverse sigmoid MF is used to shape each MF to describe this relationship, in order to produce more accurate models. This improvement of the ecological modelling is achieved through the process of evaluation results for interpretability; higher prediction accuracy and over fitting resistant. The evaluation results compared with classical classification algorithms have confirmed these findings. Based on these results, one model for each water-quality category class is presented and discussed. From ecological point of view, each model is verified with the existing diatom indicator references found in literature by the biological expert.

Keywords: Inverse sigmoid distribution, fuzzy membership function, similarity metric, water quality models.

1 Introduction

There are many potential uses of the results from clustering or partition the data into a smaller number of groups of samples with similar diatom composition or

Andreja Naumoski · Kosta Mitreski
Ss. Cyril and Methodius University in Skopje, Faculty of Electrical Engineering and Information Technologies, Skopje, Macedonia
e-mail: andrejanaumoski@gmail.com
komit@feit.ukim.edu.mk

Svetislav Krstić
Ss. Cyril and Methodius University in Skopje, Faculty of Natural Sciences and Mathematics, Institute of Biology, Skopje, Macedonia
e-mail: skrtic@pmf.ukim.edu.mk

L. Kocarev (Ed.): ICT Innovations 2011, AISC 150, pp. 207–217.
springerlink.com

environmental characteristics [1]. One example is detecting groups of samples with similar diatom composition or with similar environmental features and thus detecting indicator species for the groups or relating biologically based groups to environmental variables. These environmental features can be influenced by many factors, such as biotic and abiotic interactions, light, turbulence, resource limitation and water chemistry represented usually as water quality (WQ) or Trophic State Index (TSI) category classes. In this paper, we will focus on detecting diatom's indicator species and their property of indicating certain WQ category class.

In this domain, classical statistical approach, which gained increased interested in the compilation and analysis of small and huge diatom data sets and the necessity to detect the major patterns of variables within such data sets are different improvements of the original method of decision trees. In the paper [2] and [3] the researchers provide clear introductions of these robust techniques. Furthermore, papers like [4, 5, 6, 7] and [8] the authors discuss recent developments in classification/regression trees in terms of bagging trees, boosted trees, random forests, and multivariate adaptive regression splines in the context of ecological studies. More information in this research topic can be found in [10].

First attempt to build a classification/regression model with decision trees for detecting diatom's indicator species of one and several environment parameters for Lake Prespa, have been made by [10]. Various settings were applied to the datasets and thus different models were obtained, which later have been discussed with the biological expert. Several of the model produced, knowledge about the newly discovered the diatoms' relationships with the environment for the first-time [10, 11] with the new method of multi-target regression trees [11]. Nevertheless, the obtained models were not robust on data change, not resistant to over fitting and don't fallow the diatom-environment gradient distribution, because the structure of the algorithm implies that. These disadvantages can be overcome using fuzzy theory and several research studies considering diatom classification were conducted [12, 13] to prove that this method is more appropriate for this task.

The proposed method in this paper can be used for revealing the diatom-indicator relationship for several reasons. First of all, the proposed method is robust to over fitting, which is not the case with algorithms like C4.5, LAD Tree, NB Tree, Bagging, Boosting and, etc. Secondly, the obtained knowledge from the model is easily compared with the known ecological diatom indicator references in the literature [14]. This is vital because later, the rules produced from the models for the newly discover diatoms can be easily evaluated by the biological expert. And third, these models can achieve higher classification accuracy compared to the classical classification algorithms. One of the reasons, why this method is better compared with the previous ones, is the use of different fuzzy MFs and similarity metrics.

The rest of the paper is organized as follows: Section 2 provides the definitions of our proposed similarity metric and inverse sigmoid MF. Section 3 presents the WQ category classes and the dataset description. In section 4 we present the experimental setup and later one diatom model for selected WQ category classes based on the prediction performance, and experimental comparisons is verified by the biological expert. Finally, Section 5 concludes the paper and research direction is outlined.

2 Our Proposed Similarity Metric

The classification model method described in this section is induced by using different similarity metrics and fuzzy aggregation operators. The goal of the proposed metric is to achieve favouring the higher abundance diatoms. In this direction, the absolute distance between the abundances is multiplied by the abundance of the given diatom or the abundance of the diatoms in a given class, whichever is higher.

Let A and B be two fuzzy sets defined on the universe of discourse U. The Sim_{G_MAX} based fuzzy set similarity (Sim_{G_MAX}) of fuzzy sets A and B is computed as:

$$Sim_{G\_MAX}(A,B)=1-\left[\frac{\sum_{i=1}^{n}[|(\mu_A(x_i)-\mu_B(x_i)|*\max\{\mu_A(x_i),\mu_B(x_i)\}]}{n}\right] \quad (1)$$

where x_i, $i = 1, \ldots, n$, are the crisp values discretized in the variable domain, and $\mu_A(x_i)$ and $\mu_B(x_i)$ are the fuzzy membership values of x_i for A and B. The larger the value of Sim_{G_MAX} (A, B) is, the more similar A and B are. The $\mu_A(x_i)$, $\mu_B(x_i) \in [0, 1]$, $0 \leq Sim_{G\_MAX}(A;B) \leq 1$ holds according to (1). Note that the proposed classification method follows the same principle, if an alternative fuzzy set similarity definition such as Jaccard or RMSE used, which was done previously [14] with RMSE metric.

According to the fuzzy logic theory, the fuzzy aggregation operators are logic operators applied to fuzzy membership values or fuzzy sets. They have three subcategories, namely t-norm [15], t-conorms[15], and weighted averaging operators[16]. In our experimental setup, we use the basic operators (*Algebaric AND/OR*) that operate on two fuzzy membership values a and b, where $a, b \in [0, 1]$.

2.1 Our Inverse Sigmoid MF Definition

Many of the diatom datasets have smoothed values and nonzero points, which apply to use different MFs. This condition doesn't satisfy the straight-line MFs (triangular and trapezoidal), which has the advantage of simplicity. That's way the paper proposes the inverse sigmoid function that will follow the distribution of the diatom abundance along the environmental gradients more realistically.

The sigmoid function is defined with the equation (2). In equation 2, a and b parameters are positive constants. We also propose that the equation (2) to be modify, by taking only the mean values (μ) of the given data range into account. In this way, each fuzzy MF per attribute will fallow the increasing and decreasing of the diatom's abundance and reflect the very nature of the tested dataset evenly.

$$f(x;a;b) = \frac{1}{1+e^{-a*(x-b)}}, \quad a,b>0 \tag{2}$$

The parameter a can get two values {1 and -1}, where a=-1 representing as inverse sigmoid MF will be intensively studied in this paper. And finally when all this change is taken into account, the equation (3) mathematically represents the evenly sigmoid inverse MF as:

$$f(x;\mu;a) = \frac{1}{1+e^{+1*(x-\mu)}} \tag{3}$$

We expect that the inverse sigmoid distribution will better fallow the diatom-indicator relationship and achieve greater classification accuracy combined with the proposed similarity metric. The experimental evaluation and model verification will prove the correctness of the hypothesis.

3 Data Description and Experimental Setup

The dataset used in the experiments consist from 13 input parameters representing the TOP10 diatom's species (diatom species that exist in Lake Prespa [17]) with their abundance per sample, plus the three WQ category classes for conductivity, pH and Saturated Oxygen. The one dataset is created for each WQ category class as output class. These measurements were made as a part of the TRABOREMA project [18]. The WQ category classes were defined according to the three physical-chemical parameters: Saturated Oxygen [19], Conductivity [20] and pH [19, 20] and they are given in Table 1.

We conducted two types of experiments, which are set up as follows:

1) **Train:** A fuzzification method based on the inverse sigmoid MF, presented in this paper, for each input variable is used to transform the crisp values into fuzzy values.

2) **xVal:** Standard 10-fold cross validation is used for evaluation of the classification accuracy of the algorithm against some classical classification algorithms (C4.5, LAD Tree, NB Tree, Bagging (C4.5) and, etc.) (**Test**)

Table 1. Water quality classes for the physical-chemical parameters [20, 21]

Physico-chemical parameters	Name of the WQC	Parameter range
Saturated Oxygen	*oligosaprobous*	> 85 %
	β-mesosaprobous	70-85 %
	α-mesosaprobous	25-70 %
	α-meso / polysaprobous	10-25 %
pH	*acidobiontic*	pH < 5.5
	acidophilous	pH > 5.5
	circumneutral	pH > 6.5
	alkaliphilous	pH > 7.5
	alkalibiontic	pH > 8
	Indifferent	pH > 9
Conductivity	*fresh*	< 20 (μS·cm^{-1})
	fresh brackish	< 90 (μS·cm^{-1})
	brackish fresh	90 – 180
	brackish	180 - 900

For similarity definition, we use Sim_{G_MAX} similarity and *Algebraic AND/OR* for fuzzy aggregation metric. For evaluation purpose, we induce a simple diatom model (SDM) and general diatom model (DM) in a hieratical tree form. The simple trees consist from 1 candidate tree, 0(zero) low levels and two different depths; 5 - (SDM5) and 10 - (SDM10). While general diatom models consist from 2 candidate tree, 3 low levels trees and two different depths; 5 - (DM5) and 10 - (DM10).

4 Experimental Results

Based on the performance results, in this section we give an interpretation of several classification models and the rules derived from them. One model for the conductivity and the saturated oxygen WQ category class are discussed, and later the classification results are verified with the known diatom ecological references. The experiments are conducted with the modified bell MF with a number of MF per attribute equal to 5 for better comprehensibility. For similarity definition, we use Sim_{G_MAX} and only *Algebraic AND/OR* as a fuzzy aggregation operator.

4.1 Performance Evaluation

The Table 2 presents the highest prediction accuracy of the obtained classification models from the proposed method over two types of the classification accuracy: descriptive (Train) and the predictive (Test). Concerning the performance evaluation of the proposed similarity metric against previously used RMSE metric, this metric outperformed in 2 out of 3 diatom WQ category classes. Except for the conductivity WQ category classes - Test experimental setup, in other cases for prediction accuracy we have been increased from 1% to 3%. Based on these classification results for the predictive models, one model for two WQ category class will be presented.

Table 2. Classification accuracy per WQ category classes (in %)

Conductivity WQ category classes – Classification Accuracy								
Metric	RMSE				G_MAX			
	SDM5	SDM10	DM5	DM10	SDM5	SDM10	DM5	DM10
Train	75.69	76.15	75.69	76.15	76.15	76.15	**77.06**	**77.06**
Test	71.84	71.36	**73.61**	71.32	71.80	71.80	71.82	71.34
pH WQ category classes – Classification Accuracy								
Metric	RMSE				G_MAX			
	SDM5	SDM10	DM5	DM10	SDM5	SDM10	DM5	DM10
Train	**60.55**	60.09	**60.55**	60.09	59.63	59.63	59.63	59.63
Test	59.42	58.94	59.00	58.53	59.03	59.03	**59.48**	**59.48**
Saturated Oxygen WQ category classes – Classification Accuracy								
Metric	RMSE				G_MAX			
	SDM5	SDM10	DM5	DM10	SDM5	SDM10	DM5	DM10
Train	58.71	59.20	58.71	59.20	**59.70**	**59.70**	**59.70**	**59.70**
Test	56.50	55.50	56.50	56.00	**58.50**	**58.50**	58.00	**58.50**

4.2 Classification Models for the Water Quality Classes

We have built many classification models for each WQ category class, but due to paper constrains, we present the only one models for the Saturated Oxygen and one model for Conductivity WQ category class. The conductivity WQ category class is chosen even she didn't achieve higher classification accuracy to show that the diatom classification models with lower accuracy can obtain valuable diatom-indicator knowledge. The fuzzy aggregation operator serves as a "glue" between two input attributes of the classification models is a Fuzzy Term. To make the models easy for interpretation, each Fuzzy Term is labelled according to the value of μ {Bad, Low, Good, Very Good and Excellent} Indicator. The proposed method has generated a separate model tree for each class, which classifies the diatom into one or several classes. The diatom model shown in Fig. 1 (left) can be converted into a rule which is stated below.

Rule1: If Conductivity WQ class is ***fresh*** **THEN** ((*Cyclotella ocellata* (COCE) is **Very Good Indicator** AND *Staurosirella pinnata* (STPNN) is **Good Indicator**) AND *Cavinula scutelloides* (CSCU)) is **Excellent Indicator** OR CSCU is **Bad Indicator**. The model has highest similarity of 75.45%.

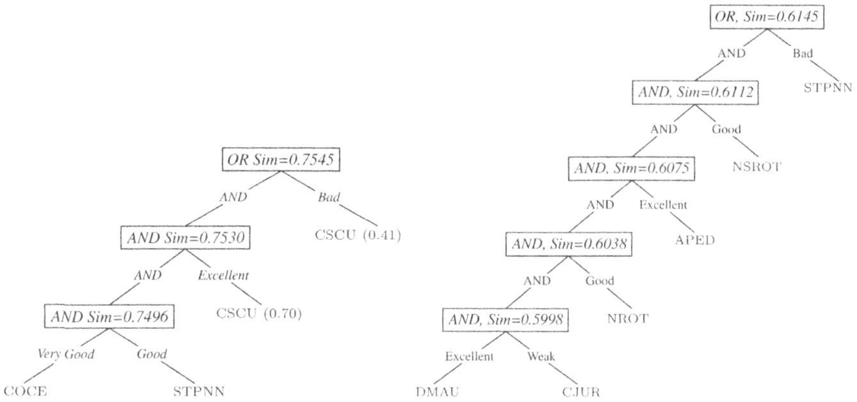

Fig. 1 Classification model generated using the proposed inverse sigmoid MF for the *fresh* Conductivity WQ class (left) and *β-mesosaprobous* Saturated Oxygen WQ class (right).

According to the diatom model given in Fig.1 - left, CSCU diatom in the same time is an excellent indicator and bad indicator. This is due to distribution of the CSCU diatoms along the Conductivity gradient. Because each input attribute is tested for similarity metric with the output class, the greater value of similarity is better. This means that the CSCU is an excellent indicator for *fresh* waters. In addition, this classification model has found that STPNN diatom is a good indicator of *fresh* waters. The COCE diatom is indicated by the model as a very good indicator of such waters. Furthermore, the classification model for *β-mesosaprobous* given with Fig.1 - right the proposed method has shown that DMAU and APED diatoms are excellent indicators of *β-mesosaprobous* waters. NROT and NSROT diatoms are found to be good indicators of such water, while the CJUR diatom is a weak indicator. STPNN diatom based on the classification model is a bad indicator of such waters. The given model is transformed into the fallowing rule:

Rule2: If Saturated Oxygen class is ***β-mesosaprobous*** **THEN** (((*Diploneis mauleri* (DMAU) is **Excellent Indicator** AND *Cyclotella jurilјii* (CJUR) is **Weak Indicator**) AND *Navicula rotunda* (NROT)) is **Good Indicator** AND *Cocconeis placentula* (CPLA) is **Excellent Indicator**) AND *Navicula subrotundata* (NSROT)) is **Good Indicator** OR STPNN is **Bad Indicator**. The model has highest similarity of 61.45%.

4.3 Verification of the Diatom Model Results

Ecological references for the TOP10 diatoms are taken from the diatom ecology publication [14]. Concerning the ecological preferences of the TOP10 dominant diatoms in the Lake Prespa, CJUR and NPRE are newly described taxa (diatoms) with no records for their ecological references in the literature. Some of the results from the classification models are the first known ecological references for certain WQ category classes.

In the relevant literature, the APED diatom is known to as an *alkaliphilous*, *fresh-brackish*, *β-mesosaprobic* and *eutrophic* (because of organic nitrogen tolerance) diatom indicator [14]. The diatom model given with Fig. 1- right has successfully verified the APED diatoms as *β-mesosaprobic* indicator. In the relevant literature, the CSCU diatom is known as an *alkalibiontic*, *freshwater* to *brackish* water taxon, being *oligosaprobic* indicators with eutrophic preferences [14]. According to the models, the CSCU diatom is *fresh* water taxon, which verifies the known ecological reference. COCE is known as *meso-eutro* taxon [14]. According to models, the COCE diatom is a very good indicator for *fresh* waters. Regarding the STPNN diatom, in the literature is known as *hyper-eutrophic* (*oligo-eutrophic*; indifferent) taxon frequently found on moist habitats [14], while the models have found that this diatom is a good indicator of *fresh* waters, but not indicator of *β-mesosaprobous*. The NSROT diatom has no ecological references in the literature, so the results of the model are the first to be known. According to the models, the NSROT diatom can be used for indicating *β-mesosaprobous* waters. Additionally, the CJUR diatom has been found by the models as a weak indicator of *β-mesosaprobous* waters. Further investigation is needed before to any conclusion is made for the newly discovered diatom indicators.

4.4 Over Fitting Comparison with C4.5, Bagging (C4.5), Boosting (C4.5), LAD Tree, NB-Tree and CLUS-MS9 Algorithms

Over fitting refers to the phenomena that a classifier may fit well to the training data but is not generalized enough to classify unseen data. The 10-fold cross validation based experiments fairly present the normal behaviour of classifiers, but it does not reveal which classifiers are prone to over-fitting. In this section, the whole data (rather than the 10-fold cross validation data) are used to train and test all the classical classification algorithms and four variants of method using the same experimental setup. The results collectively are shown in Table 3.

The classical classifiers (C4.5, Bagging (C4.5), Boosting (C4.5), LAD Tree, NB Tree and CLUS-MS9) obtained from the Weka machine learning toolkit [21] from a crisp classifier group, except the CLUS-MS9 algorithm which is implemented in CLUS system [22] available at http://www.cs.kuleuven.be/~dtai/clus/. The default settings of each classifier in the toolkit are used, excepting for the CLUS MS9 algorithms which the Maximum Size of the tree is set to 9. For example, the minimal number of instances per leaf is set to 2 for C4.5 and the number of neighbours to use is set to 1 for KNN. The LAD Tree the number of Boosting Interaction is set to 10. For each classifier, the root mean square error (RMSE) of the classification accuracy between the 10-fold cross-validation and whole data based is shown at the bottom of the table. The RMSE reveals how much improvement one classifier can gain based on the entire data experiment comparing to the 10-fold cross validation one. It is assumed that the more gain for one classifier, the more likely that the classifier is prone to over-fitting. The results collectively are shown in Table 3.

Table 3. Over fitting experimental evaluation between the classical classification and the four variants of the diatom fuzzy classification algorithm (%)

WQC Datasets	C4.5	Bagging C4.5	Boosting C4.5	LAD Tree	NB Tree	CLUS MS9	SDM5	SDM10	DM5	DM10
					Train					
WQC_Condu	83.49	88.99	95.87	81.65	82.57	72.48	76.15	76.15	77.06	77.06
WQC_pH	70.18	80.28	87.61	71.56	61.47	66.97	59.63	59.63	59.63	59.63
WQC_SatO	80.10	81.09	86.07	70.65	58.71	63.18	59.70	59.70	59.70	59.70
					Test					
WQC_Condu	65.60	63.30	63.76	72.02	66.51	71.56	71.80	71.80	71.82	71.34
WQC_pH	55.50	56.42	49.54	56.88	60.09	63.76	59.03	59.03	59.48	59.48
WQC_SatO	54.73	53.23	56.22	54.73	58.71	54.23	58.50	58.50	58.00	58.50
RMSE	34.33	44.78	58.06	23.69	16.11	9.552	**4.552**	**4.552**	5.510	5.846

The results of this test revealed that proposed method maintains the best generality compared with the classical approaches (with RMSE being 4.552) and the two remaining variants of fuzzy classification models perform slightly worse (with RMSE from 5.51 and 5.846 respectively) for Sim_{G_MAX} similarity metric. Second best classification algorithm is the CLUS MS9. Other crisp classifiers performed worse than the proposed method and the CLUS MS9 implementation. The LAD Tree and the NB Tree have shown better results compared to the C4.5 algorithm. The improvement of over fitting resistance of the proposed method is based on the construction of each rule, which is based on a portion of training data, which causes the overfitting of individual rules and in turn causes the overfitting of the whole model. This problem gets worse when a long rule is constructed; as such a rule only represents a small amount of training data rather than a meaningful pattern. In contrast, the proposed method uses the whole training data to find a rule (tree) for each class. The growth of the tree is permitted only if all training data (in contrast to a portion of training data) can be better fit into the tree. Overall conclusion is that the even complex fuzzy classification models obtained with this algorithm do not suffer from over fitting.

5 Conclusion

In this paper, we propose a fuzzy method for classifying diatom indicating properties for certain WQ category classes from diatoms' assembly data with novel similarity metric and inverse sigmoid MF. Because the diatoms are not influenced by the geographical location, but rather the physico-chemical parameters of the environment [23], the proposed method could be used for diatom classification for any ecosystem. Combined with the fuzzy theory the inverse sigmoid curve produced diatom models with easy interpretable results and high prediction accuracy compared with other classification algorithms. The experiments with the diatom WQ datasets show that the inverse sigmoid distribution models combined with Sim_{G_MAX} metric outperformed previously models, which use similarity metrics in terms of prediction accuracy. The descriptive classification accuracy of the models has achieved satisfied values compared with the models obtained with the previous similarity metrics. More important, the 10-fold cross validation used to compare the prediction performance of this algorithm with crisp algorithms, proof that we develop a similarity metric have outperformed classical classification algorithms in terms of prediction power and maintained resistance of the over fitting.

From ecological point of view, it is very important that the method is interpretable, immediately to show the results from the diatom's indicator pattern. This is achieved with the method, and the results have been verified by the biological experts with the known ecological knowledge. For example, the diatom model presented with Fig. 1 - right for *β-mesosaprobous* Saturated Oxygen WQ class clearly states that the APED diatom can be an indicator of these waters. The models have also added some new diatom indicator references, which have to be proven with more experiments before any conclusion is made.

Further research needs focus on developing more similarity metric and fuzzy operators in a process of building diatom classification models is necessary. The new similarity metrics may be more suitable for this diatom community dataset and can therefore, lead to higher accuracy. In future we plan to test more datasets based on metal parameters and thus forming datasets that can be used to discover new diatom indicators.

References

1. Legendre, P., Birks, H.J.B.: Clustering and partitioning. In: Birks, H.J.B., Lotter, A.F., Juggins, S., Smol, J.P. (eds.) Tracking Environmental Change Using Lake Sediments, Data Handling and Numerical Techniques, vol. 5, Springer, Dordrecht (2011)
2. Deáth, G., Fabricus, K.E.: Classification and regression trees: a powerful yet simple technique for ecological data analysis. Ecology 81, 3178–3192 (2000)
3. Fielding, A.H.: Cluster and Classification Techniques for the Biosciences. Cambridge University Press, Cambridge (2007)
4. Deáth, G.: Multivariate regression trees: a new technique for modeling species–environment relationships. Ecology 83, 1105–1117 (2002)
5. Deáth, G.: Boosted trees for ecological modelling and prediction. Ecology 88, 243–251 (2007)
6. Prasad, A.M., Iverson, L.R., Liaw, A.: Newer classification and regression tree techniques: bagging and random forests for ecological predictions. Ecosystems 9, 181–199 (2006)
7. Cutler, D.R., Edwards, T.C., Beard, K.H.: Random forests for classification in ecology. Ecology 88, 2783–2792 (2007)
8. Elith, J., Leathwick, J.R., Hastie, T.: A working guide to boosted regression trees. Journal of Animal Ecology 77, 802–881 (2008)
9. Stroemer, E.F., Smol, J.P.: The diatoms: Applications for the environmental and earth sciences, 2nd edn., pp. 23–47. Cambridge University Press (2010)

10. Naumoski, A., Kocev, D., Atanasova, N., Mitreski, K., Krstić, S., Džeroski, S.: Predicting chemical parameters of water quality form diatoms abundance in lake Prespa and its tributaries. In: Athanasiadis, I.N., Mitkas, P.A., Rizzoli, A.E., Gómez, J.M. (eds.) 4th International ICSC Symposium on Information Technologies in Environmental Engineering - ITEE 2009, pp. 264–277. Springer, Heidelberg (2009)
11. Kocev, D., Naumoski, A., Mitreski, K., Krstić, S., Džeroski, S.: Learning habitat models for the diatom community in Lake Prespa. Ecological Modelling 221(2), 330–337 (2010)
12. Naumoski, A., Mitreski, K.: Classifying diatoms into trophic state index classes with novel classification algorithm. Procedia Environmental Sciences 2, 1124–1138 (2010)
13. Naumoski, A., Mitreski, K.: Diatom Classification with Novel Bell Based Classification Algorithm. In: Gusev, M., Mitrevski, P. (eds.) ICT Innovations 2010. CCIS, vol. 83, pp. 245–254. Springer, Heidelberg (2011)
14. Van Dam, H., Mertens, A., Sinkeldam, J.: A coded checklist and ecological indicator values of freshwater diatoms from the Netherlands. Netherlands Journal of Aquatic Ecology 28(1), 117–133 (1994)
15. Peter, K.E., Radko, M., Endre, P.: Triangular Norms, pp. 78–98. Kluwer, Dordrecht (2000)
16. Torra, V., Narukawa, Y.: Modeling Decisions: Information Fusion and Aggregation Operators, pp. 147–175. Springer, Heidelberg (2007)
17. Levkov, Z., Krstić, S., Metzeltin, D., Nakov, T.: Diatoms of Lakes Prespa and Ohrid (Macedonia). Iconographia Diatomologica 16, 603 (2006)
18. Krstić, S.: Description of sampling sites. FP6-project TRABOREMA: Deliverable 2.2 (2005)
19. Krammer, K., Lange-Bertalot, H.: Die Ssswasserflora von Mitteleuropa 2: Bacillariophyceae. 1 Teil, p. 876. Gustav Fischer-Verlag, Stuttgart (1986)
20. Van Der Werff, A., Huls, H.: Diatomeanflora van Nederland. Abcoude - De Hoef (1974, 1957)
21. Witten, I.H., Frank, E.: Data Mining: Practical Machine Learning Tools and Techniques, 2nd edn. Morgan Kaufmann (2005)
22. Blockeel, H., Struyf, J.: Efficient algorithms for decision tree cross-validation. Journal of Machine Learning Research 3, 621–650 (2002)
23. Gold, C., Feurtet-Mazel, A., Coste, M., Boudou, A.: Field transfer of periphytic diatom communities to assess short term structural effects of metals (Cd, Zn) in rivers. Water Research 36, 3654–3664 (2002)

Analysis of the Convergence of Iterative Geophysical Inversion in Parallel Systems

Neki Frasheri and Betim Cico

Abstract. Geophysical inversion is an ill-posed problem, resolved partially mostly in 2D or with specific constrains. We analyze the convergence of iterative 3D gravity inversion based in the idea of the CLEAR algorithm in both cases of sequential and parallel processing, focusing in the time and number of iterations depending in number of discretization nodes and number of processors involved. Variations of calculation runtime and number of iterations depending on the size of the model indicate the usefulness of parallel processing for large models of 3D geophysical inversion with simple algorithms. The work is done for the FP7 project HP-SEE.

1 Introduction

Geophysical inversion [Sen and Stoffa 1995] remains a difficult problem, where physical parameters of 3D geosection have to be evaluated on basis of values measured in a 2D surface array. Such problem may be classified "ill posed" with the definition of [Hadamard 1902], and despite the progress it remains "uncertain" [Wellmann et al 2010]. Practical solutions of the problem yet are constrained for convex bodies (for example [Silva et al 2000]), regular bodies (for example [Zhou 2009]), or simplified 2D problems (for example [Zhou 2010]). Interactivity is used in many cases in order to give an initial solution and lead the process of iterations towards an optimal solution.

Except the uncertainty of extrapolation of information from a 2D array into a 3D geosection, the complication of 3D case is conditioned by the huge volume of data that representing 3D structures. In a geosection extended for kilometers different geological objects may have thin shapes of the size of one meter or less, which implies in the general case the need for a multitude of 10^9 discretization nodes. Volume of data implies huge calculation power necessary for their processing, and this is another obstacle for fast calculations required when interactive applications are used.

Neki Frasheri · Betim Cico
Polytechnic University of Tirana, Albania
e-mail: nfrasheri@fti.edu.al
betim.cico@gmail.com

L. Kocarev (Ed.): ICT Innovations 2011, AISC 150, pp. 219–226.
springerlink.com

We considered that the use of grid and parallel systems for fast calculations may help engineers to apply simple algorithms for complicated problems, balancing the complexity with the calculation power of such platforms. In this context for the FP7 project HP-SEE we proposed an application for the 3D geophysical inversion using a simple calculation methodology and tried to evaluate its timing performances when executed in a parallel system. As a first step we focused in gravity inversion as the simplest case.

The work is supported by the HP Cluster Platform Express 7000 operated by the Institute of Information and Communication Technologies, Bulgarian Academy of Sciences in framework of the project HP-SEE [http://www.hp-see.eu/].

2 Methodology of Study

We based the work using the traditional formula for the vertical gradient of gravitational potential field ([Lowrie 2007]). For the inversion we applied the idea behind the algorithm CLEAN [Högbom 1974] using an iterative process as follows:

1. Starting with a 3D geosection array representing the rocks density and initialized by zeros, and a 2D gravity anomaly array measured in the field.
2. Search the node in the geosection array, which effect offers the best least square approximation of the gravity anomaly array.
3. Increment the density of the selected node by a fixed amount (density step).
4. Ssubtract the effect of the modification of the geosection from the surface gravity anomaly.
5. Repeat the steps (2):(3):(4) until residual anomaly changes less than a fixed predefined value.

Main parameters of the geosection and an anomalous body were selected (Fig. 1):

1. Geosection with dimensions 4,000m*4,000m*2,000m with density zero,
2. Inclusion of anomalous body with dimensions 400m*400m*1,800m with density 5 g/cm3 situated at the center of geosection.

We calculated several versions of the same model using:

1. Discretization of geosection with spatial steps 400m, 200m, 100m, 50m and 25m with respectively 11*11*6, 21*21*11, 41*41*21, 81*81*41 and 161*161*81 nodes, multiplying by a spatial factor of 2.
2. Applying three cases for the density step, starting with a value of 1g/cm3 and multiplying it with factors 1/2 and 1/10 (respectively 0.5 g/cm3 and 0.1 g/cm3)
3. Discretization of surface profile of 4,000m*4,000m with respective point arrays of the same scale as in 3D.

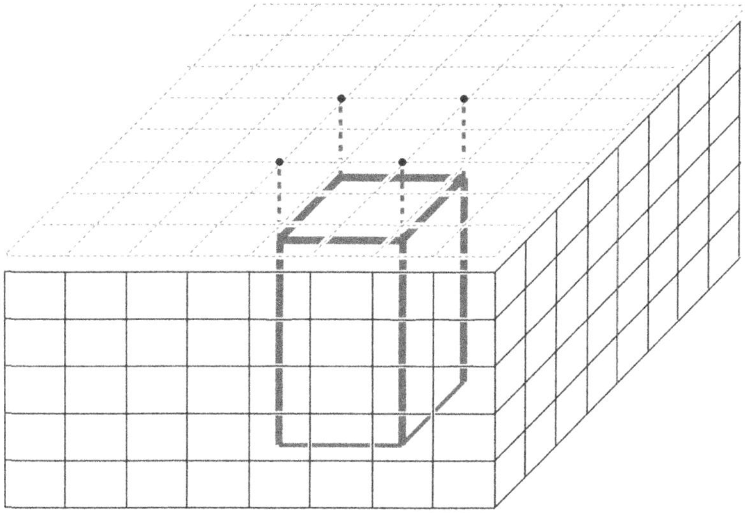

Fig. 1 The geosection model (— ~ 3D geosection array; ▪ ▪ ▪ ~ 2D surface profiles array; ▬ ~ anomalous body).

The quantity of nodes for different discretization scales is given in Table 1.

Table 1. Volume of data and calculations.

Parameter	Scale & number N of linear nodes					Factor
Spatial step (m)	400	200	100	50	25	O(1/N)
Linear nodes	11	21	41	81	161	O(N)
2D array points	121	441	1,681	6,561	25,921	$O(N^2)$
2D array (bytes)	968	3,528	13,448	52,488	207,368	$O(N^2)$
3D array nodes	726	4,851	35,301	269,001	2,099,601	$O(N^3)$
3D array (bytes)	5,808	38,808	282,408	2,152,008	16,796,808	$O(N^3)$
Elementary calculations	87,846	2,139,291	59,340,981	1,764,915,561	54,423,757,521	$O(N^5)$
Increase of number of iterations due to reduction of density step with O(N)						O(N)
Increase of number of iterations due to reduction of density contribution when spatial step decreases with O(N)						$O(N^3)$

Volume of calculations theoretically is proportional with the number of nodes. A theoretical evaluation of the increasing factor for the volume of data and calculations when the density of spatial grid is multiplied by a factor of 2 is given as well. We counted the number of elementary calculations per iteration, which is calculation of the effect of one 3D node in one 2D point in the framework of single iteration; and the increase of number of iterations for the same contribution due to variation of density step and of spatial discretization.

We succeeded to run three first models in the HPC within a reasonable time up to several hours. Increasing of density of spatial nodes beyond the schema 41*41*21 remains for the successive phase of the work.

3 Computational Results

In the first preparatory stage we solved the direct geophysical problem calculating the gravity anomaly from the inserted body for all cases of arrays in consideration. The calculation time is given in the Fig. 2, increasing by a factor of 2^5 when density of arrays is linearly duplicated.

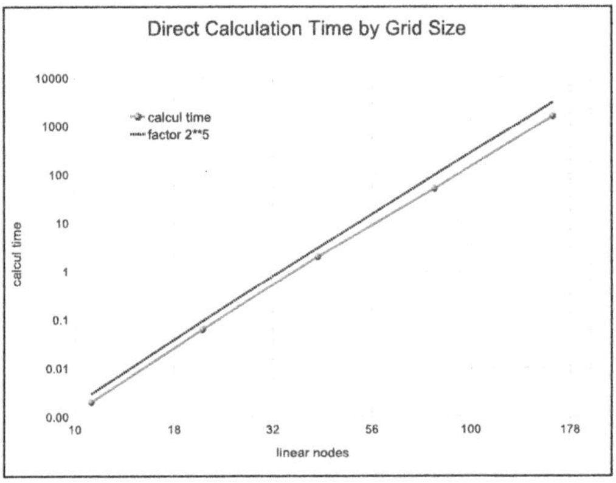

Fig. 2 Direct problem calculation time (dotted line represent increments by a factor of 2^5).

Number of iterations of the inversion problem for the first three cases 11*11*6, 21*21*11, and 41*41*21 array nodes geosection is given in Fig. 3.

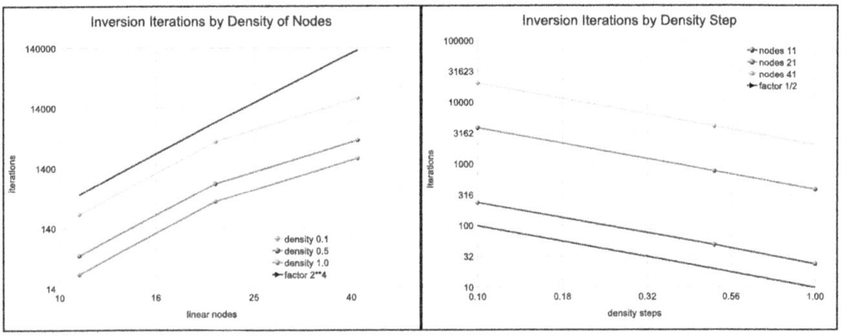

Fig. 3 Increase of Iterations for the inversed problem.

Number of iterations does not depend on parallelization, while the increasing factor for iterations when density step is inversed proportional with the density step as predicted theoretically. Instead, we have a slight decrease of the number of iterations when the spatial grid density is increased. It remains unanswered if this trend may continue for more dense arrays.

Variation of the inversion runtime in serial and parallel (16 virtual cores in 8 physical cores) mode is given in Fig. 4. In both serial and parallel cases the runtime increases with a factor that approximates 2^8=256 when the density of spatial arrays is duplicated, which corresponds with the combination of factors for the increase of elementary calculations (2^5) and of number of iterations due to decrease of density contribution (2^3).

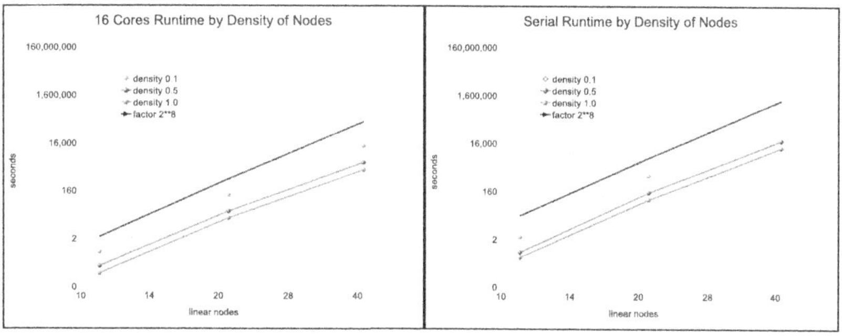

Fig. 4 Dependence of inversion runtime by spatial density of arrays.

Inversion runtime per iteration depends only on spatial parameters both in serial and parallel modes (Fig. 5). Increase factor of runtime per iteration when linear array nodes are duplicated varies between 2^4 and 2^5, apparently lower in case of parallel processing. The theoretical factor is 2^5.

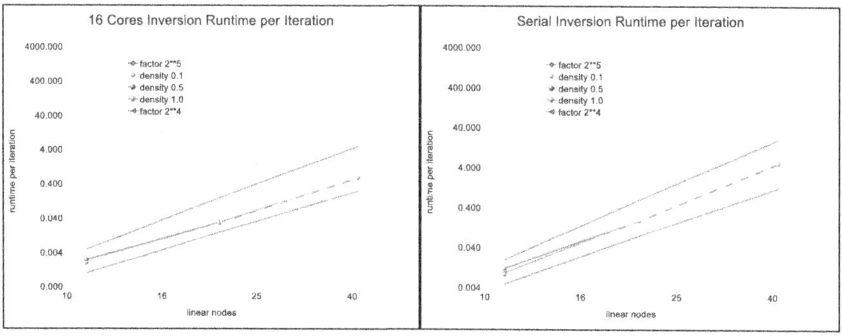

Fig. 5 Dependence of inversion runtime per iteration.

The ratio of 16 virtual cores runtime per serial runtime is presented in Fig. 6. With the increase of the density of array nodes the role of parallel calculations improves with the ratio converging towards 1/16. Least square error for both serial and 16 cores parallel running modes is presented in Fig. 7.

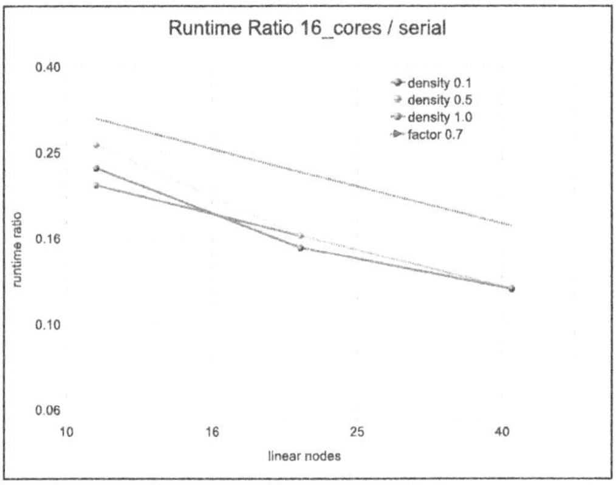

Fig. 6 Decrease of runtime in 16 virtual cores.

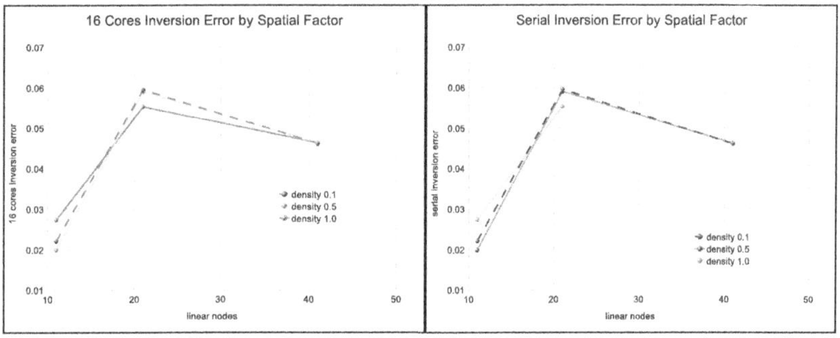

Fig. 7 Inversion problem error.

An example of the concrete solution – the vertical gravity gradient anomaly and the result of inversion for delineation of the anomalous body is given in Fig. 8.

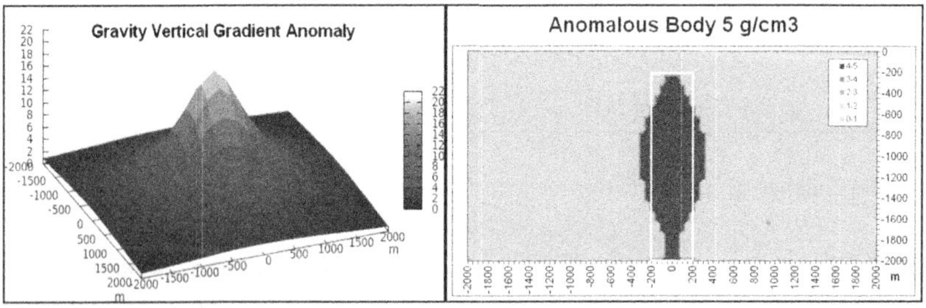

Fig. 8 Vertical gravity gradient anomaly (left) and the delineation of anomalous body (right – the central cut of geosection with the original body contour in white).

4 Conclusions

Obtained results throw light on the impact of parallel computing for the gravity inversion problem. Nevertheless the volume of calculations increases with factor $O(N^5)$ when the density N of linear nodes increases by O(N); with the modest number of 16 cores the case of a 3D array of 41*41*21 nodes and density step 0.1g/cm3 the runtime was 12,317 seconds (3.4 hours). In our models this array has the spatial step between nodes 25m; this spatial resolution may be reasonable for gravity problems but is still too low for other geophysical problems (magnetism and electrical fields) where resolutions of 1m or less may be required.

The increase of the number of iterations with a smaller factor when the density of array nodes increases may help for calculations with dense arrays, but the phenomenon has to be verified for parallel calculations with greater number of cores. A similar problem remains open for the runtime per iteration.

The runtime per iteration and the least square error obtained for the approximation of the gravity gradient do not depend on the density step. This particularity promotes the idea to avoid the use of small density steps as much as possible depending on the expected density contrasts between rocks situated in the geosection.

Delineation differences between the original anomalous body and resulting from inversion show the effect of lack of solution uniqueness.

The future work will be focused on testing of convergence and runtime hypotheses for arrays with spatial step 50m and 25m using more than 16 cores in parallel. Use of GPUs may be with interest as well, bringing the parallel system in the desktop.

Acknowledgments. This work makes use of results produced by the High-Performance Computing Infrastructure for South East Europe's Research Communities (HP-SEE), a project co-funded by the European Commission (under contract number 261499) through the Seventh Framework Programme. HP-SEE involves and addresses specific needs of a number of new multi-disciplinary international scientific communities (computational physics, computational chemistry, life sciences, etc.) and thus stimulates the use and expansion of the emerging new regional HPC infrastructure and its services. Full information is available at http://www.hp-see.eu/.

References

1. Hadamard, J.: Sur les prolemes aux derivees partielles et leur signification physique, vol. 13, pp. 1–20. Bull. Princeton Univ. (1902)
2. Högbom, J.A.: Aperture Synthesis with a Non-Regular Distribution of Interferometer Baselines. Astr. Astrophys. Suppl. 15, 417 (1974)
3. Lowrie, W.: Fundamentals of Geophysics. Cambridge University Press (2007)
4. Sen, M., Stoffa, P.: Global Optimization Methods in Geophysical Inversion. Elsevier Science B.V. (1995)
5. Silva, J., Medeiros, W.E., Barbosa, V.C.F.: Gravity inversion using convexity constraint. Geophysics 65(1), 102–112 (2000)

6. Wellmann, F.J., Horowitz, F.G., Schill, E., Regenauer-Lieb, K.: Towards incorporating uncertainty of structural data in 3D geological inversion. Elsevier Tectonophysics TECTO-124902 (2010) (accessed September 07, 2010), `http://www.elsevier.com/locate/tecto`
7. Zhou, X.: 3D vector gravity potential and line integrals for the gravity anomaly of a rectangular prism with 3D variable density contrast. Geophysics 74(6), 143–153 (2009)
8. Zhou, X.: Analytic solution of the gravity anomaly of irregular 2D masses with density contrast varying as a 2D polynomial function. Geophysics 75(2), I11–I19 (2010)

Parallel Genetic Algorithms for Finding Solution of System of Ordinary Differential Equations

Jane Jovanovski, Boro Jakimovski, and Dragan Jakimovski

Abstract. The goal of our research is to evaluate the general methods of finding solution of a system of differential equations. In this paper we investigate a novel two step genetic algorithm approach that produces an analytical solution of the system. The evaluation of the algorithm reveals its capability to solve non-trivial systems in very small number of generations. In order to find the best solution, and due to the fact that the simulations are computational intensive, we use grid genetic algorithms. Using the gLite based Grid, we propose a grid genetic solution that uses large number of computational nodes, that archives excellent performance. This research will be the basis on our goal of solving more complex research problems based around the Schrodingers equation.

1 Introduction

Almost any problem in all fields of science are expressed in terms of ordinary differential equations (ODE's), partial differential equations (PDE's) or more generally as a system of ordinary or partial differential equations (SODE or SPDE). For that reason many methods have been proposed for solving ODE's and PDE's, such as Runge Kutta, Predictor - Corrector [1], radial basis functions [2] and feedforward neural networks [3]. Also several methods based on genetic programming have also been proposed [4, 5]. The technique of genetic algorithms (GA) is an optimization process based on the evolution of a large number of candidate solutions through genetic operations such as replication, crossover and mutation [6].

Our research focuses on developing an efficient method based on genetic algorithms for finding an approximate analytical solution of a SODE. In this paper we

Jane Jovanovski · Boro Jakimovski · Dragan Jakimovski
Ss. Cyril and Methodius University in Skopje, Faculty of Natural Sciences and Mathematics, Arhimedova bb, Skopje, Macedonia
e-mail: janejovanovski@gmail.com, boro.jakimovski@finki.ukim.mk, dragan.jakimovski@gmail.com

L. Kocarev (Ed.): ICT Innovations 2011, AISC 150, pp. 227–237.
springerlink.com

propose a novel method based on two step genetic algorithm for solving SODE. This feature of two step genetic algorithms, dramatically improves the optimization process. We further investigate its parallel implementation using Parallel GA.

The rest of this article is organized as follows: in section 2 we describe the chromosome function representation, in section 3 we describe in detail the new algorithm and its implementation, in section 4 we present several experiments and in section 5 we present our conclusions and ideas for further work.

2 Function Representation and Evaluation

Search for the solution of system of ordinary differential equations using genetic algorithms requires an adequate representation of the candidate solution. Since the solution of a SODE is a set of functions, we need a chromosome that specifies a single and thus a set of functions. Computer representation of mathematical functions can be found im many different problems. The solution mainly lies in the fact that any set of correctly formed mathematical functions are based on a context-free grammar. This is the reason why the usual approach for representation or verification of a well formed mathematical function is either a push-down automate (or just a stack), a sequence of context-free rules, or a function tree.

In [5] they choose the approach from Grammatical Evolution for representation of the chromosome. Grammar Evolution is an evolutionary approach to automatic program generation, which evolves strings of binary values, and uses a BNF grammar to map those strings into programs. This mapping process involves transforming each binary individual into a string of integer values, and using those values to choose transformations from the given grammar. Even though this approach has good research background, in our opinion it is too complex and inefficient, due to the fact that the chromosomes can suffer from the "Ripple Effect" [7].

On the other hand in [4] the author opts for Revers Polish Notation for representation of the chromosomes. This approach gives the advantage of easy functional evaluation and function tree reconstruction. We find Polish Notation to be more adequate for function representation since it enables easier implementation of crossover function that we propose.

2.1 Candidate Solution

For the construction of the functions, elements of the candidate solution of the SODE, we define a grammar $G = V, \Sigma, R, S$ for the function representation. The set of non-terminals is fixed to $V = S$. The set of terminals Σ contains the building blocks of the candidate solutions. In our current experiments we chose

$$\Sigma = Variable \cup Constants \cup Functions, \tag{1}$$

where the set $Variables = \{x\}$, the set $Constants = \{c_j | j \in \{1..J\}\}$ that contains the placeholders for the constants that will appear in the solution, and the set of all accepted functions $Functions = \bigcup_k \{f_{k,l} | l \in \{1..L_k\}\}$ where k represents the arity. The set of rules is shown in equation 2.

$$R = \{S \to x, S \to c_j, S \to f_{1,l}S, S \to f_{2,l}SS, S \to f_{3,l}SSS, ...\}. \tag{2}$$

In our experiments we used the following sets and values as basis for solving the systems: number of constants J=10 and $Functions = \{sin, cos, exp, log, +, -, *, /\}$.

The chromosomes that represent the candidate solution for the SODE contain:

- one dynamic array per solution function representing a stack that holds the polish notation of the function
- one array for holding the values of the constants that appear in the system
- computed fitness value

Unlike the chromosomes that appear in Grammatical Evolution [7, 5], where chromosomes keep a lot of recess genes and their lenght is of fixed, we choose to keep the chromosomes length dynamic and containing only the functional elements.

2.2 Fitness Function

The evaluation of the fitness of the candidate solution needs to estimate the amount of difference with the correct solution. Naturally, a good estimation of the fitness is to calculate the integral of the difference between the two functions (candidate and solution), or even better, an integral of the square difference between the two functions. The fitness values are always positive, with the best fitness value 0.

In our experiments we simplify the fitness by calculating an approximation of the integral of the square difference by calculating a sum of square differences on an equidistant points. Following is a detailed description of fitness function evaluation.

Lets assume that we want to solve system of differential equations 3, where x is the differentiating variable and we assume that functions are k_i differentiable on the interval $[a,b]$. In order to find a solution, we need to define initial conditions shown in equation 4.

$$\begin{cases} f_1(x, y_1, y_1^{(1)} \dots y_1^{(k_1)} y_2^{(1)} \dots y_2^{(k_2)} \dots y_n^{(1)} \dots y_n^{(k_n)}) = 0 \\ f_2(x, y_1, y_1^{(1)} \dots y_1^{(k_1)} y_2^{(1)} \dots y_2^{(k_2)} \dots y_n^{(1)} \dots y_n^{(k_n)}) = 0 \\ \vdots \\ f_n(x, y_1, y_1^{(1)} \dots y_1^{(k_1)} y_2^{(1)} \dots y_2^{(k_2)} \dots y_n^{(1)} \dots y_n^{(k_n)}) = 0 \end{cases} \tag{3}$$

$$\begin{cases} y_1(a) = y_{1a} \\ y_1^{(1)}(a) = y_{1a}^{(1)} \\ \cdot \\ y_1^{(k_1-1)}(a) = y_{1a}^{(k_1-1)} \\ \cdot \\ \cdot \\ y_n(a) = y_{na} \\ y_n^{(1)}(a) = y_{na}^{(1)} \\ \cdot \\ y_n^{(k_n-1)}(a) = y_{na}^{(k_n-1)} \end{cases} \tag{4}$$

We start the evaluation of fitness function by choosing N equidistant points (x_i) on the interval $[a,b]$, $(x_0 = a,\ x_1,\ \ldots,\ x_{N-1} = b)$. Since the candidate solution contains n functions (stored as stacks) we will refer to these functions as $M_1,...,M_n$. The evaluation of the candidate functions is by simulating the execution of $M_i(x_i)$ using an additional stack. Once we finish the evaluation of function points, we estimate the values of the derivatives $M_j^{(k)}(x_i)$ by using the Central Difference Approximation and Ridders interpolation [8]. Computed values are used to evaluate the functions square differenece E using the equations 5. In order to give more emphasis on the initial conditions, we compute the difference from the initial condition and the candidate solution and multiply it using a emphasis factor λ. The computation of this part and the complete fitness function is shown in equations 6 and 7.

$$\begin{aligned} E(M_j) = \textstyle\sum_{l=0}^{N-1} & (f_j(x_l,\ M_1(x_l),\ M_1^{(1)}(x_l),\ldots,M_1^{(k_1)}(x_l), \\ & M_2^{(1)}(x_l),\ldots,M_2^{(k_2)}(x_l),\ldots,M_n^{(1)}(x_l),\ldots,M_n^{(k_n)}(x_l))^2, \\ E = \textstyle\sum_{j=1}^{n} & E(M_j) \end{aligned} \tag{5}$$

$$\begin{aligned} P(M_j) &= \lambda(M_j(a) - y_{ja})^2,\ \forall j = 1,\ldots n \\ P(M) &= \textstyle\sum_{j=1}^{n} P(M_j) \end{aligned} \tag{6}$$

$$u = E + P \tag{7}$$

3 Genetic Algorithm Implementation

In this section we will present the two level genetic algorithm that we use to find the approximative analytical solution of the system of differential equations. The two levels are based on the construction of the chromosomes that represent the candidate solutions for the system. The top level optimization focuses on the function structure. This means that the focus on this level are the different candidate solutions, each representing a completely different function set. On the other hand, the bottom level optimization focuses on the optimization of the constants that are introduced in the chromosomes. On this level we focus on each candidate solution, and try to find optimal values of the constants for achieving as optimal as possible fitness values.

Having this, out approach is, for each generation of the genetic algorithm, first to run the bottom optimization on the entire population, and once we have optimal constants for all chromosomes, to start the top level optimization.

Following are the principles and implementation decisions of the genetic algorithm optimization. We will start from the top and proceed to the bottom level of optimization.

3.1 Initial Population

The proper generation of an initial population of chromosomes is crucial for the success of the genetic algorithms. In our approach, we choose to build the initial population by randomly selecting rules from the chosen grammar. We use the George Marsaglia's KISS random generator built into the GNU Fortran and in each step, we first select which category of rule we will use on the first non-terminal character. By category we meen either $S \rightarrow x$, $S \rightarrow c_j$, $S \rightarrow f_{1,l}$ and $S \rightarrow f_{2,l}$, since we have only unary and binary functions. Once a category is chosen, we again randolmy choose a rule from that category, and apply the rule. The approach using categories is better than using a single random number to choose from uncategorzed rules, because if there are many constants, or many binary functions, the probability of choosing a terminating character (contant) might be much bigger than choosing a function, and the rule $S \rightarrow x$ will be chosen with very low probability. This will result with very nested or very simple functions with many constants.

3.2 Selection

The initial population is the basis of the optimization. Each generation of the algorithm starts with a selection of the population, in order to decide which chromosomes will survive to reproduce and mutate. The selection algorithm that we used is a variant of the tournament selection with two turns, quarterfinals and semifinals and at the end obtaining the finalists (reproduction set). In each turn we select a random subset of chromosomes and choose the winner that proceeds in the next turn.

3.3 Crossingover

Once the finalists are selected, we proceed with the reproduction steps using the following crossingover algorithm. The crossingoever between the chromozomes is on the level of separate functions, i.e. the first function of one chromosome exchanges material with all function of the other chromosome. this results with n^2 children from two chromosomes. For the crossing over between two functions, our goal is to select a subfunction of both functions (subtrees of the function trees), and switch the subfunctions. This process is depicted in Figure 1.

The subfunction selection process is implemented by first random selection of the top of the subfunction (random number represenets the index in the array that

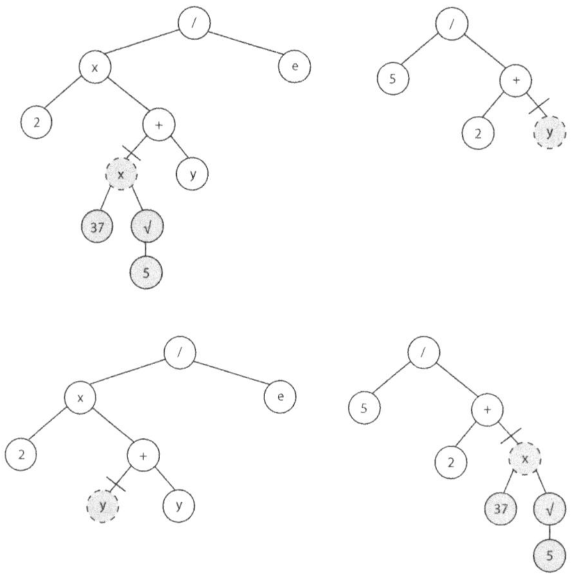

Fig. 1 Cutting and exchange of the subfunctions in the process of crossingover between two functions

holds the prefix function tree). Then we itterate towards the end of the array and try to find the closing position of the subfunction. This approach is possible since each subree is a continous segment in the array using the postfix (Polish) notation.

This approach for crossingover is much more natural when crossing two functions. A similar approach can be found in [4] where the authors search for a binary operator, and exchange the operator together with one of the subtrees. We find our solution to be more general. On the other hand, when using a Grammar Evolution approach [5] one cannot simply select a closed form subfunction. This results into additional step that need to fix the broken parts of the function.

3.4 Mutation

After the crossingover generates the new population we proceed to mutation operation. we select a relatively small number of random chromosomes, where we randomly select an element and mutate it with another elemnet from the same set (constant with constant or variable, unary function with unary function, etc.).

3.5 Constant Optimization

Since we introduced constants into the building process of the candidate functions, we need to define their values prior to solution evaluation. In similar approaches [5] predefine the values of the constants to several integer numbers. On the other hand, in [4] they define constant on every element in the function tree (a more general approach). They continue by optimizing the values of the constants by using perturbation.

In our approach, we use another genetic algorithm optimization in order to select the best values of the constants. The chromosomes represent an array of double precision values, randomly initialized. We implemented the heuristic crossover [9]. We start by selecting two chromosomes from a sorted popluation (by fitness). The first parrent u is chosen from the upper part (better fitness), and the second v from lower part of the sorted population. The offspring is generated by first selecting a random parameter $\alpha \in [0,1]$ and generate the offspring genes using the equations 8. All chromosomes enter the selection process, and the result is a new population where the upper part becomes lower part, and the newly generated chromosomes are located in the upper part of the population.

The mutation of the chromosomes is executed by selecting three chromosomes u, v, w and by selecting a single gene u_u, v_v and w_w. Then we compute the new value of $u_u = \frac{v_v + w_w}{2}$.

$$\begin{aligned} u_{inew} &= \alpha(u_i - v_i) + u_i, \quad i = 1, 2, \ldots, n \\ v_{inew} &= u_i, \quad i = 1, 2, \ldots, n \end{aligned} \tag{8}$$

Benefits of constant optimization we show in section 4. We analyze fitness value of best chromosome in population, when we exclude and include constant optimization. When we exclude constant optimization value of best chromosome convergence very slowly, need large number of generations, with large number of chromosomes, for find approximate solution, approximate solution is very long function etc. Those disadvantages are solving when we include constant optimization.

4 Evaluation

In order to evaluate the proposed two step genetic algorithm for SODE, we evaluated its performance, conversation speed and quality of solution. We tested the algorithm on several systems of differential equations with known analytical solution and show results for tree examples. Wolfram Mathematica's DSolve can not find results for first two examples. The results from the experiments are very promissing and we beleive that this approach can be used very successfully, especially its parallel implementation. The process of optimization was run using a replication rate of 15% (i.e. cross over probability 85%) and mutation rate of 5%, $\lambda = 100$, over a population of 100 chromosomes. In the tournament selection algoritm we choose the number of quaterfinalists to be 25, semifinalists to 5 and the number of finalists that are generated is $(cross\ over\ probability) * (number\ of\ chromosomes\ in\ the\ generation)$. In genetic algorithm for optimization of the constants values we set population size to 200 and number of generation to 100. When we exclude constant optimization we use 10 constant with value $\{0, 1, 2, ..., 9\}$.

4.1 Example 1

For our first experiment we used the system show in equation 9 with initial conditions 10. The solution of this system is $y_1 = \sin x$ and $y_2 = x^2$. In our search for a solution, we chose the interval $x \in [0,\ 1]$ with step 0,05. The correct solution was found in the 9th generation. The obtained solution of the SODE equation 9 is shown in Figure 2. The figure also desplays a pre-final candidate solution.

$$\begin{cases} y_1' = \cos x + y_1^2 + y_2 - x^2 - \sin^2 x \\ y_2' = 2x - x^2 \sin x + y_1 y_2 \end{cases} \tag{9}$$

$$y_1(0) = y_2(0) = 0 \tag{10}$$

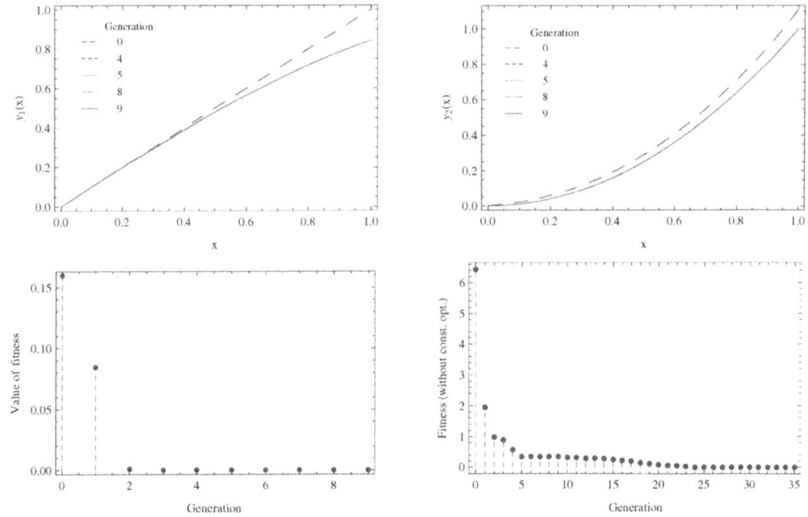

$$y1: \quad \sin x$$

$$y2: \quad -9.571 \sin\left(\log\left(\cos\left(x\left(-0.4367 + \frac{(-16.416-x)\cos\left(x+\frac{51.992+x}{16.416+x}\right)}{-801.501-34.576x+x^2}\right)\right)\right)\right)$$

$$fitness: 8.71\ 10^{-10}$$

Fig. 2 Solution convergence of the SODE equation 9 by generations

When we exclude constant optimization we find correct solution in the 24th generation.

4.2 Example 2

The second system we tested was the system show in equation 11 with initial conditions 12. The solution of this system is $y_1 = \frac{\sin x}{e^x}$ and $y_2 = e^x$. In our search for a solution, we chose the interval $x \in [0,\ 1.5]$ with step 0,05. The correct solution

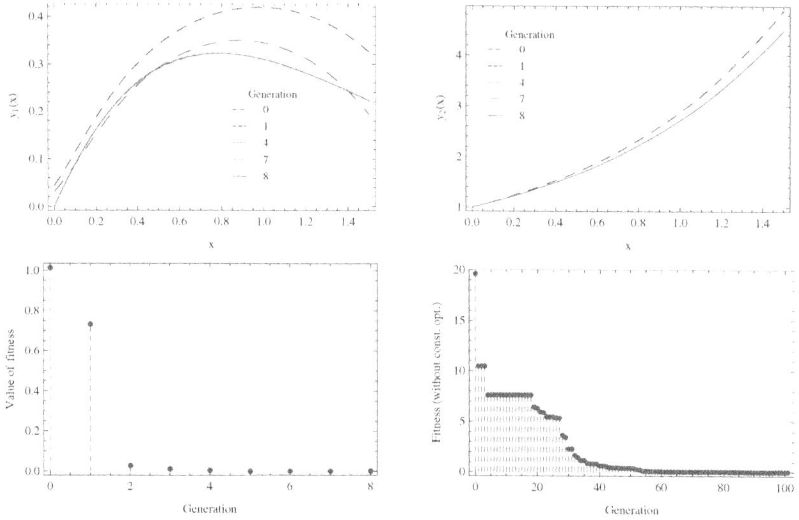

$y1$: $\cos(0.9092 - 0.417x + 0.868\cos(\cos(x)) - 0.623\log(\cos(0.74687 - x)))$
$y2$: e^x
fitness : $8.9\ 10^{-4}$

Fig. 3 Solution convergence of the SODE equation 11 by generations

was found in the 8th generation. The obtained solution of the SODE equation 11 is shown in Figure 3.

$$\begin{cases} y_1' = \frac{\cos x - \sin x}{y_2} \\ y_2' = y_1 y_2 + e^x - \sin x \end{cases} \tag{11}$$

$$y_1(0) = 0, y_2(0) = 1 \tag{12}$$

When we exclude constant optimization we find approximate solution in the 100th generation with fitness value of $3.012\ 10^{-3}$. After the 55th generation convergence is very slow.

4.3 *Example 3*

In the third experiment we used the system show in equation 13 with initial conditions 14. The solution of this system is $y_1 = \cos(e^x)$ and $y_2 = e^{-x}$. In our search for a solution, we chose the interval $x \in [0,\ 1]$ with step 0,05. The correct solution was found in the 13th generation. The obtained solution of the SODE equation 13 is shown in Figure 4.

$$\begin{cases} y_1' = -\frac{1}{y_2}\sin(e^x) \\ y_2' = -y_2 \end{cases} \tag{13}$$

$$y_1(0) = \cos(1.0), y_2(0) = 1 \tag{14}$$

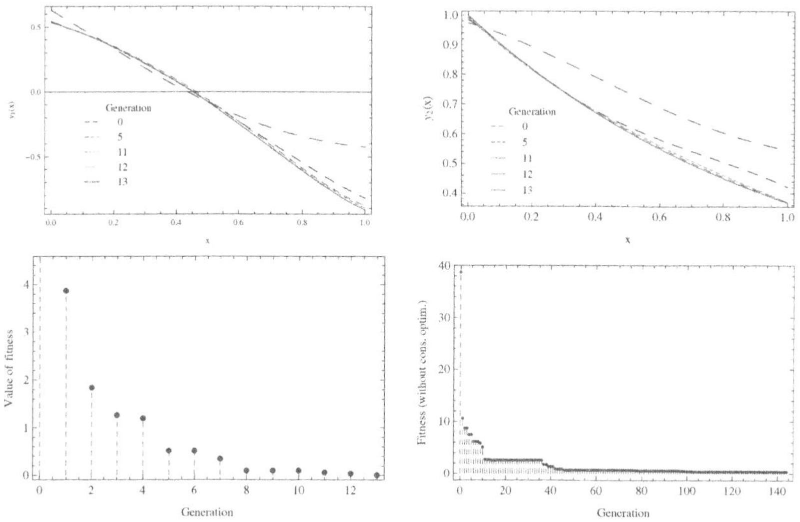

$$y1: \quad 1.05\sin\left(2.201+x+\cos\left(4.05+x^2\right)+1.05\sin\left(2.722+0.955\cos\left(3.464+x\right)\right)\right)$$

$$y2: \quad 0.210439\left(0.583738x+\frac{(-4.7394+1.55856x)\sin(4.75196+x)}{0.816161x+\sin(1.55856+x)}\right)$$

$$fitness: \ 5.578\ 10^{-2}$$

Fig. 4 Solution convergence of the SODE equation 13 by generations

When we exclude constant optimization we find the best approximate solution in the 107th generation with fitness value of 0.356. After this generation convergence is very slow.

5 Conclusion and Future Work

In this paper we presented a very successfully implementation of a novel genetic algorithm. Our two step optimization enables bigger flexibility of the solutions and makes the process of optimization very successful even with very small number of generations.

Even though we achieved great result, we will continue to optimize the algorithm and its implementation, mainly by its optimal parallel implementation on HPC clusters. We want to further investigate different initial population generation approaches, were we won't use uniform distribution to select the subset of rules. Another interesting issue is the fact that in the currently generated candidate solutions we rarely see elements such as polynomials. We believe that this problem is related to the approach for building the initial set of chromosomes and that these elements will give much better candidate solutions.

Acknowledgements. This paper is based on the work done in the framework of the HP-SEE FP7 EC funded project.

References

1. Lambert, J.D.: Numerical methods for ordinary differential systems: the initial value problem. John Wiley & Sons, Inc. (1991)
2. Fasshauer, G.E.: Solving differential equations with radial basis functions: multilevel methods and smoothing. Advances in Computational Mathematics 11, 139–159 (1999), doi:10.1023/A:1018919824891
3. Lagaris, I.E., Likas, A., Fotiadis, D.I.: Artificial neural networks for solving ordinary and partial differential equations. IEEE Transactions on Neural Networks 9, 987–1000 (1998)
4. Burgess, G.: Finding Approximate Analytic Solutions To Differential Equations Using Genetic Programming. Technical Report DSTO-TR-0838, Surveillance Systems Division, Defence Science and Technology Organisation, Australia, Salisbury, SA, 5108, Austrlia (1999), http://www.dsto.defence.gov.au/corporate/reports/DSTO-TR-0838.pdf
5. Tsoulos, I.G., Lagaris, I.E.: Solving differential equations with genetic programming. Genetic Programming and Evolvable Machines 7, 33–54 (2006)
6. Goldberg, D.E.: Genetic Algorithms in Search, Optimization and Machine Learning. Addison-Wesley Longman Publishing Co., Inc. (1989)
7. O'Neill, M., Ryan, C., Keijzer, M., Cattolico, M.: Crossover in Grammatical Evolution. Genetic Programming and Evolvable Machines 4, 67–93 (2003)
8. Ridders, C.: Accurate computation of F'(x) and F'(x) F''(x). In: Advances in Engineering Software (1978), vol. 4, pp. 75–76 (1982), doi:10.1016/S0141-1195(82)80057-0
9. Sivanandam, S.N., Deepa, S.N.: Introduction to Genetic Algorithms. Springer Publishing Company, Inc. (2007)

Enabling Large-Scale Linear Systems of Equations on Hybrid HPC Infrastructures

H. Astsatryan, V. Sahakyan, Yu. Shoukouryan, M. Daydé, and A. Hurault

Abstract. Scientific research is becoming increasingly dependent on the large-scale analysis of data using High Performance Computing (HPC) infrastructures. Scientific computing aims at constructing mathematical models and numerical solution techniques for solving problems arising in science and engineering. The solution of linear system of equations lies at the heart of most calculations in scientific computing. HPC infrastructures with many-core and graphics processing unit (GPU) challenges, Cloud and Grid technologies and e-infrastructures are currently offering interesting opportunities for solving large-scale linear system of equations. In this article, a second-generation of our Web portal for Scientific Computing is introduced based on a hybrid HPC infrastructure that provides predictable optimal execution and scales from a single resource to multiple resources. After analyzing the synergies and the complementarities of the different computing platforms, we argue for an architecture that combines the benefits of these technologies.

Keywords: Linear Algebra, P-Grade, BLAS, SGEMM, CUBLAS, GPU, Grid, Portal.

1 Introduction

Scientific computing aims at constructing mathematical models and numerical solution techniques for solving problems arising in science (including life and social

H. Astsatryan · V. Sahakyan · Yu. Shoukouryan
Institute for Informatics and Automation Problems of the National Academy of Sciences of the Republic of Armenia, 1, P. Sevak str., Yerevan, 0014, Armenia
e-mail: {hrach,svlad,shouk}@sci.am

M. Daydé · A. Hurault
Institut de Recherche en Informatique de Toulouse, Ecole Nationale Supérieure d'Electrotechnique, d'Electronique, d'Informatique, d'Hydraulique et des Télécommunications, 2, rue Charles Camichel, 31071, Toulouse, France
e-mail: {Michel.Dayde,Aurelie.Hurault}@enseeiht.fr

L. Kocarev (Ed.): ICT Innovations 2011, AISC 150, pp. 239–245.
springerlink.com

sciences) and engineering. The solution of linear system of equations lies at the heart of most calculations in scientific computing [1]. For the past twenty years, there has been a great deal of activity in the area of algorithms and software for solving linear algebra problems. Numerical simulations involving massive amounts of calculations are often executed on HPC infrastructures that include various platforms and architectures. One of the challenging advents in computer science in recent years was the fast evolution of parallel processors, specially the use of accelerators such as GPUs [2-3] that play today a major role in many computational environments. Over the recent years, GPU architectures have continued to develop a more flexible and uniform programming environment.

It is often difficult for non expert users to use such different computational infrastructures or even to use the available advanced libraries, because they often require to be familiar with middleware and tools, parallel programming techniques and packages (MPI, OpenMP, CUDA, etc.) and numerical calculation libraries (Scalapack, Lapack, Blas, etc..).

We focus on dense linear algebra in this paper but this work could be extended to other areas where scientific packages are available. Obviously the extension to sparse linear algebra is straightforward when considering sparse direct solvers such as MUMPS (MUltifrontal Massively Parallel sparse direct Solver) [4]. The environment we consider includes ScaLAPACK (Scalable LAPACK) [5] and BLAS (Basic Linear Algebra Subprograms) [6] libraries. ScaLAPACK is a library of high-performance linear algebra routines for distributed-memory message-passing MIMD (multiple instruction, multiple data) computers and networks of workstations supporting Parallel Virtual Machine and/or Message Passing Interface. It contains routines for solving systems of linear equations, least squares problems and eigenvalue problems. Furthermore, with ScaLAPACK comes an additional library called the PBLAS, which can be seen as a parallel version of the BLAS. Using the PBLAS simple matrix/vector and more complex operations can be performed in parallel.

CUBLAS [7] is an implementation of BLAS on top of the NVIDIA CUDA (compute unified device architecture) driver that is used for GPU computing. It allows access to the computational resources of NVIDIA GPUs. The library is self-contained at the API level, that is, no direct interaction with the CUDA driver is necessary. The basic model by which applications use the CUBLAS library is to create matrix and vector objects in GPU memory space, fill them with data, call a sequence of CUBLAS functions, and, finally, upload the results from GPU memory space back to the host.

The purpose of this article is to introduce an environment enabling the solution of large scale linear systems of equations on hybrid HPC infrastructures, which is the extension of the previous work [8-9]. The environment is responsible of dispatching computing jobs to remote GPU and Grid computing facilities by matching complexity of the given expression and the resource status information collected from information services.

The experiments have been done on the base of the Armenian National Grid infrastructure [10] that unifies different architectures (multicore processors, GPUs) and network topologies (myrinet, infiniband, gigabit). Particularly it consists of seven Grid sites located in the leading research and educational organizations of Armenia.

2 Experiments

Balancing the calculations over a heterogeneous computing infrastructure such as the Armenian National Grid is quite challenging.

The benchmarking of load balancing has been done mainly on the base of BLAS package, which is the standard programming interface for basic linear algebra operations such as vector and matrix multiplication. The GotoBLAS package, a tuned implementation of the BLAS, is used for CPU benchmarking, while the CUBLAS package is used for the GPU benchmarking. GEMM (General Matrix-Matrix Multiply) is a key computational kernel in all well-known libraries of linear algebra. One of the major advantage of GEMM is a computational intensively function, with a very good ratio of computations over communications which is crucial for the performance on modern architectures with memory hierarchy (on square matrices of order N, computations are in $O(N^3)$ while data accesses are in $O(N^2)$). Major hardware vendors maintain their own highly optimized GEMM routines often included into their BLAS implementation libraries.

The GPU experiments have been done using Tesla 1060 cards [11] that have 240 cores GPGPU (General-purpose computing on graphics processing units) which supports massively parallel programming through a threading model. In case of GPU, the library creates matrix and vector objects in GPU memory space, fill them with data, call a sequence of functions, and, finally, move the results from GPU memory space back to the host. The data transfers in GPU are expensive relative to the floating-point capability of the graphics processors.

The optimized GEMM function can achieve extremely high performance. Figure 1 describes a SGEMM (single precision operations) benchmarking on CPU (Dual Intel Xeon 3.06GHz) and GPU (Tesla C1060 running CUDA 2.2) architectures. All results are reported in GFLOP.

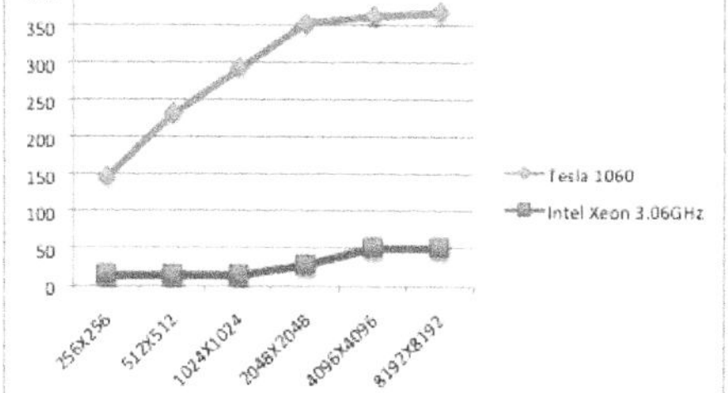

Fig. 1 BLAS Performance (CPU vs GPU): SGEMM benchmarking

Double precision is supported on many NVIDIA cards. In case of Tesla 1060 card, each of the processing units contains one double precision processor alongside the 8 single precision processors. This means that there are only 30 double precision units available. The benchmarking on DGEMM (double precision operations) function shows much less performance (see fig. 2).

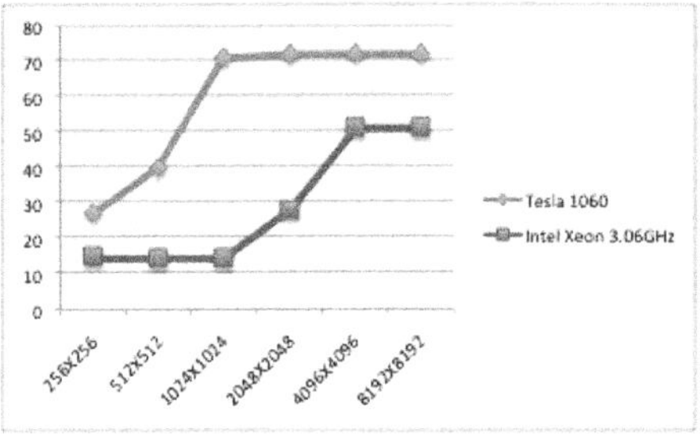

Fig. 2 BLAS Performance (CPU vs GPU): DGEMM benchmarking

3 Scientific Computing Environment

Our Scientific Computing Environment (see fig. 3) is based on the P-GRADE (Parallel Grid Run-time and Application Development Environment) [12] Grid Portal, which is a Grid portal solution that allows users to manage the whole life-cycle for executing a parallel application on the Grid. The environment provides the seamless bridge between the linear algebra calculations and the linear algebra software environment (middlewares, tools, parallel programming techniques, linear algebra libraries).

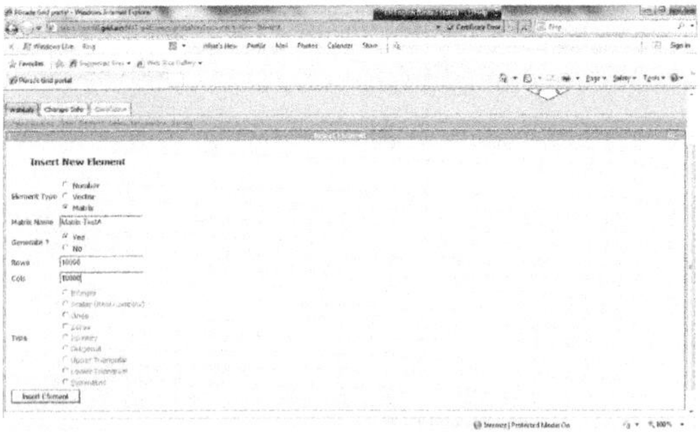

Fig. 3 Scientific Computing Environment based on P-GRADE

The rCUDA framework [13] is used for GPU remote calls that enable the concurrent usage of CUDA-compatible devices remotely. rCUDA employs the socket API for the communication between clients and servers. The portlet is compatible with the gLite middleware [14] that was developed under the umbrella of the EGEE Project.

The interface for linear algebra calculations is based on the use of linear algebra libraries. In the first stage users define and insert new numerical objects (number, vector, and matrix) that they need in their linear algebra calculations. Both uploading and generation mechanisms are available for the creation of numerical objects by specifying their types and properties: vectors (ones, zeros, integer, etc…), matrices (triangular, identity, symmetric, etc.). The interface provides users different options (download, delete or change parameters) for the uploaded numerical objects. In the next stage, users define and submit linear algebra expression. The server side processing of the portlet includes analyzer, trader and job preparation modules. The analyzer module carried out syntactic and lexical analysis by checking the main parameters and iterations of numerical objects and expressions. If the expression is correct, it converts the expression into the corresponding XML format (required as an input for the trader module) that fully defines the given expression including the parameters and properties.

The Job preparation module decides the target infrastructure taking into account the type and the complexity of the given expression (see fig. 4).

Target	4096	8192	>8192
Local			
GPU			
Grid			

Fig. 4 Target Infrastructure & Matrix size

In the second stage, the Job preparation module converts the service [15] or combination of services into the following formats of the requested target infrastructure:

- Local execution by running script
- rCUDA call
- JDL (Job Description Language) file in order to submit the expression over the Grid infrastructure. JDL attributes represent job specific information and specify in some way actions that have to be performed by the Workload Management System (WMS) to schedule the job. Some of these attributes are static and the converter automatically inserts some others.

The main benefit of such an approach is that the user has only to type a mathematical expression and to provide the required data (matrices, vectors, and numbers) in order to get the results. All given expressions and results are saved in the centralized database. The real matrixes are located in the storage element of the Grid infrastructure, which minimizes the calculation time.

4 Conclusion

The full implementation of our environment for scientific computing over the Grid will allow users, who are not familiar with the parallel programming technologies and software tools (Grid and cluster middlewares, parallel and GPU programming, queue system, linear algebra libraries), to create and submit their numerical algebra expressions over the portal that hides all details of the underlying distributed computing infrastructure. The portal minimized the execution time of the given expression by using combination of GPU, local and Grid infrastructures. The target application has been implemented based on the benchmarking results provided above. Note that it is also a nice environment for demonstrating how computations can be transparently – from the user point of view – performed over the Grid or GPU.

Acknowledgements. This work supported by the Armenian-French (Centre National de la Recherche Scientifique, France and State Committee of Science of the Republic of Armenia) joint project entitled "Development of a Second Generation Scientific Computing Infrastructure Based on P-Grade Portal".

References

1. Petitet, A., Casanova, H., Dongarra, J., Robert, Y., Whaley, R.: Parallel and Distributed Scientific Computing: A Numerical Linear Algebra Problem Solving Environment Designer's Perspective. In: Handbook on Parallel and Distributed Processing (1999)
2. Owens, J., Houston, M., Luebke, D., Green, S., Stone, J., Phillips, J.: GPU computing. Proceedings of the IEEE 96(5), 879–899 (2008)
3. General-purpose computation using graphics hardware, `http://www.gpgpu.org`
4. MUMPS: a MUltifrontal Massively Parallel sparse direct Solver, `http://mumps.enseeiht.fr`
5. Blackford, L.S., Choi, J., Cleary, A., D'Azevedo, E., Demmel, J., Dhillon, I., Dongarra, J., Hammarling, S., Henry, G., Petitet, A., Stanley, K., Walker, D., Whaley, R.C.: ScaLAPACK: A Linear Algebra Library for Message-Passing Computers. In: SIAM Conference on Parallel Processing (1997)
6. Blackford, S., Demmel, J., Dongarra, J., Duff, I., Hammarling, S., Henry, G., Heroux, M., Kaufman, L., Lumsdaine, A., Petitet, A., Pozo, R., Remington, K.: An Updated Set of Basic Linear Algebra Subprograms. ACM Trans. Math. Soft. 28(2), 135–151 (2002)
7. Barrachina, S., Castillo, M., et al.: Evaluation and tuning of the Level 3 CUBLAS for graphics processors. In: IEEE International Symposium on Parallel and Distributed Processing, pp. 1–8. IEEE (April 2008)
8. Astsatryan, H., Sahakyan, V., Shoukouryan, Y., Daydé, M., Hurault, A., Pantel, P., Caron, E.: A Grid-Aware Web Interface with Advanced Service Trading for Linear Algebra Calculations. In: High Performance Computing for Computational Science - VECPAR 2008. LNCS, vol. 4395, Springer, Toulouse (2008)

9. Astsatryan, H., Sahakyan, V., Shoukourian, Y., Srapyan, M., Daydé, M., Hurault, A., Grigoras, R.: Introduction of a Grid-Aware Portlet for Numerical Calculations. In: IEEE 1st International Conference on Parallel, Distributed and Grid Computing (PDGC 2010), Waknaghat, Solan (HP), India (2010)
10. Astsatryan, H., Shoukouryan, Y., Sahakyan, V.: Grid Activities in Armenia. In: Proceedings of the International Conference Parallel Computing Technologies (PAVT 2009), Novgorod, Russia (March 30-April 3, 2009)
11. NVIDIA® Tesla™ C1060 computing processor enables, `http://www.nvidia.com/docs/IO/43395/NV_DS_Tesla_C1060_US_Jan10_lores_r1.pdf`
12. Németh, C., Dózsa, G., Lovas, R., Kacsuk, P.: The P-GRADE Grid Portal. In: Laganá, A., Gavrilova, M.L., Kumar, V., Mun, Y., Tan, C.J.K., Gervasi, O. (eds.) ICCSA 2004. LNCS, vol. 3044, pp. 10–19. Springer, Heidelberg (2004)
13. Duato, J., Antonio, J.P., Federico, S., Rafael, M., Enrique, S.: Quintana-Ortsíi, rCUDA: Reducing the number of GPU-based accelerators in high performance clusters. doi:10.1109/HPCS.2010.5547126
14. EGEE (Enabling Grids for E-sciencE) Project, `http://glite.web.cern.ch/glite`
15. Hurault, A., Daydé, M., Pantel, M.: Advanced service trading for scientific computing over the grid. Journal of Supercomputing, Springer Netherlands 49(1), 64–83 (2008)

Faraday Waves and Collective Modes in Bose-Einstein Condensates

Alexandru I. Nicolin, Ivana Vidanović, and Antun Balaž*

Abstract. Using variational and perturbative methods we investigate the dynamics of a trapped Bose-Einstein condensate whose effective nonlinearity is harmonically modulated. We derive analytic formulae for the period of Faraday waves in low- and high-density, longitudinally-homogeneous condensates. Using the Poincaré-Lindstedt analysis, we also analytically calculate frequency shifts of collective oscillation modes of spherically and axially-symmetric condensates due to the harmonic driving of the interaction.

Keywords: Bose-Einstein condensates, Faraday waves, collective modes, perturbation theory, variational treatment.

1 Introduction

Very few physical systems are as tunable as Bose-Einstein condensates (BECs) and even fewer benefit from a well-established theoretical framework that reinforced the overwhelming majority of experimental results. This uncommon combination between experimental maneuverability and theoretical modeling, along with the intrinsically nonlinear behavior of BECs, transformed these ultra-cold quantum gases into one of the favorite playgrounds of nonlinear scientists, and catalyzed fruitful exchanges between scientists from many distinct fields such as nonlinear mechanics, quantum and nonlinear optics, nuclear and condensed matter physics [1].

Alexandru I. Nicolin
"Horia Hulubei" National Institute for Physics and Nuclear Engineering (IFIN-HH), 30 Reactorului, Magurele-Bucharest 077125, Romania
e-mail: nicolin@theory.nipne.ro

Ivana Vidanović · Antun Balaž
Scientific Computing Laboratory, Institute of Physics Belgrade, University of Belgrade, Pregrevica 118, 11080 Belgrade, Serbia
e-mail: ivanavi@ipb.ac.rs, antun@ipb.ac.rs

* Corresponding author.

L. Kocarev (Ed.): ICT Innovations 2011, AISC 150, pp. 247–256.
springerlink.com

A BEC consists of a dilute gas of magnetically and/or optically trapped bosons that, when cooled to extremely low temperatures, occupy their lowest-energy quantum state. Nowadays, condensates can be routinely manipulated in time and space. On the spatial side, one has good experimental control over the shape and the strength of the trapping potential, while on the temporal side one can modulate the aforementioned properties of the trapping potential and can tune the strength of the two-body scattering length to the extent of probing the region between attractive and repulsive condensates. The dynamics of the condensates is governed by the so-called Gross-Pitaevskii (GP) equation, a nonlinear (cubic) Schrödinger equation where the nonlinearity accounts at the mean-field level for the interatomic interactions close to absolute zero temperature [2]. Before experimental realization of BEC in 1995, the cubic Schrödinger equation has been used to describe quasi-monochromatic wave trains propagating in a weakly nonlinear dielectric (e.g. a Kerr medium, where the dielectric constant depends on the square of the electric field), the weakly nonlinear dynamics of a wave train propagating at the surface of a liquid (the so-called water-wave problem), the Langmuir oscillations (also referred to as Langmuir waves or electron plasma waves) that arise in non-magnetized or weakly magnetized plasmas, the Alfvén waves that propagate along an ambient magnetic field in a quasi-neutral plasma and so on [3].

Motivated by the experimental observation of Faraday waves in cigar-shaped condensates subject to periodic modulations of the confinement [4] and the recent investigations into the collective modes of a BEC using periodic modulations of the atomic scattering length [5], we study here using variational and perturbative means the collective dynamics of BECs subject to periodic modulations of the effective nonlinearity. The paper is organized as follows: in Section 2 we investigate by variational means the emergence of Faraday waves in low- and high-density, cigar-shaped condensates with longitudinal homogeneity, while in Section 3 we employ Poincaré-Lindstedt analysis and a Gaussian variational approach [6, 7] to study frequency shifts of collective modes of spherically and axially-symmetric BECs due to the harmonic modulation of the s-wave scattering length via a Feshbach resonance. Section 4 summarizes our main results and presents concluding remarks.

2 Faraday Waves

Faraday waves belong to the wide-ranging field of pattern formation in driven systems which now spans almost all topics from hydrodynamics to biophysics and from nonlinear optics to reaction kinetics. They can be understood as standing waves which arise through a parametric instability on the surface of a parametrically driven classical or quantum fluid layer. Depending on the geometry of the experiment and the strength of the driving field, Faraday waves can have various symmetries ranging from triangular, square and hexagonal patterns to spatio-temporal chaos [8].

The concept of such waves originated in the classical studies of M. Faraday [9] devoted to the behavior of groups of particles placed upon vibrating elastic surfaces. Faraday's experiment became a classical example of pattern formation, whereby the uniform state loses its stability against spatially modulated wave forms, whose dominant length scale is determined by the intrinsic properties of the system and is only weakly dependent on boundary or initial conditions. While Faraday already realized that the excited waves had a frequency equal to one half that of the excitation (and Rayleigh suspected that this phenomenon was related to a parametric driving [10]), Benjamin and Ursell [11] were the first to show in a rigorous manner that the linear behavior of inviscid surface waves can be understood in terms of a Mathieu oscillator. The prototypical hydrodynamic experimental setup is that of a tray filled with fluid which is vertically driven by mechanical means. Depending on the nature of the driving field (either a single- or multi-frequency drive), on the properties of the fluid (e.g., temperature and viscosity), and the specific geometry of the tray that is used (i.e. with or without significant capillary effects), one sees well-ordered structures (such as stripes, squares, hexagons, etc.), coexistence and competitions of these structures, order-disorder transitions to spatio-temporal chaos, etc.

Faraday waves are particularly appealing in the BEC community because they can be obtained experimentally by harmonically modulating either the atomic scattering length (as shown in Ref. [12] through extensive numerical simulations) or the strength of the confining potential. This latter method was used in the experiments of Engels *et al.* [4] and was subsequently reinforced theoretically in Refs. [13, 14] by analytical means and extensive numerical simulations. While the first method of obtaining Faraday waves is not currently supported by experiments, the recent results of Pollack *et al.* [5] on the collective dynamics of BECs using periodic modulations of the atomic scattering length pave the way for experimental investigations.

The starting point of our variational treatment is the well-known GP Lagrangian density (with units set to $\hbar = m = 1$):

$$\mathscr{L}(\mathbf{r},t) = \frac{i}{2}\left(\psi\frac{\partial\psi^*}{\partial t} - \psi^*\frac{\partial\psi}{\partial t}\right) + \frac{1}{2}|\nabla\psi|^2 + V(\mathbf{r},t)\,|\psi|^2 + \frac{g(t)N}{2}|\psi|^4, \qquad (1)$$

expressed in cylindrical coordinates (ρ,ϕ,z), with $\dot{\mathbf{r}} = (\rho\cos\phi,\rho\sin\phi,z)$, $\psi = \psi(\mathbf{r},t)$ and $V(\mathbf{r},t) = \Omega_\rho^2(t)\rho^2/2$. To derive a consistent set of ordinary differential equations for the dynamics of a condensate that exhibits Faraday waves, we compute the Lagrangian $L = \int d\mathbf{r}\,\mathscr{L}(\mathbf{r},t)$ in both the low- and the high-density regime. In the low-density regime we use the following ansatz:

$$\psi_{\mathrm{ld}}(\mathbf{r},t) = f_{\mathrm{ld}}\left(k,w_\rho,u,v\right) e^{-\frac{\rho^2}{2w_\rho^2}+i\rho^2\alpha}\left[1+(u+iv)\cos kz\right], \qquad (2)$$

where w_ρ, α, u, v are time-dependent parameters, and a normalization factor f_{ld} ensures that $\int d\mathbf{r}\,|\psi_{\mathrm{ld}}(\mathbf{r},t)|^2 = 1$. In the high-density regime, the ansatz has the form:

$$\psi_{\mathrm{hd}}(\mathbf{r},t) = f_{\mathrm{hd}}\left(k,w_\rho,q,u,v\right)\left(1-\frac{\rho^2(1-q)}{2w_\rho^2}\right)^{1/(1-q)}\left[1+(u+iv)\cos kz\right], \qquad (3)$$

where q is a new time-dependent parameter, and f_{hd} is normalization factor. The latter ansatz is tailored around a q-Gaussian function that was shown to describe very accurately the dynamics of the condensate in the Thomas-Fermi regime [15], actually reproducing, for $q \to -1$, the well-known hydrodynamic approach [16]. In both cases the trial wave function accounts for the Faraday wave as a surface wave, which is not grafted to a longitudinal envelope due to the homogeneity of a BEC.

Taking into account that within this variational approach a condensate extends longitudinally between $-\pi/k$ and π/k, which defines the linear density $\rho_z = kN/2\pi$, the two Lagrangians are analytically tractable for arbitrary values of variational parameters, and one gets in the low-density regime the expression:

$$L_{\mathrm{ld}}(t) = \frac{k^2}{2}\frac{u^2+v^2}{2+u^2+v^2} + \frac{1}{2w_\rho^2} + \frac{\Omega_\rho^2(t)w_\rho^2}{2} + \frac{u\dot{v}-v\dot{u}}{2+u^2+v^2} + 2w_\rho^2\alpha^2 + \frac{kNg(t)}{16\pi^2}\frac{8+3u^4+8v^2+3v^4+6u^2\left(4+v^2\right)}{w_\rho^2\left(2+u^2+v^2\right)^2}, \tag{4}$$

while in the high-density regime the Lagrangian has the form:

$$L_{\mathrm{hd}}(t) = \frac{k^2}{2}\frac{u^2+v^2}{2+u^2+v^2} + \frac{3-q}{2w_\rho^2(1+q)} + \frac{\Omega_\rho^2(t)w_\rho^2}{4-2q} + \frac{u\dot{v}-v\dot{u}}{2+u^2+v^2} - \frac{kNg(t)(q-3)^2}{16\pi^2(q-5)}\frac{8+3u^4+8v^2+3v^4+6u^2\left(4+v^2\right)}{w_\rho^2\left(2+u^2+v^2\right)^2}. \tag{5}$$

From the above expressions we can derive ensuing Euler-Lagrange equations, but they are not particularly insightful. However, to first order in $u(t)$ and $v(t)$, namely for the small-amplitude waves, one can cast the Euler-Lagrange equations in a very convenient form. For low-density BECs we obtain the system of equations:

$$w_\rho(t) = \frac{\left(4\pi^2 + kNg(t)\right)^{1/4}}{\sqrt{2\pi\Omega_\rho(t)}}, \tag{6}$$

$$\ddot{u}(t) = -u(t)k^2\left(\frac{k^2}{4} + \frac{kNg(t)}{4\pi^2 w_\rho^2(t)}\right), \tag{7}$$

while for high-density condensates we have the following system:

$$q(t) = -1 + \frac{12\pi^2}{\pi^2(3+\Omega_\rho^2(t)w_\rho^4(t)) - 2kNg(t)}, \tag{8}$$

$$w_\rho(t) = \sqrt{\frac{6}{\Omega_\rho(t)(1+q(t))}}, \tag{9}$$

$$\ddot{u}(t) = -u(t)k^2\left(\frac{k^2}{4} + \frac{kN\Omega_\rho(t)g(t)(1+q(t))}{9\pi^2}\right). \tag{10}$$

Note that in both cases the equations describing the dynamics of the transverse envelope of the condensate $w_r(t)$ are algebraic and do not depend on $u(t)$ and $v(t)$, meaning that, to first order in $u(t)$ and $v(t)$, the dynamics of the bulk of the condensate is unperturbed by the Faraday waves.

If we consider the experimentally interesting case of harmonic modulation of the trapping frequency $\Omega_\rho(t) = \Omega_{\rho 0}(1 + \varepsilon \sin \omega t)$, with a small modulation amplitude $\varepsilon \ll 1$, Eqs. (7) and (10) can be easily recast to the form

$$\ddot{u}(t) + (a(k,\omega) + b(k,\omega)\sin 2t)\, u(t) = 0, \tag{11}$$

where $b(k,\omega)$ is proportional to ε and is therefore much smaller than 1. Using the Floquet-Bloch theorem one sees that solutions of the previous equation are of the form $h(t)\exp(i\mu t)$, where $h(t)$ has the same period as $\sin 2t$, and μ is in general complex-valued. Faraday waves correspond to the most unstable solution of Eq. (11), and it has been shown [13] that, for small positive values of $b(k,\omega)$, they correspond to $a(k,\omega) = 1$. This leads to the following dispersion relation for Faraday waves in the low-density regime:

$$k = \frac{\omega\left(4\pi^2 + 2\pi\rho_z g\right)^{1/4}}{2\sqrt{\rho_z g \Omega_{\rho 0}}}, \tag{12}$$

while for the high-density regime Faraday waves correspond to:

$$k = \frac{3\omega\sqrt{\pi}}{2\sqrt{2\rho_z g \Omega_{\rho 0}(1+q)}}. \tag{13}$$

3 Driving of Collective Modes

Usually, collective modes in BECs are induced by a modulation of the external trapping potential. In the recent paper [5], however, an alternative way of a condensate excitation has been experimentally realized. A broad Feshbach resonance of ^{7}Li allows to modulate the atomic s-wave scattering length by tuning an external magnetic field. Based on this property, a harmonic modulation of the s-wave scattering length has been achieved

$$a(t) \simeq a_{\rm av} + \delta_a \cos \Omega t, \tag{14}$$

with $a_{\rm av} \approx 3a_0 > \delta_a \approx 2a_0$, where a_0 is the Bohr radius, yielding a time-dependent interaction among atoms.

On the mean-field level at zero temperature, the BEC dynamics of such system is captured by the GP equation for the condensate wavefunction and the time-dependent interaction leads to a time-dependent nonlinearity. Depending on the closeness of the external modulation frequency Ω to one of condensate's eigenmodes, a qualitatively different dynamical behavior emerges. In the non-resonant case, we have small-amplitude oscillations of the condensate size around the equilibrium widths, and we are in the regime of linear response. However, as Ω

approaches an eigenmode, we expect a resonant behavior which is characterized by large amplitude oscillations. In this case it is clear that a linear response analysis does not provide a qualitatively good description of system's dynamics.

In order to obtain analytical insight into the condensate dynamics induced in this way, we use the Gaussian approximation from Refs. [6, 7]. To this end we assume that the condensate wave function with contact interaction has the same Gaussian form as in the non-interacting case, just with renormalized parameters. Thus, we use a time-dependent variational method based on a Gaussian ansatz, which reads for an axially symmetric trap:

$$\psi_{\mathrm{G}}(\rho,z,t) = \mathcal{N}(t)\exp\left[-\frac{1}{2}\frac{\rho^2}{u_\rho(t)^2} + i\rho^2\phi_\rho(t) - \frac{1}{2}\frac{z^2}{u_z(t)^2} + iz^2\phi_z(t)\right], \qquad (15)$$

where $\mathcal{N}(t) = \pi^{-\frac{3}{4}}u_\rho(t)^{-1}u_z(t)^{-\frac{1}{2}}$ is a time-dependent normalization, while $u_\rho(t)$, $u_z(t)$, $\phi_z(t)$ and $\phi_\rho(t)$ are variational parameters. They have straightforward interpretation: $u_\rho(t)$ and $u_z(t)$ correspond to the radial and to the axial condensate width, while $\phi_\rho(t)$ an $\phi_z(t)$ represent the corresponding phases. The above ansatz therefore describes the dynamics of the condensate in terms of the time-dependent widths and phases and does not incorporate center-of-mass motion due to the symmetry of the system.

Following the variational approach introduced in Ref. [6], we insert ansatz (15) to the action yielding the GP equation, and minimize it with respect to variational parameters. In this way we obtain a system of ordinary differential equations that govern the BEC dynamics, which we refer to as a Gaussian approximation:

$$\ddot{u}_\rho(t) + u_\rho(t) - \frac{1}{u_\rho(t)^3} - \frac{p(t)}{u_\rho(t)^3 u_z(t)} = 0, \qquad (16)$$

$$\ddot{u}_z(t) + \lambda^2 u_z(t) - \frac{1}{u_z(t)^3} - \frac{p(t)}{u_\rho(t)^2 u_z(t)^2} = 0. \qquad (17)$$

In the previous set of equations and in all equations that follow, we express all lengths in the units of the characteristic harmonic oscillator length $l = \sqrt{\hbar/m\omega_\rho}$ and time in units of ω_ρ^{-1}. The dimensionless interaction parameter $p(t)$ is given by $p(t) = \sqrt{2/\pi}Na(t)/l$. Taking into account Eq. (14), we have:

$$p(t) = p + q\cos\Omega t, \qquad (18)$$

where $p = \sqrt{2/\pi}Na_{\mathrm{av}}/l$ denotes the average interaction, $q = \sqrt{2/\pi}N\delta_a/l$ is a modulation amplitude, and Ω represents the modulation or driving frequency. By linearizing Eqs. (16) and (17) around the equilibrium widths $u_{\rho 0}$ and u_{z0}, we can obtain frequencies of both the breathing mode ω_{B0} and of the quadrupole ω_{Q0}. For the repulsive interaction, the quadrupole mode has a lower frequency and is characterized by out-of-phase radial and axial oscillations, while in-phase oscillations

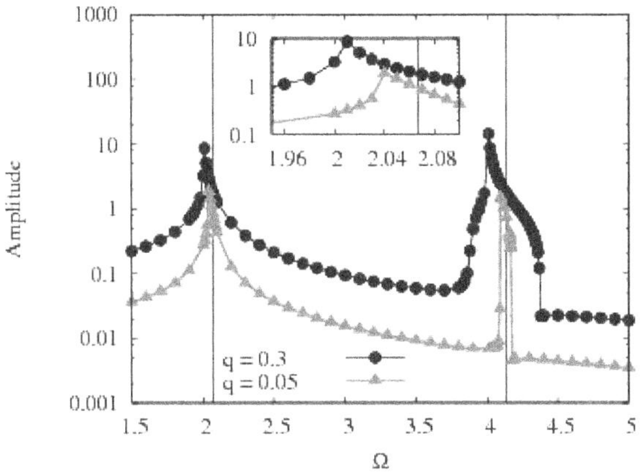

Fig. 1 Oscillation amplitude $(u_{\max} - u_{\min})/2$ versus driving frequency Ω for $p = 0.4$. In the inset, we zoom to the first peak to emphasize that the shape and value of a resonance occur at a driving frequency Ω which differs from ω_0 depends on the modulation amplitude q. The solid vertical lines correspond to ω_0 and $2\omega_0$.

correspond to the breathing mode. In the case of the experiment [5], their values are $\omega_{Q0} = 0.035375$, $\omega_{B0} = 2.00002$.

The main feature of the modulation induced dynamics is that it is strongly depending on the value of the driving frequency Ω. To illustrate this, we first study the spherically symmetric case. We set $p = 0.4$, $q = 0.1$ and solve the corresponding variational equation for different values of Ω. From the linear response theory, we have $u_0 = 1.08183$, $\omega_0 = 2.06638$ and we assume that the condensate is initially in equilibrium, i.e. $u(0) = u_0, \dot{u}(0) = 0$. Large amplitude oscillations and beating phenomena are present for both $\Omega \approx \omega_0$ and for $\Omega \approx 2\omega_0$.

This phenomenology is shown in Fig. 1, where we plot the oscillation amplitude defined as $(u_{\max} - u_{\min})/2$ versus the driving frequency Ω. A resonant behavior becomes apparent for both $\Omega \approx \omega_0$ and $\Omega \approx 2\omega_0$. In the same figure we also show the expected positions of resonances calculated using the linear stability analysis. Clearly, the prominent peaks exhibit shifts with respect to the solid vertical lines representing ω_0 and $2\omega_0$. As expected, a stronger modulation amplitude leads to a larger frequency shift, as can be seen from the inset.

To provide an analytical description of the observed frequency shift, we use the standard Poincaré-Lindstedt analysis. Linearizing the variational equation around the time-independent solution u_0 for vanishing driving $q = 0$, we obtain the zeroth-order approximation for the collective mode $\omega = \omega_0$. To calculate the collective mode to higher orders, we explicitly introduce the sought-after eigenfrequency ω into the calculation by rescaling the time from t to $s = \omega t$, yielding the equation:

$$\omega^2 \ddot{u}(s) + u(s) - \frac{1}{u(s)^3} - \frac{p}{u(s)^4} - \frac{q}{u(s)^4} \cos \frac{\Omega s}{\omega} = 0 . \qquad (19)$$

In the next step, we assume the following perturbative expansions in the modulation amplitude q:

$$u(s) = u_0 + q u_1(s) + q^2 u_2(s) + q^3 u_3(s) + \ldots , \tag{20}$$

$$\omega = \omega_0 + q\,\omega_1 + q^2\,\omega_2 + q^3\,\omega_3 + \ldots , \tag{21}$$

where we expand ω around ω_0 and introduce frequency shifts ω_1, ω_2, ... for each order in the expansion in q. By inserting the above expansions into the Eq. (19) and collecting terms of the same order in q, we obtain a hierarchical system of linear differential equations. Although the calculation is straightforward, it easily becomes tedious for higher orders of perturbation theory. Note that it is necessary to perform the calculation to at least the third order since it turns out to be the lowest-order solution where secular terms appear and where a nontrivial frequency shift, quadratic in q, is obtained:

$$\omega = \omega_0 + \frac{q^2}{12u_0^{20}\omega_0^3}\,\frac{\mathrm{P}(\Omega)}{(\Omega^2-\omega_0^2)^2\,(\Omega^2-4\omega_0^2)} + \ldots , \tag{22}$$

where the polynomial $\mathrm{P}(\Omega)$ is explicitly calculated in Ref. [7].

The result given by Eq. (22) is the main achievement of our analytical analysis. It is obtained within a perturbative approach up to the second order in q and it describes the breathing mode frequency dependence on Ω and q as a result of nonlinear effects. Due to the underlying perturbative expansion, we do not expect Eq. (22) to be meaningful at the precise position of the resonances. However, by comparison with numerical results based on the variational equation, we find that Eq. (22) represents a reasonable approximation even close to the resonant region. To illustrate this, in Fig. 2 we present the comparison for the parameter set $p = 0.4$ and $q = 0.1$, where we observe significant frequency shifts in the narrow resonant

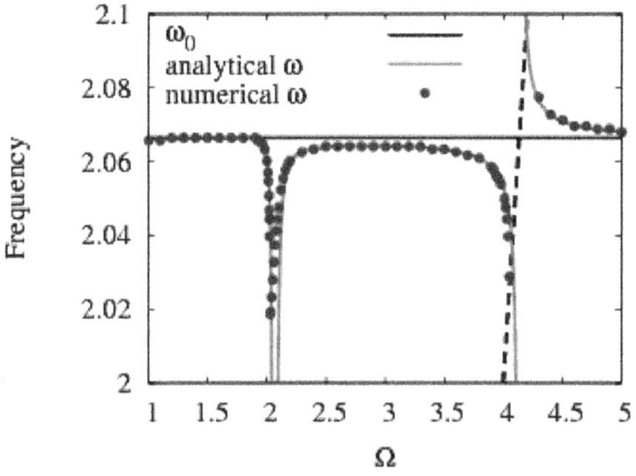

Fig. 2 Frequency of the breathing mode versus the driving frequency Ω for $p = 0.4$ and $q = 0.1$. The dashed line represents $\Omega/2$ and is given to guide the eye.

regions. We notice an excellent agreement of the numerical values with the analytical result given by Eq. (22).

A similar analysis can be also done for the axially-symmetric case, where one can calculate frequency shifts for both the quadrupole and the breathing mode. The analytic expressions are cumbersome and can be found in Ref. [7]. In Fig. 3 we present a comparison of analytical and numerical results for the parameter set $p = 1$, $q = 0.2$, and $\lambda = 0.3$.

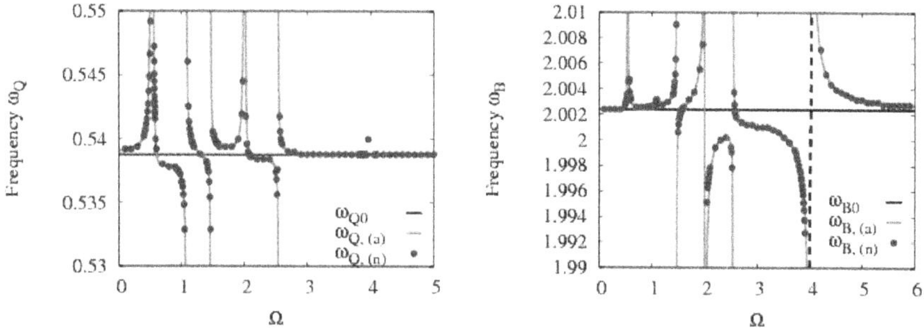

Fig. 3 Frequency of the quadrupole mode ω_Q (left) and breathing mode ω_B (right) versus driving frequency Ω for $p = 1$, $q = 0.2$, and $\lambda = 0.3$. We plot linear response result, second-order analytical result $\omega_{(a)}$ and numerical values $\omega_{(n)}$ for each case.

4 Conclusions

In this paper we have derived by variational means a set of effective equations for the dynamics of Faraday waves in low- and high-density longitudinally-homogeneous BECs. Focusing on the experimentally attainable spacings of the Faraday waves, we have obtained Eqs. (12) and (13), which describe the experimental conditions in Ref. [4]. Our predicted spacings of the Faraday waves are in good qualitative agreement with the ones observed experimentally. The small discrepancies are due to the longitudinal inhomogeneities of the condensate, which are not accounted for in the current framework [17].

We have also considered the case of spherically and axially symmetric condensates with harmonically modulated interactions. The external driving leads to resonances in the BEC dynamics, to mode coupling, and to the shifts of frequencies of collective modes. These effects are studied analytically, using the Poincaré-Lindstedt method, and numerically, using the Gaussian variational ansatz, and by solving the GP equation. Analytically derived formulae for frequency shifts of collective oscillation modes are found to be in good agreement with the numerics.

Acknowledgements. The authors would like to thank Mihnea Dulea, Mihaela Carina Raportaru, Axel Pelster, and Hamid Al-Jibbouri for stimulating discussions. This work was supported in part by the European Commission under EU FP7 project HP-SEE (contract No. 261499). A. I. N. was supported by CNCS-UEFISCDI through the postdoctoral grant

PD122 (contract No. 35/28.07.2010). I. V. and A. B. were supported by the Serbian Ministry of Education and Science under projects No. ON171017 and NAD-BEC, by DAAD - German Academic and Exchange Service under project NAD-BEC, and by the European Commission under EU FP7 projects PRACE-1IP and EGI-InSPIRE.

References

1. Kevrekidis, P.G., Frantzeskakis, D.J., Carretero-González, R. (eds.): Emergent Nonlinear Phenomena in Bose-Einstein Condensates. Springer, New York (2008)
2. Pethick, C.J., Smith, H.: Bose-Einstein Condensation in Dilute Gases. Cambridge University Press, Cambridge (2008)
3. Sulem, C., Sulem, P.-L.: The Nonlinear Schrödinger Equation. Springer, New York (1999)
4. Engels, P., Atherton, C., Hoefer, M.A.: Observation of Faraday Waves in a Bose-Einstein Condensate. Phys. Rev. Lett. 98, 095301 (2007)
5. Pollack, S.E., Dries, D., Hulet, R.G., et al.: Collective Excitation of a Bose-Einstein Condensate by Modulation of the Atomic Scattering Length. Phys. Rev. A 81, 53627 (2010)
6. Pérez-García, V.M., Michinel, H., Cirac, J.I., Lewenstein, M., Zoller, P.: Low Energy Excitations of a Bose-Einstein Condensate: A Time-Dependent Variational Analysis. Phys. Rev. Lett. 77, 5320 (1996)
7. Vidanović, I., Balaž, A., Al-Jibbouri, H., Pelster, A.: Nonlinear Bose-Einstein-Condensate Dynamics Induced by a Harmonic Modulation of the *s*-wave Scattering Length. Phys. Rev. A 84, 013618 (2011)
8. Cross, M.C., Hehenberg, P.C.: Pattern Formation Outside of Equilibrium. Rev. Mod. Phys. 65, 851 (1993)
9. Faraday, M.: On a Peculiar Class of Acoustical Figures; and on Certain Forms Assumed by a Group of Particles Upon Vibrating Elastic Surfaces. Philos. Trans. R. Soc. London 121, 299 (1831)
10. Rayleigh, L.: On Maintained Vibrations. Phil. Mag. 15, 229 (1883)
11. Benjamin, T.B., Ursell, P.: The Stability of the Plane Free Surface of a Liquid in a Vertical Periodic Motion. Proc. R. Soc. Lond. A 225, 505 (1954)
12. Staliunas, K., Longhi, S., de Valcárcel, G.J.: Faraday Patterns in Bose-Einstein Condensates. Phys. Rev. Lett. 89, 210406 (2002)
13. Nicolin, A.I., Carretero-Gonzáles, R., Kevrekidis, P.G.: Faraday Waves in Bose-Einstein condensates. Phys. Rev. A 76, 063609 (2007)
14. Nicolin, A.I., Raportaru, M.C.: Faraday Waves in High-Density Cigar-Shaped Bose-Einstein Condensates. Physica A 389, 4663 (2010)
15. Nicolin, A.I., Carretero-Gonzáles, R.: Nonlinear Dynamics of Bose-Condensed Gases by Means of a *q*-Gaussian Variational Approach. Physica A 387, 6032 (2008)
16. Dalfovo, F., Minniti, C., Stringari, S., Pitaevskii, L.: Nonlinear Dynamics of a Bose Condensed Gas. Phys. Lett. A 227, 259 (1997)
17. Nicolin, A.I.: Variational Treatment of Faraday Waves in Inhomogeneous Bose-Einstein Condensates. Physica A 391, 1062 (2012)

Monte Carlo Methods Using New Class of Congruential Generators

T. Gurov, S. Ivanovska, A. Karaivanova, and N. Manev

Abstract. In this paper we propose a new class of congruential pseudo random number generator based on sequences generating permutations. These sequences have been developed for other applications but our analysis and experiments show that they are appropriate for approximation of multiple integrals and integral equations.

1 Introduction

Stochastic simulation methods need sequences of numbers whose behavior corresponds to the statistical properties on which these methods are based. In particular Monte Carlo methods use *deterministic algorithms* that produce *independent and identically distributed* random variables having the uniform distribution over the interval $(0,1)$ (*i.i.d.* $U(0,1)$ *variables*). If it is necessary, these variables are later transformed to random variables (vectors) with the desired distribution.

Real-valued random variables $\{u_j\}_{j=0}^{\infty}$ are i.i.d. $U(0,1)$, if for all integers $i \geq 0$ and $t > 0$, the vectors $(u_i, u_{i+1}, \ldots, u_{i+t-1})$ are uniformly distributed over the t-dimensional hypercube $(0,1)^t$.

Such sequences of random variables are generated by so called Pseudo Random Number Generators (PRNG). A short introduction to them is given in Section 2.

In section 3 we describe in brief a new class of PRNGs based on a special linear recursions modulo prime power. These recursions have been originally constructed

T. Gurov · S. Ivanovska · A. Karaivanova
Institute of Information and Communication Technologies, BAS
Acad. G. Bonchev St., Bl.25A, 1113 Sofia, Bulgaria
e-mail: gurov@parallel.bas.bg, sofia@parallel.bas.bg,
anet@parallel.bas.bg

N. Manev
Institute of Mathematics and Informatics, BAS, Acad. G. Bonchev St., Bl. 8,
1113 Sofia, Bulgaria
e-mail: nlmanev@math.bas.bg

L. Kocarev (Ed.): ICT Innovations 2011, AISC 150, pp. 257–267.
springerlink.com

and studied in order to be applied to areas far from stochastic simulations, but their statistical properties have inspired us to test whether these recursions can be utilized by Monte Carlo methods.

The numerical experiments that we have carried out by using two generators from the proposed new class are: computing multidimensional integrals, and solving integral equations. They are described in Section 4. The obtained results are compared with the results of integrations based on using Mersenne Twister random number generator.

2 Preliminaries

Definition 1 (L'Ecuyer [10]). A PRNG is a structure (S, μ, f, U, g), where

- S is a finite set called **the state space**;
- μ is the probability distribution on S;
- $f : S \rightarrow S$ is a function called **transition function**; starting with a given initial state s_0, which is selected according to the distribution μ, all elements of S are generated according to $s_i = f(s_{i-1})$.
- U is the **output space**;
- $g : S \rightarrow U$ is the **output function**: $u_i = g(s_i)$.

The above definition corresponds to the PRNGs used in Monte Carlo simulations. In most such PRNGs we have

- $S = \mathbb{Z}_m^k$, where $\mathbb{Z}_m$ is the ring of integers modulo m,
- the output set U is i.i.d. $U(0,1)$ (or $U(0,1)^t$),
- $u_i = s_i/m$.

To some extend, the quality requirements for PRNGs depend on applications to which they are applied. But there are some properties that are relevant to and required by any application. The most important such properties are listed below.

Efficiency

The generator has to be implementable by a deterministic polynomial-time algorithm, i.e., to run in time bounded by a polynomial of the length of the initial state. The implementation has to be realized by as few as possible arithmetical operations and use little memory.

Long period

The period T of the generator has to be a square, or sometimes a cube of the required number of points, that is, even for modest applications we need $T > 10^{18}$.

Repeatability and Portability

These properties guarantee the ability to exactly generate the same sequence of random numbers on different machines and at different times. For the purposes of testing and development these properties are very important.

Uniformity and Independence

They are relevant to the ability to generate i.i.d. $U(0,1)$ sequences.

Ability "to skip ahead"

This property characterizes the ability of calculating u_k for large k without generating all values $u_0, u_1, \ldots, u_{k-1}$. It is a property important for parallel realizations.

Pseudorandomness

This is very important characteristic for random number generators. Informally pseudorandomness means "The generators output has to look random". This is quite vague statement and different trends in its understanding can be observed. Indeed there are three main approaches to formalization of pseudorandomness:

- ***Probabilistic*** (Shannon): Shannon's information theory considers perfect randomness as the extreme case and it is associated with a unique distribution, the uniform one.
- ***Computational Complexity*** (Kolmogorov, Chaitin, Solomonov): This approach is based on the Kolmogorov's computational complexity [9], [13].
- ***Computational Indistinguishability*** (Blum, Goldwasser, Micali,Yao, Goldreich): A distribution is pseudorandom if no efficient procedure can distinguish it from the uniform distribution [6], [7].

We will not enter into details since the discussion on such topics is far from the goals of this paper. Such considerations are of greater concern in cryptography. For Monte Carlo integration, for example, it is not so important if the next generated value is unpredictable. We refer the interested reader to the cited literature.

As examples of generator often used in Monte Carlo applications we give the next PRNGs which are included in the GNU Library:

1. Fifth-order multiple recursive with period $\approx 10^{46}$:
$x_n = 107374182 x_{n-1} + 104480 x_{n-5} \pmod{2^{31}-1}$.
2. Combined multiple recursive $z_n = x_n - y_n \pmod{2^{31}-1}$, where $\{x_n\}$ and $\{y_n\}$ are 3rd order linear recurrent sequences modulo $2^{31}-1$ and 2145483479, respectively. Its period is $\approx 2^{185}$.
3. Generalized (lagged) Fibonacci: $y_n = y_{n-s} \theta y_{n-r}$, $r > s$, where θ is $+, -, *$ modulo m, or xor. If θ is the addition modulo 2^k, then the period is $(2^r - 1)2^{k-1}$.
4. Mersenne Twister generator. It is equi-distributed in 623 dimensions and has period $2^{19937} - 1 \approx 10^{6000}$.

The last generator has recently become popular for simulation and it has been installed as the default PRNG for the most used mathematical packages. That is why we have chosen to compare our random number generator with the Mersenne Twister generator.

3 A New Class of Congruential Generators

Definition 2 ([14]). Let $\mathbf{S} = \{s_n\}_{n\geq 0}$ be a sequence with terms in a finite ring R. The sequence $\mathbf{S}$ is called **strictly balanced (in short SB) sequence**, if it is periodic and each element of R occurs equal number of times in one period of the sequence. If each element of R appears exactly once in a period, the sequence $\mathbf{S}$ is called **sequence generating permutation (in short SGP)**.

The period of an SGP sequence is equal to the cardinality $|R|$ of the ring.

Both from algebraic and practical points of view, the most important case is $R = \mathbb{Z}_{p^m}$.

Recall that k^{th}-*order homogeneous linear recurrence sequence*, $\mathbf{S} = \{s_n\}$, *with constant terms in* R is defined by the recursion

$$s_{n+k} = a_{k-1}s_{n+k-1} + a_{k-2}s_{n+k-2} + \cdots + a_1 s_1 + a_0 s_n \tag{1}$$

and *initial terms* $s_0, s_1, \ldots, s_{k-1}$. It is obvious that any such sequence over a finite ring is periodic.

Theorem 1 ([14]). *Let* $\{s_n\}$ *be a second-order sequence defined by*

$$s_{n+2} = as_{n+1} + bs_n, \quad (\text{mod } p^m). \tag{2}$$

It is an SGP sequence if and only if

$$\mu(x^2 - ax - b) = (x-1)^2, \tag{3}$$

where $\mu : \mathbb{Z}_{p^m}[x] \to \mathbb{Z}_p[x]$ *for* $p > 2$ *and* $\mu : \mathbb{Z}_{2^m}[x] \to \mathbb{Z}_4[x]$ *for* $p = 2$.
In the case $p = 3$ *the condition*

$$u + v \not\equiv 2 \pmod 3, \qquad \text{where } a = 2 + 3u, \quad b = -1 + 3v,$$

has to be added to (3) in order the theorem to be true.

The higher order case is more complicated and a necessary and sufficient condition cannot be formulated in a simple form. But if $f(x) = x^k - a_{k-1}x^{k-1} - a_{k-2}x^{k-2} - \cdots - a_1 x - a_0$ is the minimal polynomial of the recursion of order $k \geq 3$ with $M = p^m$, the following conditions are sufficient for generating an SGP sequence:

- $f(x) \equiv (x-1)^k \pmod p$
- $s_0, s_1, \ldots, s_{k-1}$ must be different modulo p, e.g., $0, 1, 2, \ldots, k-1$.

Using the Chinese Reminder Theorem we can construct an SGP sequence for any $M = p_1^{e_1} p_2^{e_2} \ldots p_k^{e_k}$.

Given an SGP sequence $\{s_n\}$ over $\mathbb{Z}_M$ we can transform it into a sequence of numbers in $[0,1)$ dividing each element by M. Hence the resulting sequence of M elements is a permutation of the numbers

$$\left\{ \frac{i}{M} \mid i = 0, 1, 2, \ldots, M-1 \right\}. \tag{4}$$

Examples

Generator M1
Consider the second-order sequence with terms in $\mathbb{Z}_M$, where $M = 5^{19}$ defined by

$$s_{n+2} = (5^{10} + 2)s_{n+1} - 5556 s_n$$

We start with $s_0 = 0$ and $s_1 = 1$ (or any $s_1 \equiv 1 \pmod 5$).

Similarly, the coefficients $a_1 = 5^{10} + 2$ and $a_0 = -5556$ can be replaced with any $a_1 \equiv 2,\ a_0 \equiv -1 \pmod 5$.

Generator M2
Consider the third-order sequence with terms in $\mathbb{Z}_M$ ($M = 2^{61}$) defined by

$$s_{n+3} = (2^{10} - 1)s_{n+2} + (2^8 + 2^6 + 1)s_{n+1} + s_n \pmod{2^{61}}.$$

Any elements s_0, s_1, s_2 of $\mathbb{Z}_M$, such that $s_0 \equiv 0,\ s_1 \equiv 1,\ s_2 \equiv 2 \pmod 4$ are suitable for initial parameters. The coefficients a_2, a_1, a_0 can be replaced with any $a_2 \equiv -1,\ a_1 \equiv 1$, and $a_0 \equiv 1 \pmod 4$, but they should be taken with large absolute values in order to improve uniformity of the distribution of segments $(u_{n+1}, u_{n+2} \ldots, u_{n+t})$.

Bellow some features which characterize the proposed class of PRNGs are listed:

- ***Good lattice structure can be arranged***
  We can make the lower bound $\left(\sum_{i=0}^{k-1} a_i^2\right)^{-1/2}$ for the minimal distance between hyperplanes, d_t, sufficiently small by choosing large coefficients.
- ***The generators modulo 2^m are very efficient***
  For instance, the implementation of Example 2 gives a two times faster generator than Mersenne Twister. The generation is realized by shifting and addition and only in the last step of the algorithm a multiplication by $1/2^m$ is used.
- ***There is no theoretical limit for the period***
  But in order to keep the advantage in speed and low complexity of the considered class of PRNGs the period should be less than 2^{128} in practical implementation. This is relatively short period in comparison to one of Mersenne Twister but enough long for many applications. Indeed we can lengthen the period without enlarging M by permitting the repetition of the elements of (4).
- ***The proposed class of PRNGs is significantly different*** from algebraic point of view. In contrast to other congruential generators the minimal polynomial of each generator is purely inseparable polynomial (this corresponds to purely inseparable extensions of the basic field).

4 Computational Experiments

4.1 Study Case 1: Monte Carlo Integration

We have carried out our computational experiments with the following d-dimensional test integrals:

$$\mathrm{I}_1 = \int_{(0,1)^d} F_1(\mathbf{x})d\mathbf{x} \qquad \text{and} \qquad \mathrm{I}_2 = \int_{(0,1)^d} F_2(\mathbf{x})d\mathbf{x},$$

where $\mathbf{x} = (x_1, x_2, \ldots, x_d)$ and

$$F_1(\mathbf{x}) = \prod_{i=1}^{d} (x_i^3 + 0.75) \qquad \text{and} \qquad F_2(\mathbf{x}) = \prod_{i=1}^{d} |4x_i - 2|.$$

The values of I_1 and I_2 are both equal to 1.

The first of these test functions is taken from Schmid and Uhl [17]. I_2 is known as Roos and Arnold's example and it is suggested as a test function by Owen [16].

It is straightforward to calculate the variances σ_1^2 and σ_2^2 of F_1 and F_2:

$$\sigma_1 = \sigma[F_1] = \sqrt{\left(\frac{121}{112}\right)^d - 1} \qquad \text{and} \qquad \sigma_2 = \sigma[F_2] = \sqrt{\left(\frac{4}{3}\right)^d - 1} \tag{5}$$

In particular we have

$$d = 10: \quad \sigma_1 = 1.07984949546134, \quad \sigma_2 = 4.09362023566092$$

$$d = 20: \quad \sigma_1 = 1.92142671333385; \quad \sigma_2 = 17.72954751823117;$$

$$d = 30: \quad \sigma_1 = 3.02703898016620; \quad \sigma_2 = 74.82423185191648;$$

As it is well known the error of integration tends asymptotical to

$$e_N[F_i] \approx \frac{\sigma_i}{\sqrt{N}} \nu,$$

where ν is a standard normal ($N(0,1)$) random variable and σ_i is the square root of the variance of F_i.

Here are in brief what we have tested in our experiments:

- Tested generators: *the Mersenne Twister, Example 1, and Example 2.*
- Dimensions: $d = 10,\ 20,\ 30$.
- Number of points: $N = 2^m$, where $m = 10, 11, \ldots, 20$.
- 200 calculations have been done for each generator and for each pair (d, m). The presented value of the error is the average over these 200 calculations of the absolute values of the error for each pair (d, m).

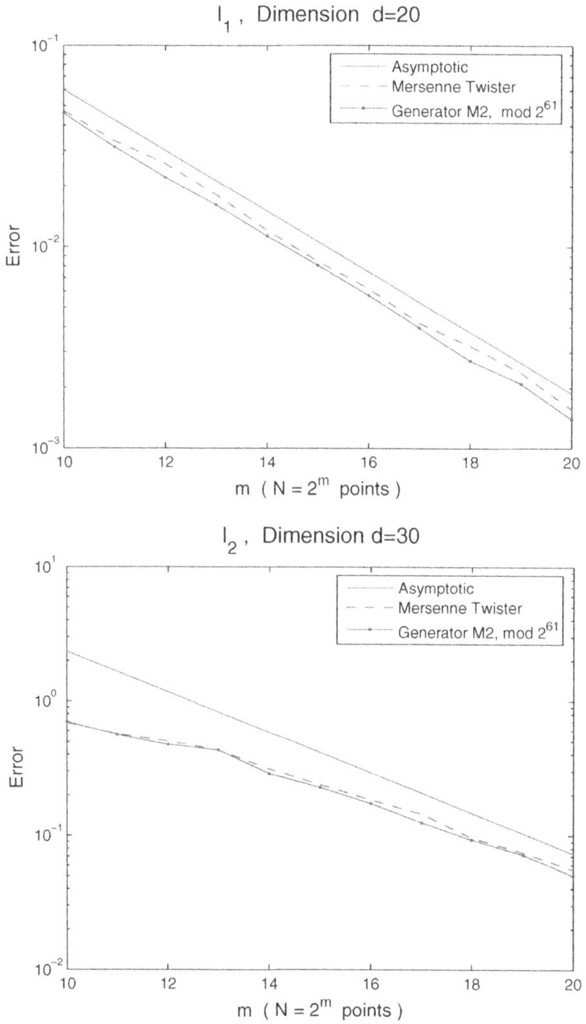

Fig. 1 Graphical presentations of the experimental results

The described experiments can be considered as a continuation of the research given in [5].

Obtained results

Two of the obtained results are graphically represented in Figure 1. The term "asymptotic" is used for the graph of $\sigma_i/\sqrt{N}$ ($N = 2^m$). The experiments show that generators from the new class demonstrate even modest better behavior than the Mersenne Twister. The generators (Examples 1 and 2 and many others not described here) have been chosen at random. Hence we believe that their behavior is intrinsic to all class.

4.2 Study Case 2: Simulation of Electron Transport

As a second study case we consider the quantum kinetic equation describing a electron-phonon interactions in presence of applied electric field [15]. This equation can be written in the following integral form [15]:

$$f(\mathbf{k},t) = \phi(\mathbf{k}) + \int_0^t dt'' \int_G d\mathbf{k}' K(\mathbf{k},\mathbf{k}') \times \tag{6}$$
$$\left\{ \int_{t''}^t dt' S_1(\mathbf{k},\mathbf{k}',\mathbf{F},t',t'') f(\mathbf{k}',t'') + \int_{t''}^t dt' S_2(\mathbf{k},\mathbf{k}',\mathbf{F},t',t'') f(\mathbf{k},t'') \right\},$$

where the kernel is separated in two terms:

$$K(\mathbf{k},\mathbf{k}') = \frac{2V}{2\pi^3\hbar^2}|g(\mathbf{q})|^2, \quad \text{and} \tag{7}$$

$$S_1(\mathbf{k},\mathbf{k}',\mathbf{F},t',t'') = -S_2(\mathbf{k}',\mathbf{k},\mathbf{F},t',t'') = exp(-\Gamma(t'-t'')) \times \tag{8}$$
$$\left[(n_{\mathbf{q}}+1)\cos\left(\frac{\varepsilon(\mathbf{k}) - \varepsilon(\mathbf{k}') + \hbar\omega_{\mathbf{q}}}{\hbar}(t'-t'') - \frac{\hbar}{2m}(\mathbf{k}'-\mathbf{k})\cdot\mathbf{F}(t'^2 - t''^2) \right) \right.$$
$$\left. + n_{\mathbf{q}}\cos\left(\frac{\varepsilon(\mathbf{k}) - \varepsilon(\mathbf{k}') - \hbar\omega_{\mathbf{q}}}{\hbar}(t'-t'') - \frac{\hbar}{2m}(\mathbf{k}'-\mathbf{k})\cdot\mathbf{F}(t'^2 - t''^2) \right) \right].$$

Here, $\mathbf{k} = (k_x,k_y,k_z)$ and t are the wave vector and the evolution time, respectively. $f(\mathbf{k},t)$ is the distribution function. $\phi(\mathbf{k})$ is the initial electron distribution function. $\mathbf{F} = e\mathbf{E}/\hbar$, where $\mathbf{E}$ is the applied electric field. $n_{\mathbf{q}} = 1/(exp(\hbar\omega_{\mathbf{q}}/\mathcal{K}T) - 1)$ is the Bose function, where $\mathcal{K}$ is the Boltzmann constant and T is the temperature of the crystal, corresponds to an equilibrium distributed phonon bath. $\hbar\omega_{\mathbf{q}}$ is the phonon energy which generally depends on $\mathbf{q} = \mathbf{k}' - \mathbf{k}$, and $\varepsilon(\mathbf{k}) = (\hbar^2\mathbf{k}^2)/2m$ is the electron energy. A Fröhlich coupling is considered

$$g(\mathbf{q}) = -i\left[\frac{2\pi e^2\hbar\omega_{\mathbf{q}}}{V}\left(\frac{1}{\varepsilon_\infty} - \frac{1}{\varepsilon_s} \right)\frac{1}{(\mathbf{q})^2} \right]^{\frac{1}{2}},$$

where (ε_∞) and (ε_s) are the optical and static dielectric constants. The damping factor Γ is considered independent of the electron states $\mathbf{k}$ and $\mathbf{k}'$. This is reasonable since Γ weakly depends on $\mathbf{k}$ and $\mathbf{k}'$ for states in the energy region above the phonon threshold, where the majority of the electrons reside due to the action of the electric field. The solution of the quantum kinetic equation (6) is evaluated by a Monte Carlo algorithm suggested in [8] and using the generator described in *Example 2*.

The numerical tests have been performed on the High Performance cluster deployed at the Institute of Information and Communication Technologies of the Bulgarian Academy of Sciences (IICT-BAS). This cluster has the following hardware: HP Cluster Platform Express 7000 enclosures with 36 blades BL 280c with dual Intel Xeon X5560 @ 2.8Ghz (total 576 cores), 24 GB RAM ; 8 controlling nodes HP DL 380 G6 with dual Intel X5560 @ 2.8 Ghz, 32 GB RAM; non-blocking low-latency 20Gpbs DDR interconnection via Voltaire Grid director 2004 switch; two SAN switches for redundant access to storage; and MSA2312fc with 96 TB SAN storage, available under /home and /gscratch filesystems.

Obtained results

The numerical results presented in the Figure 2 are obtained for zero temperature and $GaAs$ material parameters: the electron effective mass is $0.063m_e$, the optimal phonon energy is $36meV$, the static and optical dielectric constants are $\varepsilon_s = 10.92$ and $\varepsilon_\infty = 12.9$. The initial condition at $t = 0$ is given by a function which is Gaussian in energy, ($\phi(k) = exp(-(b_1 k^2 - b_2)^2)$, $b_1 = 96$ and $b_2 = 24$), scaled in a way to ensure, that the peak value is equal to unity.

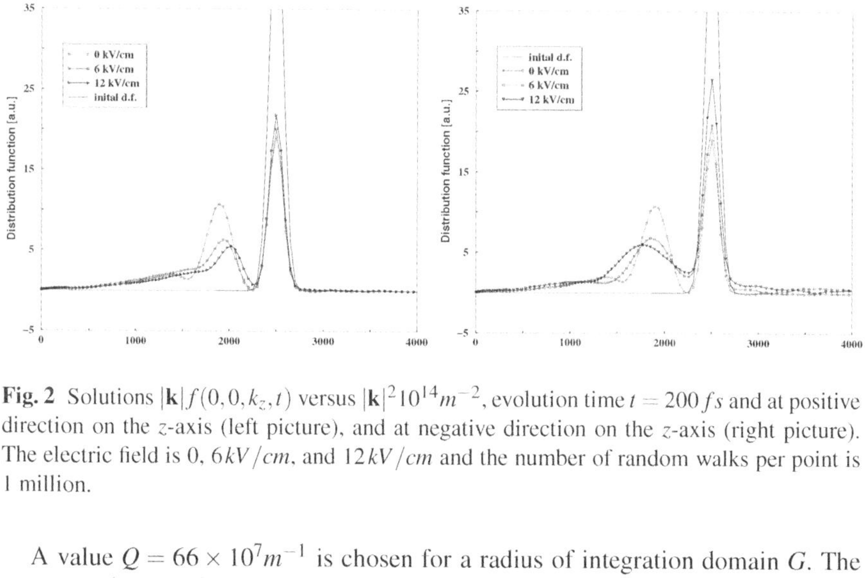

Fig. 2 Solutions $|\mathbf{k}|f(0,0,k_z,t)$ versus $|\mathbf{k}|^2 10^{14} m^{-2}$, evolution time $t = 200 fs$ and at positive direction on the z-axis (left picture), and at negative direction on the z-axis (right picture). The electric field is 0, $6kV/cm$, and $12kV/cm$ and the number of random walks per point is 1 million.

A value $Q = 66 \times 10^7 m^{-1}$ is chosen for a radius of integration domain G. The solution $f(0,0,k_z,t)$ is estimated in 2×96 points that are symmetrically located on z-axes, the direction of applied field. The truncation parameter $\varepsilon = 0.001$. The quantity presented on the y-axes in all figures is $|\mathbf{k}| * f(0,0,k_z,t)$, i.e. it is proportional to the distribution function multiplied by the density of states. It is given in arbitrary units. The quantity $\mathbf{k}^2$, given on the x-axes in units of $10^{14}/m^2$, is proportional to the electron energy.

The numerical results show that this class of RNGs can be used in order to investigate the quantum kinetic equation under consideration.

5 Conclusion

The study of the consider class of PRNGs is at the initial stage but the obtained results encourage us to continue. We are looking for PRNGs in the proposed class that produced sequences of points that demonstrate high level of uniform distribution in high dimensional hypercubes. We cannot recommend yet concrete parameters (i.e., a concrete generators) since we have not enough knowledge about discrepancy and lattice structure of the generated sequences. This is the subject of a further research.

Acknowledgements. This work makes use of results produced by the High-Performance Computing Infrastructure for South East Europe's Research Communities (HP-SEE), a project co-funded by the European Commission (under contract number 261499) through the Seventh Framework Programme. HP-SEE involves and addresses specific needs of a number of new multi-disciplinary international scientific communities (computational physics, computational chemistry, life sciences, etc.) and thus stimulates the use and expansion of the emerging new regional HPC infrastructure and its services. Full information is available at http://www.hp-see.eu.

This work was partially supported by the Bulgarian National Science Fund under Contracts DO02-146/2008 and DO02-215/2008.

References

1. Broy, M.: Software engineering — from auxiliary to key technologies. In: Broy, M., Dener, E. (eds.) Software Pioneers, pp. 10–13. Springer, Heidelberg (2002)
2. Dod, J.: Effective substances. In: The Dictionary of Substances and Their Effects. Royal Society of Chemistry (1999), DIALOG, `http://www.rsc.org/dose/titleofsubordinatedocument`. (cited January 15, 1999)
3. Geddes, K.O., Czapor, S.R., Labahn, G.: Algorithms for Computer Algebra. Kluwer, Boston (1992)
4. Hamburger, C.: Quasimonotonicity, regularity and duality for nonlinear systems of partial differential equations. Ann. Mat. Pura. Appl. 169, 321–354 (1995)
5. Atanassov, E., Karaivanova, A., Gurov, T., Ivanovska, S., Durchova, M., Dimitrov, D.: Quasi-Monte Carlo integration on the grid for sensevity studies. Earth Sci. Inform. 3, 289–296 (2010)
6. Blum, M., Micali, S.: How to Generate Cryptographically Strong Sequences of Pseudo-Random Bits. SIAM J. Comput. 13, 850–864 (1984)
7. Goldreich, O.: Pseudorandomness. Notices 46(10), 1209–1216 (1999)
8. Gurov, T.V., Dimov, I.T.: A Parallel Monte Carlo Method for Electron Quantum Kinetic Equation. In: Lirkov, I., Margenov, S., Waśniewski, J., Yalamov, P. (eds.) LSSC 2003. LNCS, vol. 2907, pp. 153–161. Springer, Heidelberg (2004)
9. Kolmogorov, A.: Three approaches to the concept of the amount of information. Probl. Inf. Transm. 1(1), 1–7 (1965)
10. L'Ecuyer, P.: Uniform random number generation. Annals of Operations Research 53, 77–120 (1994)
11. L'Ecuyer, P.: Uniform random number generation. In: Henderson, S.G., Nelson, B.L. (eds.) Ch. 3, Simulation, Handbooks in Operations Research and Management Science, pp. 55–81. Elsevier, Amsterdam (2006)
12. L'Ecuyer, P.: Pseudorandom Number Generators, in Encyclopedia of Quantitative Finance. In: Cont, R. (ed.) Simulation Methods in Financial Engineering, pp. 1431–1437. John Wiley, Chichester (2010)
13. Li, M., Vitanyi, P.: An Introduction to Kolmogorov Complexity and Its Applications. Springer, New York (1993)

14. Manev, N.: Sequences generating permutations. Applied Mathematics and Computation 216, 708–718 (2010)
15. Nedjalkov, M., Gurov, T.V., Kosina, H., Whitlock, P.A.: Statistical Algorithms for Simulation of Electron Quantum Kinetics in Semiconductors - Part II. In: Margenov, S., Waśniewski, J., Yalamov, P. (eds.) LSSC 2001. LNCS, vol. 2179, pp. 183–190. Springer, Heidelberg (2001)
16. Owen, A.: The dimension distribution and quadrature test functions. Stat. Sin. 13, 1–17 (2003)
17. Schmid, W., Uhl, A.: Techniques for parallel quasi-Monte Carlo integration with digital sequences and associated problems. Math. Comp. Sim. 55, 249–257 (2001)

Implementation of Hybrid Monte Carlo (Molecular Dynamics) Quantum Mechanical Methodology for Modeling of Condensed Phases on High Performance Computing Environment*

Anastas Misev, Dragan Sahpaski, and Ljupco Pejov

Abstract. The overall objective of the present work is to develop and implement a novel multi-step general computational methodology for modeling of complex condensed-phase systems on high-performance computing environments. First, molecular dynamics (MD) or Monte Carlo (MC) simulations of the free interacting clusters, as well as of clusters microsolvated by several molecules from the medium (solvent) are performed. MD or MC simulations are carried out applying either classical empirical interaction potentials, or implementing quantum mechanical MD or MC methodologies. Quantum mechanical MD simulations are carried out with the Born-Oppenheimer approach (BOMD), the Car-Parrinello (CPMD) approach, or using the atom-centered density matrix propagation scheme (ADMP). Sequential to this step, a series of suitably chosen configurations from the statistical physics simulations corresponding to the equilibrated system, which are mutually statistically independent, are subjected to further more rigorous quantum mechanical analysis. In this way, a realistic simulation of complex physico-chemical systems is enabled, in real computational time, without loosing, in statistical sense, any relevant information about the system. Due to the complexity of the algorithms which are used for this hybrid approach, it is of crucial importance to be able to implement the computational strategy on high-performance computing environment. Often, the overall CPU time which is required is very high. Therefore, achieving good parallel

Anastas Misev · Dragan Sahpaski
University Sts Cyril and Methodius, Faculty of Natural Sciences & Mathematics Institute of Informatics, Skopje, Macedonia
e-mail: anastas@ii.edu.mk, dragans@ii.edu.mk

Ljupco Pejov
University Sts Cyril and Methodius, Faculty of Natural Sciences & Mathematics Institute of Chemistry, Skopje, Macedonia
e-mail: ljupcop@pmf.ukim.mk

* This paper is based on the work done in the framework of the SEE-GRID-SCI FP7 EC funded project, with partial support from NSFB grant D002 - 146/2008.

L. Kocarev (Ed.): ICT Innovations 2011, AISC 150, pp. 269–277.
springerlink.com

efficiency for calculations of such type is far from a trivial task without the use of high-performance low-latency MPI interconnect.

1 Introduction

To study a vast variety of properties of condensed phases, liquids (such as e.g. solutions of ions and various molecular systems in molecular liquids), solids (including small molecular systems adsorbed on surfaces), it is often essential to carry out complex numerical simulations.

In order to describe exactly the fluctuating nature of the liquid phases in particular, to analyze the influence of the intra and intermolecular dynamics on the energetic and spectroscopic characteristics of the systems, to account explicitly for the temperature effects, as well as to enable theoretical modeling of other parameters which can be measured by experimental techniques (such as, e.g. the half-widths of the spectral bands), sequential statistical physics quantum mechanical methodologies have been developed for theoretical treatment of condensed phases [1, 2, 3, 4]. However, most of these methodologies have been non-standardized, as it is often required to make a balance between the quality of the implemented computational method and the resulting computational cost. While usage of more sophisticated methodology is usually a prerequisite to obtain higher-quality results, such methodology also often requires vast computational resources. Due to their complexity and sequential character, the above-mentioned methodologies are inherently multi-step ones. In the first step, molecular dynamics (MD) or Monte Carlo (MC) simulations of the "free" interacting clusters, clusters microsolvated by several molecules from the medium (solvent), or of *nD* periodic systems are performed. MD or MC simulations have often been carried our applying classical empirical interaction potentials [1]. Sequential to this step, a series of suitably chosen configurations from the statistical physics simulations corresponding to the equilibrated system, which are mutually statistically independent, are subjected to further more rigorous quantum mechanical analysis. Classical MC or MD simulations are rather quick, in comparison to their quantum mechanical counterparts, at least for modest-size systems. However, their inherent drawbacks arising from the empirical character of the interaction potentials have been well-documented. The previous argument particularly refers to the simplest two-body potentials which are often used for classical statistical physics simulations. The main aim of the present work is to implement a more advanced hybrid MC (MD) quantum mechanical methodology which would enable a more realistic simulation of complex physico-chemical systems, in real computational time, without loosing any relevant information about the system. Although we do not disregard the possibility to use classical empirical interaction potentials in the first phase, especially if those have been derived on the basis of careful explorations of potential energy hypersurfaces of interacting species, the emphasis in this study is put on the *ab initio* molecular dynamic techniques preceding the quantum mechanical computation sequence.

The algorithms that are used for such hybrid and complex computational approach are rather involved, and it is therefore of crucial importance to be able to implement the computational strategy on high-performance computing environment. Is not easy to predict the scalability of this computational approach, as its various subsequent steps scale rather differently with the number of CPUs. Often, the overall CPU time which is required is very high. Therefore, achieving good parallel efficiency for calculations of such type is far from a trivial task without the use of high-performance low-latency MPI interconnect.

2 Computational Details and Algorithms

As the main emphasis in the present work is put on the *ab initio* molecular dynamics (or MC) techniques preceding the quantum mechanical part, and we have also described in more details some particular cases in which classical MC or MD simulations precede the QM phase [5], we hereby focus on the former algorithms.

2.1 Born-Oppenheimer Molecular Dynamics Simulations

In BOMD, the total Hamiltonian of the system is given by [6, 7]:

$$\hat{H} = \sum_{i>j} \frac{|\hat{\vec{P_i}}|^2}{2M_i} + \frac{1}{4\pi\varepsilon_0} \sum_{i>j} \frac{Z_i Z_j e^2}{|\vec{R_i} - \vec{R_j}|} + E_{elec}(\{\vec{R_i}; \vec{r}\}) \tag{1}$$

In (1), the coordinates represented by lower-case letters refer to electrons, while the upper-case ones refer to nuclei. M_i is the mass of i-th atom (i.e. nucleus), while Z_i and Z_j are the effective relative core charges for atoms i and j respectively; E_{elec} is the total electronic energy of the system. The equations of motions (EOM) for the nuclei (within the classical nuclei approximation) are written as:

$$M_i \frac{d^2 R_i^2}{dt^2} = -\nabla_{\vec{R_i}} H \tag{2}$$

Eq. (2) can be solved numerically, e.g. using the predictor-corrector algorithm, as implemented in the Gaussian series of codes [8].

2.2 Atom-Centered Density Matrix Propagation (ADMP) Scheme

ADMP methodology belongs to the group of extended Lagrangian approaches to molecular dynamics, where Gaussian basis functions are used, propagating the density matrix [9, 10, 11].

The extended Lagrangian for the system can be written as:

$$L = \frac{1}{2} Tr(V^T M V) + \frac{1}{2} \mu Tr(WW) - E(R,P) - Tr[\Lambda(PP - P)] \tag{3}$$

where M, R and V are the nuclear masses, positions and velocities, respectively. P, W and μ are the density matrix, density matrix velocity and the fictitious mass for the electronic degrees of freedom, while Λ is a Lagrangian multiplier matrix, used to impose the constraints on the total number of electrons in the system, as well as on the idempotency of the density matrix. Applying the principle of stationary action, the Euler-Lagrange equations for propagation of the density and the nuclei may be written in the following form:

$$\mu \frac{d^2P}{dt^2} = -\left[\left. \frac{\partial E(R,P)}{\partial P} \right|_R + \Lambda P + P\Lambda - \Lambda \right] \tag{4}$$

$$M \frac{d^2R}{dt^2} = - \left. \frac{\partial E(R,P)}{\partial R} \right|_P \tag{5}$$

Equations (4) and (5) were integrated in the present study using the velocity Verlet algorithm, with a time step of 0.2 fs. Thermostat was applied to maintain a temperature of 300 1K. We havent implemented the periodic boundary conditions in ADMP simulations within the present study, however, as is often done in the case of systems that have an overall net charge or dipole moment.

2.3 *The Car-Parrinello Approach*

The basic idea behind the Car-Parrinello (CP) approach is to exploit the different time scales of electronic and nuclear motions in a molecular system [12].

The (molecular) two component quantum-classical problem is redefined as a two-dimensional purely classical one, with two separate energy scales. By this mapping, however, the explicit time-dependence of the quantum subsystem dynamics is in a sense lost. In practice, in CP approach, the quantum-classical classical-classical mapping is enabled by considering the extended Kohn-Sham energy functional EKS. The derivative of the Lagrangian with respect to nuclear coordinates in classical mechanics gives the forces acting on the nuclei. By analogy, a functional derivative with respect to orbitals, interpreted as classical fields, can yield the forces on orbitals. Using the EKS, Car and Parrinello have postulated Lagrangian of the form:

$$L_{cp} = \left[\vec{R}^N, \dot{\vec{R}}^N, \{\Phi_i\}, \{\dot{\Phi}_i\} \right] = \sum_I \frac{1}{2} M_I \dot{\vec{R}}_I^2 + \sum_i \mu \langle \dot{\Phi}_i | \dot{\Phi}_i \rangle - E^{KS} \langle \{\dot{\Phi}_i\}, \vec{R}^N \rangle \tag{6}$$

The corresponding equations of motions are further obtained from the associated Euler-Lagrange equations, but this time for both nuclear positions and orbitals. Finally, one arrives at the following Car-Parrinello equations of motion for nuclei and orbitals respectively:

$$M_I \ddot{\vec{R}}_I = \frac{\partial E^{KS}}{\partial \vec{R}_I} + \sum_{ij} \Lambda_{ij} \frac{\partial}{\partial \vec{R}_I} \langle \Phi_i | \Phi_j \rangle \tag{7}$$

$$\mu|\ddot{\Phi}_i\rangle = -\frac{\partial E^{KS}}{\partial\langle\Phi_I|} + \sum_j \Lambda_{ij}|\Phi_j\rangle \tag{8}$$

The parameter which appears in (8) is a "fictitious mass", i.e. an inertia parameter assigned to the orbital degrees of freedom.

Choice of initial geometries for *ab initio* molecular dynamics simulations. The starting geometries for *ab initio* molecular dynamics simulations were chosen in the following way. As a sample system, we have chosen the aqueous hydroxide ion. First, Monte-Carlo simulations of one hydroxide ion plus 83 water molecules placed in a cubic box with side length determined from the data on the density of liquid water at ambient conditions (of 0.9966 gcm^{-3}), were carried out imposing 3D periodic boundary conditions. All MC simulations were performed in the isothermal-isobaric (NPT) ensemble, implementing the Metropolis sampling algorithm, at T = 298 K, P = 1 atm, using the statistical mechanics code DICE [13]. In all MC simulations carried out in the present study, intermolecular interactions were described by a sum of Lennard-Jones 12-6 site-site interaction energies plus Coulomb terms:

$$U_{ab} = \sum_i^a \sum_j^b 4\varepsilon_{ij}\left[\left(\frac{\sigma_{ij}}{r_{ij}}\right)^{12} \left(\frac{\sigma_{ij}}{r_{ij}}\right)^{6}\right] + \frac{q_i q_j e^2}{4\pi\varepsilon_0 r_{ij}} \tag{9}$$

where i and j are sites in interacting molecular systems a and b, r_{ij} is the interatomic distance between sites i and j, while e is the elementary charge. The following combination rules were used to generate two-site Lennard-Jones parameters ε_{ij} and σ_{ij} from the single-site ones:

$$\varepsilon_{ij} = \sqrt{\varepsilon_i \varepsilon_j} \tag{10}$$

$$\sigma_{ij} = \sqrt{\sigma_i \sigma_j} \tag{11}$$

For water, we have used the SPC model potential parameters [14], while the charge distribution in the case of hydroxide anion was described in two ways: as a set of two point charges with positions coinciding with the centers of O and H nuclei (simple charge description - SC), and implementing a somewhat modified charged-ring (CR) description, proposed in [15]. Subsequently, several snapshots from the equilibrated MC simulations were chosen, and finite clusters consisting of the central OH^- ion and all water molecules the oxygen atoms of which reside within spheres with radius of 5$\AA$ were cut-out. These clusters were subsequently used as starting points for the *ab initio* molecular dynamics simulations.

3 Results and Discussion

3.1 ADMP Molecular Dynamics Trajectory and the Stability of the Simulation

All ADMP simulations in the present study have been carried out on a HPC cluster, using up to 20 processors for parallel computations. B3LYP level of theory was used, with the $6-31+G(d,p)$ basis set for orbital expansion. The total angular momentum was conserved to better than $10^{-10}\hbar$ in the present study, since projection methods have been implemented to remove the residual angular forces. The idempotency of the density matrix was maintained to 10^{-12} or better throughout the simulations. After the "initial heating" of the electronic degrees of freedom, the drift of the fictitious kinetic energy of the density matrix (i.e. the parameter which actually characterizes the adiabaticity) was conserved to better than 10^{-3} Hartrees. The initial heating of the electronic degrees of freedom is due to the choice of non-zero density matrix velocities at $t=0$ (Fig. 1). Subsequent evolution may be regarded as a sort of "equilibration process", in which the density matrix elements "catch-up" with the nuclei. A typical snapshot from the simulations is shown in Fig. 2.

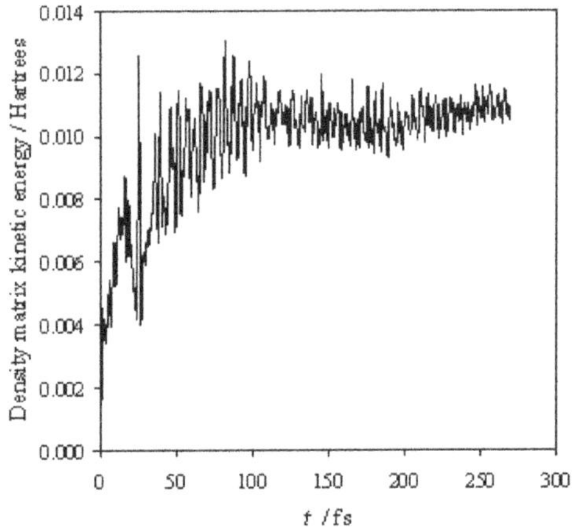

Fig. 1 Evolution of the electronic kinetic energy (i.e. the density matrix kinetic energy) with time in the ADMP molecular dynamics simulation

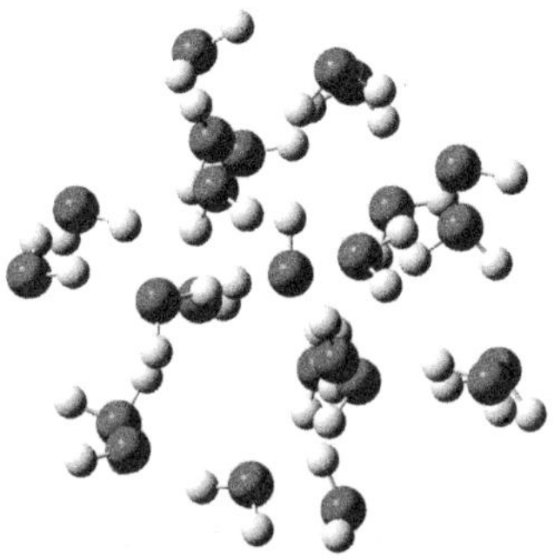

Fig. 2 A typical snapshot from the ADMP simulations, with the hydroxide ion emphasized, in the center of the finite cluster

3.2 *Subsequent Quantum-Mechanical Calculations of the Vibrational States of Hydrated Hydroxide Anion*

Since all X-H vibrations are known to be large-amplitude motions, and therefore, inherently considerably anharmonic, to compute the corresponding frequencies with a sufficient accuracy it is necessary to go beyond the widely used harmonic approximation in quantum chemistry.

Harmonic vibrational frequencies are easily and straightforwardly obtained by the available quantum chemistry codes by diagonalization of the Hessian matrices, a procedure which has become almost a routine. However, it has been shown in the literature that theoretically calculated *harmonic* X-H vibrational frequency shifts may in some cases even be in fortuitous excellent agreement with the experimental data, due to cancellation of errors [16]. Anharmonic contributions to the overall observed vibrational frequency shifts, on the other hand, may be as high as 30 - 40 %. To account for the anharmonicty effects (and other systematic deficiencies of computational methods), various variants of the scaled quantum mechanical force fields methods have been developed in the literature [17]. This approach, is, however, based on system-specific "magic scaling factors, which are by any means artificial. In the last years, also the vibrational self-consistent field methodology (VSCF) [18, 19] has become widely available and implemented in computed codes. This methodology is, however, much more computationally demanding. In our approach, instead of using an arbitrary scaling procedure, we accounted explicitly for the anharmonicity of the O-H stretching motion, using a computationally feasible approach. To obtain the vibrational potential energy function ($V = f(r_{OH})$) for an OH oscillator in a particular aqueous environment, a series of 20 pointwise HF, DFT or MP2 energy calculations were performed for each O-H oscillator, varying the O-H distances from 0.850 to 1.325 $\AA$ with a step of 0.025 $\AA$. As mentioned before, the nuclear displacements corresponding to the O-H stretching vibration were generated keeping the center-of-mass of the vibrating hydroxide ion fixed. The obtained energies were least-squares fitted to a fifth-order polynomial in $\Delta r_{OH} (\Delta r = r - r_e)$:

$$V = V_0 + k_2 \Delta r^2 + k_3 \Delta r^3 + k_4 \Delta r^4 + k_5 \Delta r^5 \tag{12}$$

The resulting potential energy functions were subsequently cut after fourth order and transformed into Simons-Parr-Finlan (SPF) type coordinates [20]:

$$\rho = 1 - r_{OH,e}/r_{OH} \tag{13}$$

(where $r_{OH,e}$ is the equilibrium, i.e. the lowest-energy, value). The one-dimensional vibrational Schrdinger equation was solved variationally. Usage of only 15 harmonic oscillator eigenfunctions as a basis was shown to lead to excellent convergence of the computed vibrational frequencies. Superiority of the SPF-type coordinates over the "ordinary" bond stretch ones when a variational solution of the vibrational Schrdinger equation is sought has been well established, as they allow for a faster convergence (for the number of basis functions used) and a greatly extended region of convergence [21]. The fundamental *anharmonic* O-H stretching frequency was computed from the energy difference between the ground and first excited vibrational states. To account for the influence of the bulk solvent molecules on the clusters energies, these were treated as a polarizable continuum, within the SCRF model of Onsager [22]. The computed anharmonic OH and OD vibrational stretching frequencies for 5 snapshots from the ADMP simulations, together with the vibrational potential energy parameters are given in Tables 1 and 2. It is seen that the fluctuating in-liquid environment affects significantly the hydroxide ion OH stretching frequency, though the range of values is not very wide. This is in agreement with the appearance of a very sharp peak due to this vibrational mode in the IR and Raman spectra of aqueous hydroxides. However, much more anharmonic vibrational frequency calculations, for numerous snapshots from the ADMP trajectories are required in order to be able to make direct comparisons with the experimentally measured half-width of the corresponding stretching bands.

Table 1 The computed *anharmonic* OH and OD stretching frequencies for the five snapshots selected from the equilibrated ADMP run

Snapshot No.	$v(OH)/cm^{-1}$	$v(OH)/cm^{-1}$	$r_e/Å$
1	3633.4	2880.2	0.9678
2	3624.8	2873.4	0.9686
3	3622.4	2871.4	0.9688
4	3622.1	2871.2	0.9688
5	3620.3	2869.8	0.9690

Table 2 Anharmonic vibrational potential energy parameters for the OH(D) oscillators for the five snapshots selected from the equilibrated ADMP run

Snapshot No.	$k_2/(mdyn/Å^{-1})$	$k_3/(mdyn/Å^{-2})$	$k_4/(mdyn/Å^{-3})$	$k_5/(mdyn/Å^{-4})$
1	4.1016	-9.670	13.93	-10.1
2	4.0822	-9.622	13.88	-10.1
3	4.0768	-9.608	13.87	-10.1
4	4.0761	-9.607	13.87	-10.1
5	4.0722	-9.597	13.86	-10.1

4 Conclusions and Future Works

In the present work, we have implemented a complex hybrid quantum mechanical (*ab initio*) molecular dynamics quantum mechanical methodology which is aimed to be used for simulation of complex condensed phase systems. The methodology is inherently a multistep one, and certain steps are excellently scalable on parallel high-performance clusters. Judging the scalability of the overall algorithm is a more complex task, as it is difficult to fully automatize the overall procedure. However, such efforts are in progress, and once this aim is achieved, we believe that this quite non-standard methodology will be available for more standardized use within the community of physicists, chemists, and engineering scientists.

References

1. Coutinho, K., Canuto, S.: Adv. Quantum Chem. 28, 89 (1997)
2. Pejov, L., Spångberg, D., Hermansson, K.: J. Chem. Phys. 133, 174513 (2010)
3. Hermansson, K., Bopp, P.A., Spångberg, D., Pejov, L., Bakó, I., Mitev, P.D.: Chem. Phys. Lett. (Frontier Article) (in press)
4. Pejov, L., Spångberg, D., Hermansson, K.: J. Phys. Chem. A 109, 5144 (2005)
5. Sahpaski, D., Pejov, L., Misev, A.: LNCS (in press)
6. Helgaker, T., Uggerud, E., Jensen, H.J.A.: Chem. Phys. Lett. 173, 145 (1990)
7. Uggerud, E., Helgaker, T.: J. Am. Chem. Soc. 114, 4265 (1992)
8. Frisch, M.J., Trucks, G.W., Schlegel, H.B., Scuseria, G.E., Robb, M.A., Cheeseman, J.R., Zakrzewski, V.G., Montgomery, J.A., Stratmann, R.E., Burant, J.C., Dapprich, S., Millam, J.M., Daniels, A.D., Kudin, K.N., Strain, M.C., Farkas, O., Tomasi, J., Barone, V., Cossi, M., Cammi, R., Mennucci, B., Pomelli, C., Adamo, C., Clifford, S., Ochterski, J., Petersson, G.A., Ayala, P.Y., Cui, Q., Morokuma, K., Malick, D.K., Rabuck, A.D., Raghavachari, K., Foresman, J.B., Cioslowski, J., Ortiz, J.V., Stefanov, B.B., Liu, G., Liashenko, A., Piskorz, P., Komaromi, I., Gomperts, R., Martin, R.L., Fox, D.J., Keith, T., Al-Laham, M.A., Peng, C.Y., Nanayakkara, A., Gonzalez, C., Challacombe, M., Gill, P.M.W., Johnson, B.G., Chen, W., Wong, M.W., Andres, J.L., Head-Gordon, M., Replogle, E.S., Pople, J.A.: Gaussian 1998 (Revision A.11). Gaussian, Inc., Pittsburgh PA (1998)
9. Schlegel, H.B., Iyengar, S.S., Li, X., Millam, J.M., Voth, G.A., Scuseria, G.E., Frisch, M.J.: J. Chem. Phys. 117, 8694 (2002)
10. Schlegel, H.B., Millam, J.M., Iyengar, S.S., Voth, G.A., Daniels, A.D., Scuseria, G.E., Frisch, M.J.: J. Chem. Phys. 114, 9758 (2001)
11. Iyengar, S.S., Schlegel, H.B., Millam, J.M., Voth, G.A., Scuseria, G.E., Frisch, M.J.: J. Chem. Phys. 115, 10291 (2001)
12. Car, R., Parrinello, M.: Phys. Rev. Lett. 55, 2471 (1985)
13. Coutinho, K., Canuto, S.: DICE: a Monte Carlo Program for Molecular Liquid Simulation. University of São Paulo, São Paulo (2003)
14. Berendsen, H.J.C., Grigera, J.R., Straatsma, T.P.: J. Phys. Chem. 91, 6269 (1987)
15. Ufimtsev, I.S., Kalinichev, A.G., Martinez, T.J., James Kirkpatrick, R.: Chem. Phys. Lett. 442, 128 (2007)
16. Kocevski, V., Pejov, L.: J. Phys. Chem. A 114, 4354 (2010)
17. Rauhot, G., Pulay, P.: J. Phys. Chem. 99, 14572 (1995)
18. Chaban, C.M., Jung, J.O., Gerber, R.B.: J. Chem. Phys. 111, 1823 (1999)
19. Yagi, K., Hirao, K., Taketsugu, T., Schmidt, M.W., Gordon, M.S.: J. Chem. Phys. 121, 1383 (2004)
20. Simons, G., Parr, R.G., Finlan, J.M.: J. Chem. Phys. 59, 3229 (1973)
21. Carney, D.G., Curtiss, L.A., Langhoff, S.R.: J. Mol. Spectrosc. 61, 371 (1976)
22. Wong, M.W., Frisch, M.J., Wiberg, K.B.: J. Am. Chem. Soc. 113, 4776 (1991)

Parallelization of Dynamic Programming in Nussinov RNA Folding Algorithm on the CUDA GPU

Marina Zaharieva Stojanovski, Dejan Gjorgjevikj, and Gjorgji Madjarov*

Abstract. When an RNA primary sequence is folded back on itself, forming complementary base-pairs, a form called RNA secondary structure is created. The first solution for the RNA secondary structure prediction problem was the Nussinov dynamic programming algorithm developed in 1978 which is still an irreplaceable base that all other approaches rely on. In this work, the Nussinov algorithm is analyzed but from the CUDA GPU programming perspective. The algorithm is radically redesigned in order to utilize the highly parallel NUMA architecture of the GPU. The implementation of the Nussinov algorithm on CUDA architecture for NVidia GeForce 8500 GT graphic card results with substantial acceleration compared with the sequential executed algorithm.

1 Introduction

The tertiary structure of the RNA molecule plays the most important role in understanding and analyzing the RNA molecule function. However, since prediction of this structure can be done only with some expensive and time-consuming methods, the easier solution is predicting the RNA secondary structure which can be later used for the tertiary structure prediction. As a result of this, although the form in which the RNA occurs in the real organisms is its tertiary structure, the research in this domain has been primarily focused on the RNA secondary structure prediction. This is a problem which has been researched for more than 30 years, but is

Marina Zaharieva Stojanovski
Formal Methods and Tools Group, Faculty EEMCS, University of Twente, The Netherland
e-mail: m.zaharieva@utwente.nl

Dejan Gjorgjevikj · Gjorgji Madjarov
Faculty of Computer Science and Engineering, Ss. Cyril and Methodius University, Skopje, Macedonia
e-mail: dejan@finki.ukim.mk, gjorgji.madjarov@finki.ukim.mk

L. Kocarev (Ed.): ICT Innovations 2011, AISC 150, pp. 279–289.
springerlink.com

still very popular in the field of bioinformatics and there is still no single solution considered as the most valuable one.

The Nussinov algorithm [1] is the base and carries the logic which is used in most of the other algorithms. This paper focuses on implementing the Nussinov algorithm on CUDA GPU architecture. The results obtained with this implementation would not be as accurate as those from the other later developed more advanced algorithms, but the key point with this work is accelerating the Nussinov algorithm as a base for the other algorithms. The positive result found with this implementation would lead to reusing the base idea from this work for efficiency improvement of others advanced RNA folding algorithms. The choice for the Nussinov algorithm to be the target of this research was predominantly because of its simplicity.

1.1 RNA Secondary Structure

In mathematical term, the RNA primary structure can be considered as a sequence of symbols contained in the set S ={A,C,G,U} where each of the symbols is representing one of the four possible bases: Adenine, Cytosine, Guanine, and Uracil, respectively. This long sequence can be folded [6] in a way that the symbols are paired with each other forming the well-known Watson-Crick pairs (A-U, C-G) or the Wooble pair (G-U). Thus, the RNA secondary structure can be defined as a set of ordered pairs

$$S = \{(i, j) | 1 \le i < j \le n, \text{where } n \text{ is the length of the sequence } \} \qquad (1)$$

The pair (i, j) represents that a pairing has occurred between symbols at position i and position j. The following rules must be satisfied:

1. The distance between two paired symbols must be greater than 3;
2. A symbol can be included in at most 1 pair.

The secondary structure prediction problem can be defined as: **Given the RNA primary structure, the secondary structure should be predicted.**

1.2 Nussinov Algorithm for RNA Folding

The first solution for the RNA secondary structure prediction problem is the Nussinov algorithm [1] developed in 1978. The base idea for Nussinov algorithm approach is the assumption that the secondary structure is the form in which the number of paired bases is maximized. Since the number of possible secondary structures grows exponentially as the sequence length increases, the most naive solution (finding all possible structures and selecting the one with the largest number of pairs) would have no sense. The most obvious technique that can deal with this problem while achieving an efficient solution is the dynamic programming.

The core of each dynamic programming algorithm is a set of recurrent relations that define the additional algorithm steps as well as the performance and memory complexity. The relations for the Nussinov algorithm are displayed below.

$$N(i,j) = \max\begin{cases} N(i+1,j) \\ N(i,j-1) \\ N(i+1,j-1)+\delta(i,j) \\ \max_{i<k<j}[N(i,k)+N(k+1,j)] \end{cases} \quad (2)$$

where

$$\delta = \begin{cases} 1, \text{if } x_i \text{ and } x_j \text{ may pair} \\ 0, \text{else} \end{cases} \quad (3)$$

The equations defined show that the performance complexity of the Nussinov algorithm is cubic $O(n^3)$ with linear complexity of computing the value of one element. The memory complexity is $O(n^2)$, since the computed values are stored in a two-dimensional $n \times n$ array.

The cubic performance complexity results with very inefficient and time-consuming execution of the Nussinov algorithm considering the fact that the experiments are usually made with a sequence of several thousand bases. This is where the main motivation of this paper came from. The acceleration of the Nussinov algorithm is done by implementing the algorithm on CUDA GPU architecture [2, 3] in order to achieve massive parallelization and to substantially decrease the execution time.

2 CUDA Implementation

2.1 General Overview

The starting point of designing a parallel algorithm is making a good analysis of the sequential algorithm and getting the basic idea of how this algorithm can be divided into smaller tasks that can run in parallel. Analyzing the equation (2), enough information can be extracted for the Nussinov algorithm procedure. A two-dimensional array (n x n) is used for storing the computed values in this dynamic programming algorithm (Fig. 1 left). The values in the main diagonal are initialized with zeros and the lower triangular part of the matrix remains unused. Additionally the values for the next three matrix diagonals above the main diagonal are set to zero which comes from the fact that the distance between two paired bases must be greater than 3. The diagonals above the main diagonal are iterated and the values of the elements in the diagonal are computed in each iteration. The last diagonal contains only one element (the upper right corner) with the result value indicating the number of pairs contained in the secondary structure for the current RNA sequence.

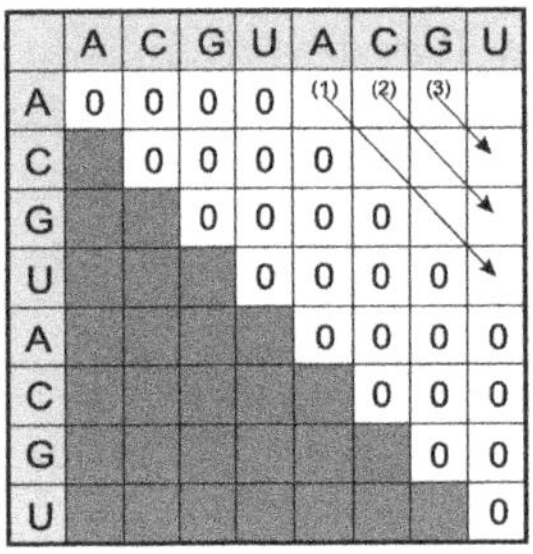

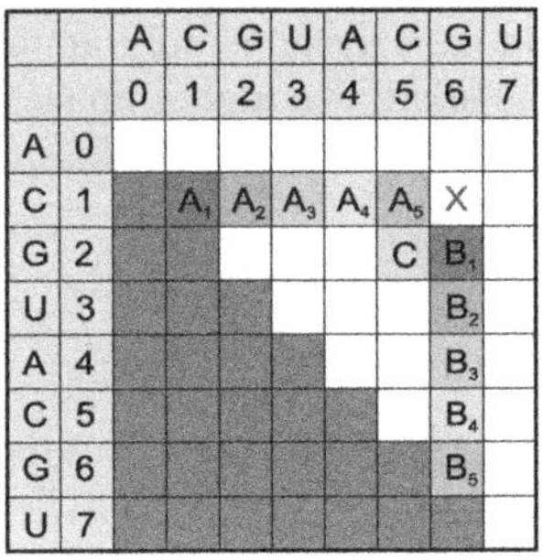

Fig. 1 Nussinov Matrix (left) and Computing the value of an element (right)

Considering the fact that one matrix element can be found as a computation of the already computed values in the same row and the same column and the adjacent element (down-left) (Fig. 1 right), the following conclusion can be defined:

All elements in a same diagonal are independent of each other and their computing can run in parallel.

The GPU device provides registers and local memory for each thread, a shared memory for each block, and a global memory for the entire grid of blocks of threads. Although all threads execute the same GPU kernel function, a thread is aware of its own identity through its block and thread indices, and thus a thread can be assigned a specific portion of the data on which it can perform computation. The shared memory for a block of threads is fast, yet it is limited in size. One strategy to attain high performance is for the threads in the same block to collaborate on loading data that they all need from the global memory to the shared memory.

A logical consequence of these facts is the general overview of the algorithm design defined through the following steps: 1) The 2-dimensional array is stored into CPU initialized with all values equal to zero; 2) The array is copied to the GPU global memory; 3) All diagonals above the main one are iterated and a kernel function is called for each of them. This design provides parallelism of computing the values for elements from one diagonal, while their independency ensures that the computed values would always be correct.

2.2 *Kernel Design*

Designing the kernel function means defining subtasks and mapping them to number of threads in a way that they would all execute the preferred function – computing the values of the elements for a specific diagonal. The first question is: "How many threads would be responsible for computing the value of one element?". The proposed solutions are: 1) a thread; 2) a block of threads; 3) a grid of threads.

If we consider the solution (1), one thread per element would mean that the number of threads included in the kernel would not be large enough in order to fully utilize the GPU device capacity. The maximum number of threads in the grid would be equal to the primary sequence length decreased by 4 (that is the length of the first iterated diagonal). Another disadvantage for this solution is the fact that computing the value for one element would remain sequential with linear complexity.

These negative sides would be avoided in case the solution (2) is used. However, in this case additional effort is needed for defining the parallel execution of the block of threads. The threads will have to be synchronized so that the computation of one element value can run in parallel. Shared memory would be used so that the threads would communicate with each other.

We avoid the solution (3) since if the whole grid is responsible for computing the value of single element, the latent global memory should be used and achieving efficiency would be more complicated. A possible solution is to use this kernel function for diagonals containing small number of elements, but this is not the case in the implementation we propose.

Following these analyses, in the current implementation we have chosen the solution (2). However, since the diagonals that are closer to the main diagonal contain elements whose computations are simple and depend of a small number of elements, we create two different kernel functions. The first one would be defined according to solution (1) and will be called only for diagonals which are close to the main diagonal, and the second one would follow the solution (2) and would be called for the remaining diagonals.

Kernel 1 function. The base idea for this kernel function is that the task of computing the values for one diagonal is divided into smaller subtasks – computing the value for one element and each of these subtasks is mapped to a specific thread. The diagonal is divided into a number of segments each of them assigned to a specific block of threads (Fig. 2). The kernel configuration parameters would be (*blocksize*, [*n*/*blocksize*]) where *blocksize* is the number of threads per block which we define to be equal to 128 and n is the number of elements in the specified diagonal.

Fig. 2 Computing values for a matrix diagonal with kernel 1 function

The logic of this kernel function is simple: First the index of the thread is computed (using the build-in variables) and then it is used to identify the element the thread should work on. There is no need to use the shared memory since during the execution of the kernel for diagonal d, a value of the element (x, y) from the global memory can be accessed at most 3 times (while computing: the value of the element at row x contained in d; the value of the element at column y contained in d; the value of the element $(x+1, y-1)$ if it is contained in d.

Kernel 2 function. This kernel function is designed in a way that the subtask of computing the value of one element is executed by a block of threads. The kernel is called with the following configuration parameters (*blocksize*, n) where *blocksize* is set to be 128, and n is the number of elements in the appropriate diagonal. Fig. 1 shows the process of computing the value in one cell. The value X is computed as:

$$X = \max\{A_1 + B_1, A_2 + B_2, A_3 + B_3, A_4 + B_4, A_5 + B_5, C + \delta(G,C)\} \quad (4)$$

The most suitable solution would be the following: The sum $A_i + B_i$ $(i=1..5)$ is computed by the thread i. The number of sums (in this case 5) will usually be greater than the number of threads contained in the block, thus the thread i would be responsible for computing more than one sum, $A_i + B_i$, $A_{i+blocksize} + B_{i+blocksize}$, $A_{i+2*blocksize} + B_{i+2*blocksize}$... and finding the max value from all sums. Each computed maximum value would be stored in the shared memory.

This procedure will result with an array of elements with length equal to *blocksize* (the number of threads per block) stored in the shared memory. The next step is finding the maximum of this array which can be seen as a typical reduction pattern problem. Thus, the reduction pattern is used in a way that a thread is computing the maximum value of two array elements, then after the threads are synchronized the same step is repeated. With every iteration the array size would be divided by two and after $\log_2 N$ operations the final result would be found, the maximum value of all elements in the array.

Global memory access pattern impact on performance. This solution seems valuable, but it still does not give sensible results. The problem that must be handled is the global memory access that gives a great negative influence to the algorithm execution performance. When loading the values from the global memory, the accesses must be coalesced in order to avoid the degradation that may happen because of the highly latent global memory. In order to achieve the coalesced global memory access the following solution is proposed:

Since the lower matrix triangular half part does not store any values, the values from the upper part are copied symmetrically to the main diagonal and shifted one position left. This way, when a thread accesses the two matrix elements which sum should be computed, both of the elements would be stored in the same matrix column.

In order to maximize the memory bandwidth while accessing the elements they should be $16k$ elements apart, requesting the matrix dimension to be $n = 16k$ $(k \in N)$. This insures that the OFFSET value while accessing the global memory will be factor of 16, that provides most effective bandwidth. All matrix values are initially set to 0. When new value for a matrix element is computed, it is stored both in the original matrix cell (above the main diagonal) and in its mirrored copy (below the main diagonal).

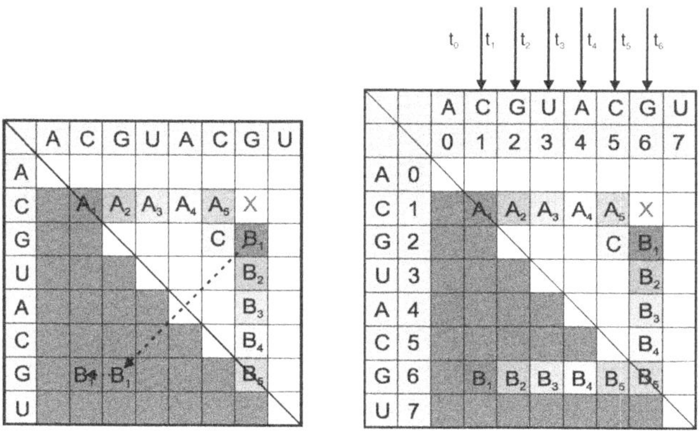

Fig. 3 Copying matrix element and distributing the threads accessing global memory

The Traceback Stage In the second part of the Nussinov algorithm, the traceback stage is executed after all kernel functions have finished. The matrix with all updated values is copied back to the host, where the traceback stage is executed sequentially. The solution is recursive with $O(n^2)$ performance complexity.

3 Results

The efficiency of the proposed Nussinov algorithm is tested with a set of experiments using RNA primary sequences with different lengths. The CUDA implementation is compared with the sequential Nussinov algorithm implemented in C++ and Java and substantial results were found.

The experiments were made on a Windows PC with Intel Core2Duo E4500 running at 2.2GHz and 1GB of RAM. The parallel CUDA algorithm was executed on NVidia GeForce 8500 GT architecture (2 multiprocessors and 16 cores). All the tables below present the execution time in seconds. It should be noticed that these results do not include the traceback stage but this would not influence the comparison since this part of the algorithm is executed sequentially on the CPU host both in the original sequential and in the parallel CUDA algorithm.

The test sequences are randomly generated using the Markovian model having order 0 (Bernoulli model) where the frequency of occurrence of the symbols A, C, G, U is equal to 0.25. The GenRGenS software [4] is used.

3.1 Sequential and Parallel CUDA Algorithm Comparison

The experiments include execution of three different implementations of the Nussinov algorithm on different sequence lengths. The three implementations include: sequential implementation in JAVA (JRE 1.6), sequential implementation in C++ (Visual C++ 2008 Express Edition, Maximize Speed Optimization) and parallel implementation in CUDA C (CUDA kernel functions called from JAVA using JCUDA [5]).

The running times of the three different implementations on RNA sequences with different lengths are given in Table 1.

Results in Table 1 shows that as the sequence length grows, the acceleration is greater, thus for sequences with small length, there is even no acceleration, but for sequence with length equal 7860, the acceleration is 18 times compared to the sequential execution of the C++ implementation, and 31 times compared to the sequential execution of the JAVA implementation. Fig. 4 shows these results graphically.

Table 1 Sequential and parallel CUDA Nussinov algorithm execution results

Sequence length	GPU (Cuda)	CPU (C++)	CPU (JAVA)
640	0.6	0.5	0.6
1280	1.8	3	4.3
1920	2.23	10	17
2560	3.98	31	42.1
3200	6.72	68	86.9
3840	10.47	119	160.2
4480	15.55	198	270.8
5120	21.9	312	438
5760	30.53	452	690.9
6400	41.3	660	1081.8
7040	54.6	895	1511.7
7860	68.5	1242	2119.2

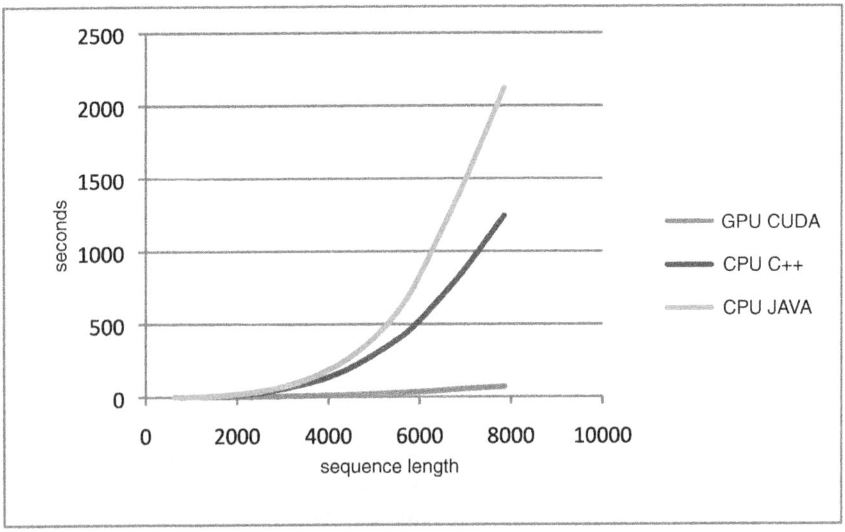

Fig. 4 Nussinov algorithm in C++, Java and CUDA

3.2 Conditions for Achieving Acceleration

The following experiments are done in order to show the key points in the algorithm that contribute the most to acceleration. If we avoid the second kernel function and use only the kernel 1 function for all diagonals which means sequentially computing the value of element matrix, one even gets increase in the execution time compared to the sequential execution. Other experiments are done when no symmetrical copying of the elements is done. The result shows that the uncoalesced global memory accesses degrade the algorithm efficiency and the execution time is even longer compared to the algorithm with only one kernel. Fig. 5 displays these results graphically.

Table 2 Comparison between: CUDA using only kernel 1, CUDA with uncoalesced global memory accesses and CUDA proposed solution

Sequence length	CUDA with kernel 1 only	CUDA within coalesced accesses	CUDA proposed solution
640	1	1.3	0.6
1280	3.4	4.2	1.8
1920	13.5	15.6	2.23
2560	38.5	41.9	3.98
3200	93.5	102.5	6.72
3840	192.3	201.2	10.47

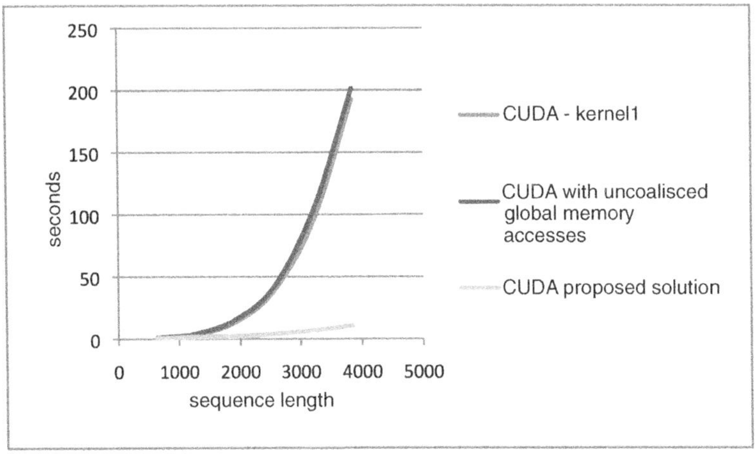

Fig. 5 CUDA using only kernel1, CUDA with uncoalesced global memory accesses and CUDA proposed solution

As a conclusion, the key point of the algorithm acceleration is a proper design of the kernel 2 function and coalesced global memory access which in the proposed solution is achieved by symmetrically copying the matrix elements.

3.3 Border between Kernel 1 and Kernel 2

The moment when the kernel 2 is called instead of kernel 1 is after 32 iterations, which means that the sums of pairs that should be computed would be 32 (Table 3). The basic idea for this is that if we call kernel 2 function earlier, not all threads in a warp will be active when computing a value for an element.

Table 3 Choosing the border between kernel 1 and kernel 2 function.

Sequence length	0 iterations (kernel 2 only)	32 Iterations	48 iterations	64 iterations	96 Iterations
1920	2.27	2.23	2.27	2.26	2.32
2560	4	3.98	4.02	4.05	4.06
3200	6.72	6.72	6.73	6.74	6.82
3840	10.48	10.47	10.55	10.53	10.59
4480	15.59	15.55	15.6	15.68	15.93
5120	22.1	21.9	22.03	22.02	22.17
5760	30.52	30.50	30.51	30.63	30.8
6400	41.5	41.3	41.4	41.9	42.1
7040	55	54.6	54.9	54.8	55.9
7860	69.1	68.5	68.7	69.5	69.8

Results show that the value 32 is the most appropriate to be the border between kernel 1 and kernel 2. The differences are small, and even if kernel 1 is totally avoided the results are similar with those of the proposed solution.

4 Conclusions

In this work, the Nussinov algorithm is redesigned in order to utilize the highly parallel NUMA architecture of the GPU. The implementation of the Nussinov algorithm on CUDA architecture for NVidia GeForce 8500 GT graphic card results with substantial acceleration compared with the sequential executed algorithm. The positive result in this implementation would lead to reusing the base idea from this work for efficiency improvement of others advanced RNA folding algorithms.

References

1. Nussinov, R., Pieczenik, G., Griggs, J.R., Kleitman, D.J.: Algorithm for Loop Matching. SIAM Journal on Applied Mathematics 35(1), 68–82 (1978)
2. NVIDIA.: CUDA C Best Practices Guide, Version 3.2. (August 2010)
3. NVIDIA.: CUDA C Programming Guide, Version 3.2. (October 2010)
4. Ponty, Y., Termier, M., Denise, A.: GenRGenS: software for generating random genomic sequences and structures. Bioinformatics 22(12), 1534–1535 (2006)
5. Yan, Y., Grossman, M., Sarkar, V.: JCUDA: A Programmer-Friendly Interface for Accelerating Java Programs with CUDA. In: Sips, H., Epema, D., Lin, H.-X. (eds.) Euro-Par 2009. LNCS, vol. 5704, pp. 887–899. Springer, Heidelberg (2009)
6. Fontana, W., Stadler, P.F., Bornberg-Bauer, E.G., Griesmacher, T., Hofacker, I.L., Tacker, M., Tarazona, P., Weinberger, E.D., Schuster, P.: RNA folding and combinatory landscapes. Phys. Rev. E 47(3), 2083–2099 (1993)

Towards Improvement of the Study Programme Quality: Alumni Tracking Information System

Danijel Mijic and Dragan Jankovic

Abstract. The quality of study programmes in higher education is essential for the education of graduates with a high level of knowledge in the subject field. Many factors influence the quality of study programmes, such as curriculum design, available human and material resources, and continual improvement and innovation according to the recent advancements in the field. According to standards for quality assurance in higher education area, improvement of study programme quality should involve all relevant stakeholders, especially graduated students, employers and organizations from the professional field. Establishing and maintaining connections to the stakeholders is a challenging job for higher education institutions. This paper describes a possible solution to information system for tracking information about graduated students - alumni, and its possible contribution to improvement and innovation of study programmes.

1 Introduction

Higher education institutions pay much attention to the quality of study programmes. Continual improvements and innovations in accordance to advancement in the subject field are essential for improving the quality of study programmes. This is especially significant in the engineering education, where the new technologies are frequently introduced. An important part of the study programme quality assurance is involvement of all relevant stakeholders in the quality assurance activities. European standards for internal quality assurance in higher education area demand active role of students, graduated students, employers, labour market and other relevant stakeholders in the process of monitoring and revision of study programmes [1]. Criteria for accreditation of higher education institutions in Bosnia and Herzegovina (BiH) are adopted in July 2010 [2]. They conform to the

Danijel Mijic
University of East Sarajevo, Faculty of Electrical Engineering, East Sarajevo, Bosnia and Herzegovina

Dragan Jankovic
University of Nis, Faculty of Electronic Engineering, Nis, Serbia

L. Kocarev (Ed.): ICT Innovations 2011, AISC 150, pp. 291–300.
springerlink.com

European standards and guidelines and they will be used for accreditation of higher education institutions in BiH. Standards for quality assurance of study programmes in BiH are proposed in the framework of a TEMPUS project ESABIH [3] during 2010. The adoption of these standards is expected by the end of 2011. Accreditation of study programmes at higher education institutions in BiH will be conducted according to the adopted standards at national level. Like the European standards, the proposed standards require involvement of all relevant stakeholders in the quality assurance activities. Establishing and maintaining connections to the stakeholders enables getting valuable feedback that should be used for improvements of study programmes.

This paper presents an information system for tracking information about graduated students and their employers, which are important stakeholders in the education process. The motivation for development of such system is the need for reestablishing and maintaining connections to graduated students – alumni, and their employers, which is currently a weak point of almost all higher education institutions in Bosnia and Herzegovina. Information collected using this system will be used for evaluation and improvements of study programmes. Information about alumni employers and job positions is also a significant indicator of higher education institution success and quality.

2 Alumni Tracking Systems

Tracking information about alumni usually becomes important when relevant accreditation institutions adopt standards and accreditation criteria that involves alumni in process of continuous assessment and evaluation of quality of study programmes. This doesn't happen at the same time at all countries around the world. Higher education institutions in USA started with getting feedback from alumni more than 20 years ago [4,5], while this process is at the beginning in Bosnia and Herzegovina. Tracking alumni information is usually closely related to alumni associations. Establishing these associations enables creating and maintaining alumni connections with universities. These connections are later used to involve alumni in many activities related to improvements of study programmes.

Universities in Europe, USA and Australia have a long tradition of alumni associations. They have very strong connections with alumni community, featuring many benefits for alumni members. Graduated students become alumni members after graduation at one of the university departments, colleges or schools. After registering as alumni member, alumni are able to access additional information about their colleagues, classmates, access contact information, job opportunity postings, news and events. Most alumni associations use some kind of information system for tracking information about their graduated students. There are in-house solutions developed by universities, as well as commercially available applications [6-13]. Alumni tracking systems differ in complexity and functionalities, but at the same time have many similar features. The most common features are: alumni directory, registering new alumni member, updating personal and employment information, joining alumni networks, searching alumni database, viewing alumni profiles, reading and posting job vacancies, donating to alumni association, etc.

Almost all of the mentioned features are available to registered alumni users, giving very small amount of information to the guests, for the purpose of personal data protection.

The situation regarding alumni associations and alumni tracking systems is significantly different in Bosnia and Herzegovina (BiH) and the neighboring countries. Only a few higher education institutions in BiH have registered alumni associations or are in the process of establishing such associations. It is expected that most of the universities in BiH will initiate activities for establishment of alumni associations and connections to former graduates in the near future. This is primary to the fact that Criteria for accreditation of higher education institutions in BiH demands collecting information about graduated students, their employment and carrier development. The Criteria also demands strong connection to the alumni and the work field for the purpose of getting valuable feedback in process of revisions and improvements of study programmes. Similar situation regarding alumni associations and alumni tracking systems is in the BiH neighboring countries, Serbia, Croatia and Montenegro. According to web sites of leading universities in these countries, only several institutions established alumni associations and have active connections to alumni. University of Nis have established alumni center at the university level [14]. A few institutions are using own alumni tracking systems [15,16]. There is no evidence of any university in the region that has established alumni tracking information system at the university level.

3 System Overview

The proposed system for tracking information about alumni at University of East Sarajevo is called ALTRIS (Alumni Tracking Information System). University of East Sarajevo (UoES) is a public higher educational institution comprised of 17 organizational units geographically distributed over the eastern part of BiH. The UoES is an integrated university, meaning that it is the only legal body representing all faculties, academies and institutes in the role of organizational units. The integration of the UoES is done several years ago, according to the national Law on Higher Education. Before the integration, all organizational units were independent legal bodies and they all used the same information system for supporting the teaching process. Since the UoES is geographically distributed, many functions of the university are also distributed. Information systems on organizational units of the UoES are still running independently, without some kind of integration at the university level. ALTRIS is designed to track information about alumni from all organizational units of the university. For that purpose the integration of relevant data from individual information systems will be required. At the time of writing the system is in use only at Faculty of Electrical Engineering (FoEE) with basic features for tracking alumni information, such as alumni list, alumni search, alumni profiles, alumni employment rates and listing of alumni job positions. The development of ALTRIS is still in progress. After adding additional features the system will be used at the university level, which is expected to happen by the end of 2011.

The alumni database currently contains 290 alumni profiles for the FoEE graduates which have graduated beginning from 1994. Alumni profiles are imported from the FoEE information system (IS) database and they initially contain accurate information about graduates at the time of graduation. The import procedure is currently done by the help of separate software that is developed for the purpose of initial creation of alumni profiles. This functionality will later be replaced with REST based web services that will be used to periodically import information about new graduates from the IS databases of all organizational units.

There is no need for alumni to manually enter information such as name, date of birth, graduation date, study programme, or other information that is already contained in the UoES IS databases. Alumni are only required to activate their profile, update personal information that is changed in the meantime and insert information about their current and past employments. The process of activation of alumni profile is easy and can be done in three ways:

- Clicking on activation link in email message containing random code associated with the alumni profile. The email message is sent to alumni email address listed in the UoES IS database. In case the email address in no longer valid or used, the email with activation link will never reach the intended recipient and the next two options are available.
- Visiting ALTRIS registration page and entering personal citizen identification number (unique number for each person in BiH, similar to social security number) which is also registered in the UoES IS database and is associated with alumni profile. In case this method fails (because of missing personal identification number for older graduates), the next option is available.
- Visiting ALTRIS registration page and entering other personal student's information from the UoES IS database, which is available in alumni profile. This information includes first name, last name, date of birth, date of graduation, study programme, organizational unit and the student's ID number.

After activation of the user profile, registered alumni can update personal contact information and add information about current and past employments. Information about alumni place of residence is stored in the database for each alumni. When entering contact address information, Google Maps Geocoding API [17] is used to determine location of the residence and placing a marker on the Google Map that is displayed inside the web page. User can move the marker to the exact position of the residence and save the coordinates in the profile. These coordinates are later used for displaying a map with markers on locations of residence of all registered alumni. When adding information about employments, users are entering all relevant data about the job which is later used for generation of various reports. Locations of employments are saved in the same way as locations of residence and could be also presented on the map. Users can select the type of employment (full-time, part-time, etc), industry, starting and ending dates, and other data that is identified as useful for generating employment statistics reports. Information that is the most interesting from the aspect of evaluation and quality assurance of study programmes is employment ratio and average time of waiting for the first employment after the graduation. This information could be reported at various levels like the university level, organization unit level, study programme level, year of enrolment, year of graduation, etc.

Collecting alumni contact data and alumni employment records is significant for establishing and maintaining connections to alumni and their employers. However, this process depends on the alumni readiness and willing to participate. By motivating alumni to register and regularly update their profiles, higher education institutions can retain permanent connections to these important stakeholders. After establishing connections to alumni and employers it is possible to introduce various mechanisms for getting valuable feedback that is necessary for evaluation and improvement of the quality of study programmes. This includes periodical surveys, meetings, participation in bodies for improvements and innovations in the study programmes, decision making bodies.

3.1 Users

Users of ALTRIS are former students, quality assurance staff and management staff at university and organizational unit levels, system administrator, and other persons in the role of guests. Access to personal data is protected using username and password. Unregistered users have limited access to data. Registered users are able to access more information, depending on the role and the privacy settings in the alumni profiles.

3.2 Architecture

The system architecture diagram is presented in Figure 1. The application server is at the location of the university computing center. Users can access the system from the Internet. Integration with information systems of organizational units is done using RESTful web services. Integration is necessary for importing new alumni profiles.

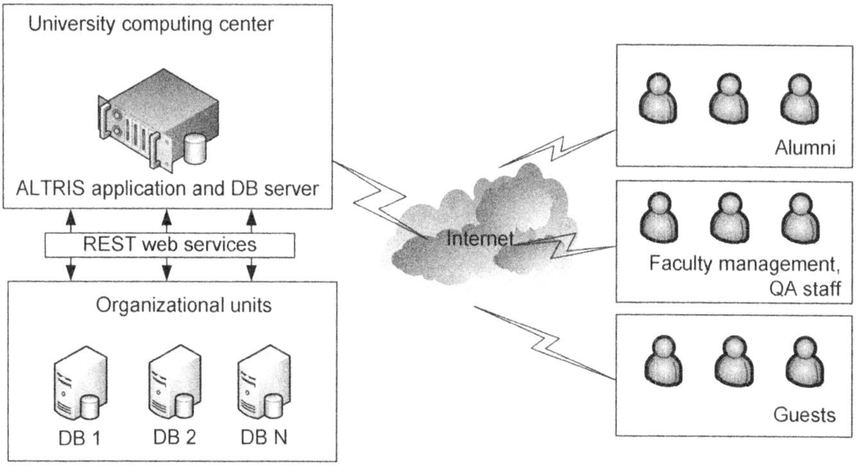

Fig. 1 ALTRIS architecture

4 System Implementation

ALTRIS is implemented on a LAMP (Linux, Apache, MySQL and PHP) platform using CakePHP rapid application development framework [18]. The application server has following features: IA64 architecture, operating system Debian Linux, Apache version 2.2.5 with PHP 5.1 module, and MySQL 5.0 database server. The CakePHP framework is used for development of web applications based on MVC (Model-View-Controller) programming pattern. Web applications designed using CakePHP framework are separated into the layers responsible for data persistence (models), business logic (controllers) and presentation (views). Following some naming conventions during the design of web applications in CakePHP framework, developers can make applications more efficiently as the framework itself enables automation of common tasks such as implementing basic database operations for creating, reading, updating and deleting data (CRUD actions). By using CakePHP shell scripts it is possible to interactively generate models, controllers and views based on the underlying database, which is a very good starting point for further system implementation. When developing ALTRIS, we used *cake bake* shell script for generation of models, controllers and views with basic CRUD functionality, and after that we manually implemented all other functionalities and made some customizations of the existing ones. Basic ALTRIS components are presented in Figure 2.

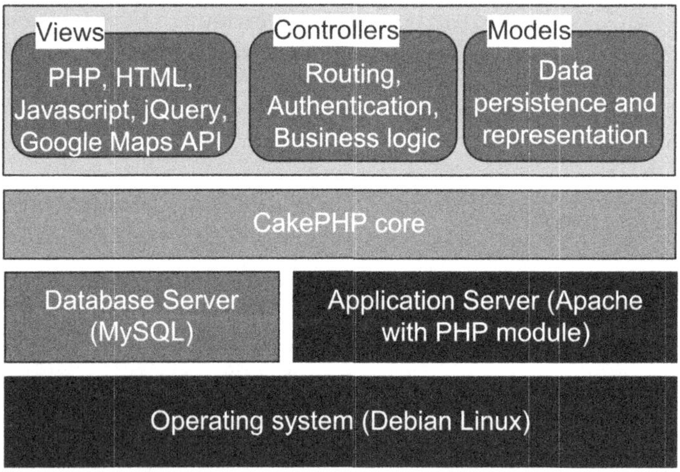

Fig. 2 ALTRIS components

4.1 Models

Models are responsible for data persistence in MVC applications developed using the CakePHP framework. For each table in the database there is a model class implementing methods for common database operations. Relations between tables are automatically detected if the naming conventions are used in applications developed using CakePHP framework. Foreign keys are represented by columns having name of the referenced table, suffixed with "_id". For example, in ALTRIS database, tables *students* and *jobs* are related in such a way that one job belongs to one student and one student can have many jobs. This is represented by column *student_id* in the *jobs* table. The type of a relationship between tables (M:N, 1:N, 1:1) can be specified in the model class.

4.2 Controllers

Controllers are responsible for processing user actions and implementing business logic. The methods implemented in controller classes are called actions. Controllers usually work with one model, but it is also possible to use more than one model.

All user defined controllers in ALTRIS inherit from application-wide *AppController*, which in turn inherits from CakePHP library class *Controller*. *AppController* can be used to implement actions used by all controllers that inherit from it. Authentication, authorization and common functions used in all other controllers are implemented in ALTRIS *AppController* class.

4.3 Views

The role of the view is parsing data and preparing it for presentation to the user. Views can be implemented as HTML documents or dynamically generated HTML code, or could be in other format according to the specific needs, like XML of PDF. Most of the views in ALTRIS are dynamically generated HTML documents. Data to be displayed in the view consists of static HTML and dynamic data that is handed over from the controller. By using standard PHP coding in the views, the data can be incorporated into appropriate HTML elements and sent to the browser.

Additional elements contained in views are client-side scripts. These scripts are used for validation of input on the client-side and for improving look-and-feel of the user interface to the system. We used publically available jQuery Javascript library [19] for handling events and improving functionality of some user interface elements, like calendars and auto complete drop-down lists. Another important component that is used at the view level is Google Maps Javascript API [20]. Views that display maps with locations of alumni residences and locations of alumni jobs are using Google Maps Javascript API for displaying Google Map inside the page, placing overlays (markers and info windows) at desired locations and for handling user generated events on the maps. An example of the view displaying a map of alumni residences (a screenshot of UoES FoEE alumni portal, http://alumni.etf.unssa.rs.ba) is shown in the Figure 3. Each marker represents a location of residence of a registered alumni. Click on the marker shows info window with additional information about alumni. Registered users can see the name and full address of alumni, while unregistered users can only see the name of the town or place at the location of the marker.

Fig. 3 Alumni residence map

5 Comparison to Similar Systems

The main goals of ALTRIS are derived from the standards and criteria for quality assurance and accreditation of study programmes in BiH. Therefore, functionalities that are implemented so far are in the first place targeting the needs for establishing and maintaining connections to alumni and employers. Another important function of the system is evaluation of study programme quality through analysis of various elements of employment data such as employments rates, average time for waiting for employment, job industry, country, town, etc. This functionality is not seen in other alumni tracking systems. Future development of ALTRIS will include development of a special module, based on business intelligence technologies, that will provide fast, multidimensional analysis of employment data based on different criteria and displaying results of analysis in the web browser.

Many features of ALTRIS are common with similar alumni tracking systems, such as alumni directory, alumni search, updating personal contact information, etc. On the other hand, some of the functionalities are rarely seen in other systems. This includes all alumni employment records (past and current) with interactive Google Map of job locations and a map of alumni residence locations, list of all alumni job positions and employers. When compared to another systems, especially commercial solutions [12-14], one of the advantages of ALTRIS is usage of existing data about graduated students through integration with the university information system.

6 Conclusion

This paper presented a solution to information system for tracking alumni related information. The significance of the system is in its usability and benefits for higher education institutions in BiH, which are currently in the process of accreditation. As mentioned earlier in this paper, connections to alumni are not existing at most of universities in BiH. By using ALTRIS, the connections could be efficiently established, which is enough to fulfill one of the accreditation requirements. Employment information provided by alumni is useful for analyzing quality of study programmes and to justify their existence. Employment rates also depend on the other economical factors in the country, but could also be used to identify study programmes whose graduates are more needed or not at all needed on the labour market.

Establishing connections to alumni by asking them to register or activate profile is not enough. Institutions should work more on establishing alumni associations and providing various benefits to registered alumni in order to motivate them to more actively participate in the activities related to improvements of study programme quality. Organizing social events, providing discounts on university services, providing usage of resources and equipment are only a few examples of measures that could be undertaken to increase alumni feedback. This is the next step after establishing connections, to maintain and strengthen connections to alumni in order to provide active alumni participation in quality improvement activities.

Besides the alumni, important benefit of using the presented system is establishing connections to the employers. Employers could provide useful information about the level of knowledge of newly graduated students, give suggestions for innovations in the structure and content of the study programmes, provide means for practical work, students research, projects, publish job vacancies, etc.

Finally, a nice presentation of alumni information on institution web site is additional indicator of commitment of the institution to all segments of quality improvement and it could be used as a good marketing tool for attracting future students and business partners.

References

1. Standards and Guidelines for Quality Assurance in the European Higher Education Area, European Association for Quality Assurance in Higher Education (2009)
2. Decision on Criteria for Accreditation of Higher Education Institutions in Bosnia and Herzegovina, Agency for Development of Higher Education and Quality Assurance of BiH, http://hea.gov.ba/Kvalitet/dokumenti/?id=1315 (accessed June 2011)
3. ESABIH TEMPUS project, http://www.esabih.ba/en/about_esabih (accessed June 2011)
4. Schneider, S.C., Niederjohn, R.J.: Assessing Student Learning Outcomes Using Graduating Senior Exit Surveys and Alumni Surveys. Frontiers in Education Conference (1995)

5. McGourty, J., Besterjeld-Sucre, M., Shuman, L.J., Wolfe, H.: Improving Academic Programs by Capitalizing On Alumni's Perceptions and Experiences. Frontiers in Education Conference (1999)
6. University of Leeds Alumni Assocciation, https://alumni.leeds.ac.uk/Page.aspx?pid=183 (accessed June 2011)
7. University of York Development and Alumni Relations Office, http://www.york.ac.uk/admin/dao/ (accessed June 2011)
8. University of Sydney Alumni and Friends, http://sydney.edu.au/alumni/ (accessed June 2011)
9. University of Michigan Alumni Association, http://alumni.umich.edu/ (accessed June 2011)
10. Carnegie Mellon University Alumni Association, http://www.alumniconnections.com/olc/pub/CMU/ (accessed June 2011)
11. Alumni Network, Versation Inc., http://www.versation.com/products/alumni (accessed June 2011)
12. Alumni Management and Contact Application, AMIXA LLC, http://www.amixa.com/alumniapp.asp (accessed June 2011)
13. Alumni Software, Rainier Software Inc., http://www.alumnisoftware.com/Platform-Features.aspx (accessed June 2011)
14. University of Nis Alumni Center, http://ni.ac.rs/images/dokumenti/alumni_publikacija_2011.pdf (accessed June 2011)
15. University of Belgrade, Faculty of Electrical Engineering, Alumni Association, http://alumni.etf.rs/ (accessed June 2011)
16. University of Zagreb, Faculty of Electrical Engineering and Computing, Alumni Association, http://www.amac.fer.hr/ (accessed June 2011)
17. Google Maps Geocoding API, http://code.google.com/apis/maps/documentation/geocoding/ (accessed June 2011)
18. CakePHP: the rapid development php framework, http://cakephp.org (accessed June 2011)
19. jQuery Javascript Library, http://jquery.com/ (accessed June 2011)
20. Google Maps Javascript API, http://code.google.com/apis/maps/documentation/javascript/ (accessed June 2011)

Computer-Added C-V Measurement and Analysis of Metal/High-κ/Si Structures

Lihnida Stojanovska-Georgievska and Nenad Novkovski

Abstract. High-κ dielectrics are intensively studied nowadays, because of their use as gate dielectrics in MOS capacitors and especially in DRAMs, is implicated by technology roadmaps. Their electrical characterization in terms of capacitance measurements is common practice for determination of parameters, reliability behaviour and for modelling the performance of devices. In our case, for measurements we use Visual Engineering Environment Programming (VEE Pro) software, a graphical dataflow programming software environment from Agilent Technologies for automated test, measurement, data analysis and reporting. VEE Pro interface is very suitable for simultaneous and succesive measurements of different parameters, that need swiching on/off various instruments and devices involved in the measurement. The obtained data were graphicaly represented and analyzed with OriginPro 8.1, data analysis and graphing software. For calculation of the value of capacitance in accumulation C_{acc} Kar's method for extraction is used, and then the flatband voltage V_{FB} and fixed oxide charge Q_{ox} are calculated.

Keywords: VEE programing, capacitance measurement, high-κ dielectric.

1 Introduction

The silicon industry has been scaling SiO_2 aggressively for the past decades for low-power, high-performance CMOS logic applications. SiO_2 as thin as 1.2 nm (physical T_{ox}) has already been successfully implemented in the 90 nm logic technology node [1]. However, continual gate dielectric scaling will require high-κ material, as SiO_2 is eventually running out of atoms for further scaling. Most of the high-κ gate dielectrics investigated are Hf, Zr and Ta-based [2-4]. Common

Lihnida Stojanovska-Georgievska
Laboratory of Physics, Faculty of Electrical Engineering and Information Technologies, Skopje, Macedonia

Nenad Novkovski
Institute of Physics, Faculty of Natural Sciences and Mathematics, Skopje, Macedonia

L. Kocarev (Ed.): ICT Innovations 2011, AISC 150, pp. 301–310.
springerlink.com

observation for most high-κ dielectrics deposited on Si substrate is the formation of intermediate SiO_2, which is inevitably grown on the interface of Si and high-κ dielectric [5]. Both polycrystalline Si and metals are being evaluated as gate electrodes for the high-κ dielectrics [6, 7]. Characterization of MOS/MIS capacitors containing high-κ gate dielectric involves determination of equivalent oxide thickness, dielectric parameters, preexisting and stress generated electrical charges, leakage current, flatband and threshold voltages and other characteristics that determine reliability behaviour of the devices. The measurements are realized by multiple interconnected instruments and devices, which demand simultaneous or successive switching on/off, adjusting and ramping of the values etc. The motivation in this work is to find the most adequate way to arrange these actions, by using programmed software for controlling and governing the measurement process. Virtual Engineering Environment Programming (VEE Pro) is a graphical dataflow programming software development environment from Agilent Technologies for automated test, measurement, data analysis and reporting. In this work we have performed electrical characterization in terms of capacitance measurements of MIS capacitors containing high-κ dielectric, using VEE Pro software, which is already proven to be a suitable tool for data acquisition and processing [8].

2 Virtual Engineering Environment Programming

Agilent VEE is a graphical dataflow programming software development environment from Agilent Technologies for automated test, measurement, data analysis and reporting. VEE originally stood for Visual Engineering Environment and developed by HP designated as HP VEE; it has since been officially renamed to Agilent VEE. Agilent VEE has been widely used in various industries, serving the entire stage of a product lifecycle, from design, validation to manufacturing. It is optimized in instrument control and automation with test and measurement devices such as data acquisition instruments like digital voltmeters and oscilloscopes, and source devices like signal generators and programmable power supplies. A VEE program consists of multiple connected VEE objects (sometimes called devices). Each VEE object consists of different types of pins, namely data pins, sequence pins, execute pins (XEQ), control pins and error pins [9]. Data pins govern the data flow propagation while sequence pins determine object execution order.

Within a VEE program, there are multiple connections between objects and data flows through object from left to right while sequence flows from top to bottom. When an object executes, it uses the input pin's value to perform an operation. When it finishes, the result is placed on the output pin. The output pin value placed is then propagated to any input pins that are connected to it. A sequence pin is used to specify some object execution order.

Agilent VEE can interact with other programming languages using the built-in ActiveX Automation Server. Other software development programs such as Visual Basic, C/C++, Visual C# and all .NET compliant languages can call Agilent VEE User Functions [10]. Access to over 2500 MATLAB analysis and visualization functions is made possible with the built-in MATLAB® Signal Processing

Toolbox. The built-in Microsoft Excel library provides direct access to save, retrieve and generate reports in spreadsheets.

3 Measurement and Analysis of *C-V* Measurements of High-*κ* Dielectrics

3.1 Capacitance Measurements

Many electrical techniques have been developed over the years to characterize gate dielectric quality [11, 12]. In particular, various current-voltage (*I-V*) characterization methods have been developed, including the following:

- Time-Dependent Dielectric Breakdown (TDDB), Charge to Breakdown (QBD)
- Tunneling Current (Fowler-Nordheim or Direct Tunneling
- Stress-Induced Leakage Current (SILC) testing

However, the most commonly used tool for studying gate-oxide quality in detail is the capacitance-voltage (*C-V*) technique. *C-V* test results offer a wealth of device and process information, including bulk and interface charges and many MOS-device parameters.

For our study we performed capacitance-voltage (*C-V*) measurements of the samples using VEE Programming software. The samples were metal/Ta_2O_5/Si structures, obtained by thermal oxidation of Ta target and deposition over Si substrate. MOS structures were obtained by evaporating top electrodes with different areas and common backside Al electrode (Fig. 1).

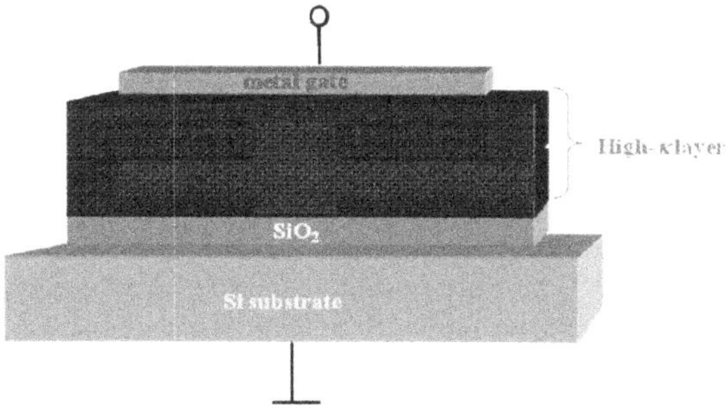

Fig. 1 Schematic representation of metal/high-*κ*/Si structures

The studied samples had three different gate electrodes Al, W and Au. They we studied in terms of capacitance-voltage measurements, performed at different frequencies, in the range from 10 kHz to 1 MHz.

The interface of the program is presented at Fig. 2, where the main view is shown. Here all the settings of instruments included in the measurements and adjustments of initial parameters, are done, in an interactive, user friendly manner.

The programming with VEE Pro software is step by step process, reliably guided through the successive stages, like imputing the range of voltages, number and length of steps, frequency range etc. Using VEE Pro software reading on program window and graphing can be automatically recorded into the data worksheet, wherefrom parameters as oxide thickness, flat band capacitance and flat band voltage, threshold voltage and metal-semiconductor work function difference can be automatically calculated using standard formulas.

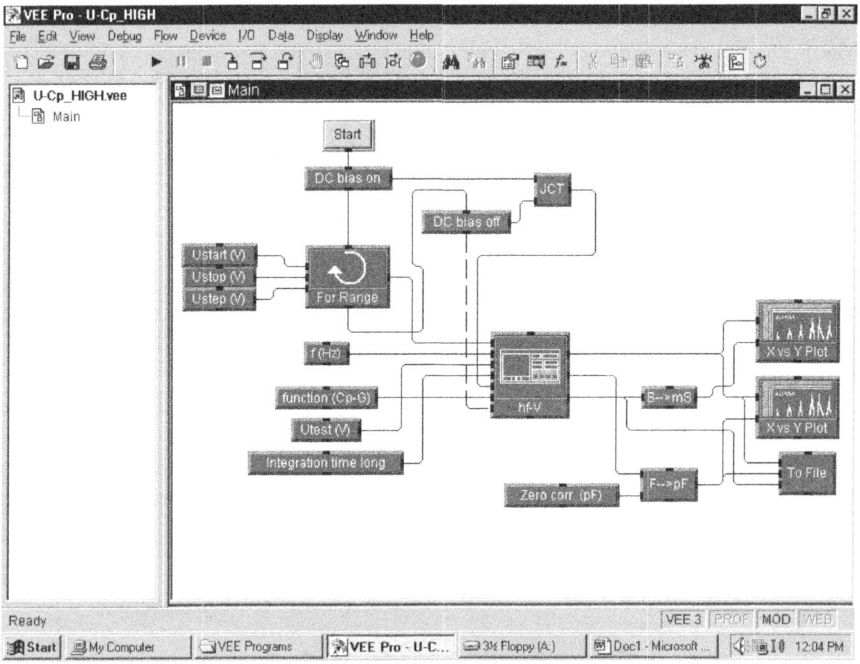

Fig. 2 Interface of VEE Pro software

The appearance of the interface during measurement of capacitance-voltage curves is shown on Fig. 3. This curve refers to p-type Si, where three characteristic regions of *C-V* curve can be noticed: accumulation, depletion and inversion. For p-type semiconductor capacitance in accumulation is measured at negative polarities. *C-V* curve for n-type samples is mirror image of the previous one. The real importance of *C-V* measurement techniques is that a large number of device parameters can be extracted from a high-frequency and quasistatic *C-V* curve: oxide thickness, flatband capacitance and flatband voltage, threshold voltage, effective and total bulk oxide charge. On the Fig. 3 we can notice that the main window shows the voltage dependence of two different physical quantities: capacitance and conductance. This figure is an example of one of the most frequently used features of VEE Pro programming.

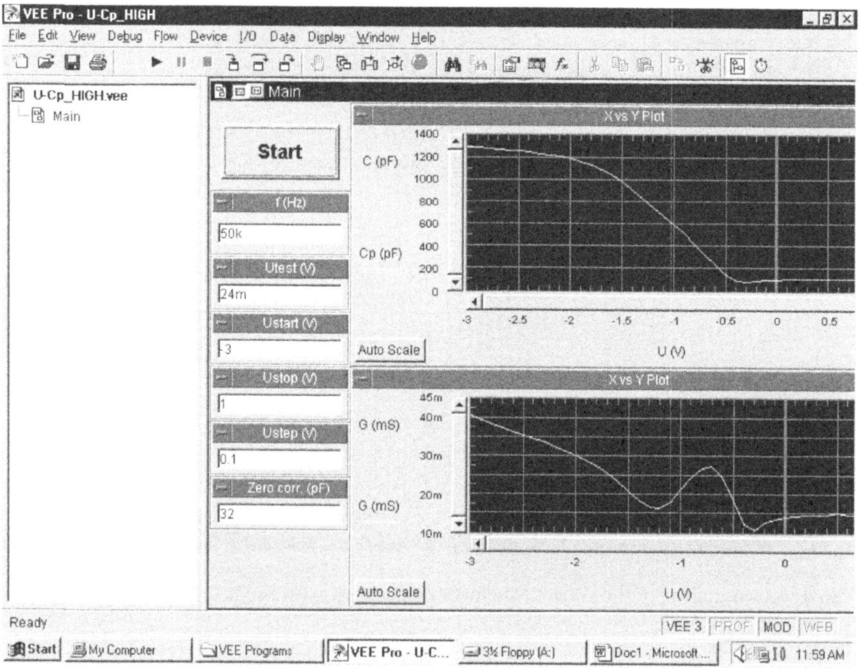

Fig. 3 Measuring *C-V* curves via VEE Pro software

3.2 Extracting the Capacitance in Accumulation C_{acc}

In the case of high-κ layers, it is an often situation, that accumulation cannot be achieved during the measurements, especially when considering ultrathin layers, when higher voltages are threatening to degrade and cause breakdown of the films. There is a technique for extraction of capacitance in accumulation regime developed by Samares Kar [13]. This method for extraction of the capacitance of ultrathin high-κ gate dielectrics is based on the assumption that the space charge and the interface trap capacitances are exponential function of the surface potential. A very useful feature of the proposed technique is that the linear fit (Fig. 4) to data points obtained by the measurement of high frequency *C-V* plots (the extraction plot) for any part of the accumulation regime has the same intercept and slope:

$$\left|\frac{1}{C}\frac{dC}{dV}\right|^{\frac{1}{2}} = \frac{|\beta|^{\frac{1}{2}}}{C_{acc}}(C_{acc} - C) \tag{1}$$

Here the intercept with the *C*-axis yields the gate dielectric capacitance, C_{acc}. *C-V* curves were measured for the sample with Au gate electrode [14].

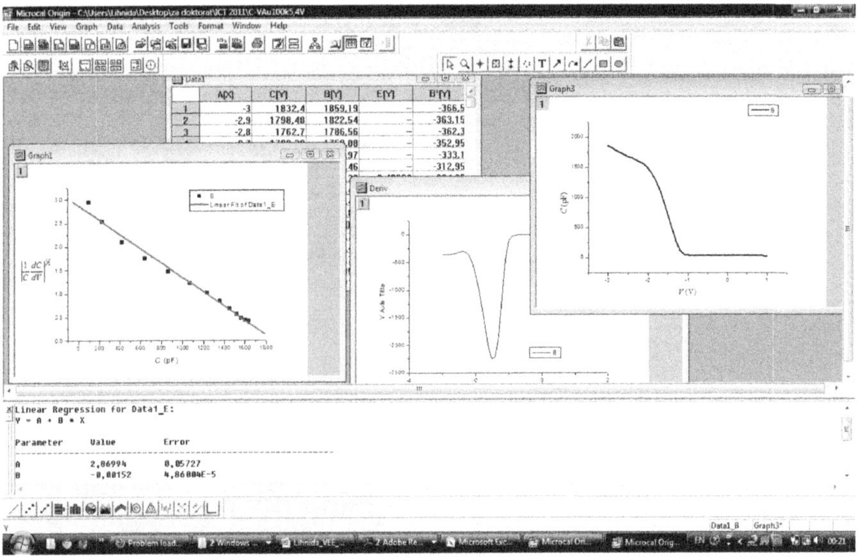

Fig. 4 Determination of the capacitance in accumulation with Kar's technique

Graphing of the data from the VEE Pro data worksheet, we performed using OriginPro 8.1, a software application with tools for data analysis, publication-quality graphing, and programming. OriginPro 8.1 contains powerful tools for all analytic needs, including peak analysis, curve fitting, statistics, and signal processing. To make data analysis more efficient, Origin supports many common formats for importing data, and exporting results. The interface of the program is as it is shown on Fig. 4. We can see the data sheet, where the value can be imported by external sources or calculated using sophisticated and comprehensive mathematical operations, than graph windows, which can be arranged and exported in appropriate format, and also the active window for results from the analysis performed.

3.3 Extracting of Flatband Voltage V_{FB}

Application of a certain gate voltage, the flatband voltage (V_{FB}) results in the disappearance of band bending. At this point, known as the flatband condition, the semiconductor band is said to become flat [15]. Because the band is flat, the surface potential is zero (with the reference potential being taken as the bulk potential deep in the semiconductor). Flatband voltage and its shift are widely used to extract other device parameters, such as oxide charges. We can identify V_{FB} from the *C-V* curve. One way is to use the flatband capacitance method. In this method, the ideal value of the flatband capacitance (C_{FB}) is calculated using equations

$$C_{FB} = \frac{\frac{C_{acc}\varepsilon_s A}{\lambda}}{C_{acc} + \frac{\varepsilon_s A}{\lambda}} \tag{2}$$

where C_{FB} is the flatband capacitance, C_{acc} is the oxide capacitance, ε_s is the permittivity of the substrate material, A is the gate area and λ is the extrinsic Debye length, as calculated:

$$\lambda = \sqrt{\frac{\varepsilon_s kT}{q^2 N_x}} \tag{3}$$

where kT is the thermal energy at room temperature, q is the electron charge and $N_x = N$ at 90% W_{max} [16].

Once the value of C_{FB} is known, the value of V_{FB} can be obtained from the *C-V* curve data, by interpolating between the closest gate-to-substrate (V_{GS}) values.

3.4 Calculation of Oxide Charges Q_{eff}

The effective oxide charge (Q_{eff}) represents the sum of the oxide fixed charge (Q_f), the mobile charge (Q_m), and the oxide trapped charge (Q_{ot}). The calculation of Q_{eff} is based on the assumption that the charge is located in a sheet at the silicon-to-silicon dioxide interface. From Nicollian and Brews [15], we have:

$$V_{FB} - W_{MS} = -\frac{Q_{eff}}{C_{acc}} \tag{4}$$

where V_{FB} is the flatband voltage, W_{MS} is the metal-semiconductor work function, and C_{acc} is the oxide capacitance.

During voltage stress of the MOS capacitor, a shift of the *C-V* curves (and flatband voltage) along the voltage axis is observed [17]. This shift is due to the oxide charges generated with the stress. The value of the flatband voltage shift ΔV is given with the following expression:

$$\Delta V = \frac{\Delta Q_{ox}}{C_{acc}} \tag{5}$$

where ΔQ_{ox} is oxide charge generated during the stress and C_{acc} is the capacitance in accumulation. High-frequency *C-V* characteristics obtained when sweeping the voltage from accumulation to inversion, from -3 V to +1 V, after consequent CVS are showing the expected shift along the voltage axis (Fig. 5). Using the equation (5) and the value of the voltage shift from high-frequency *C-V* characteristics, we have calculated the generated oxide charge in the dielectrics (Fig. 6).

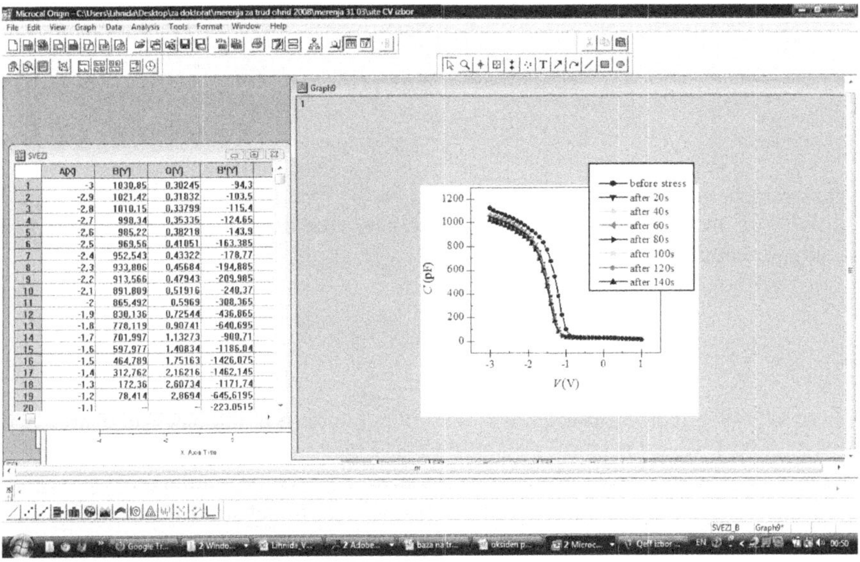

Fig. 5 Voltage shift during consequent CVS

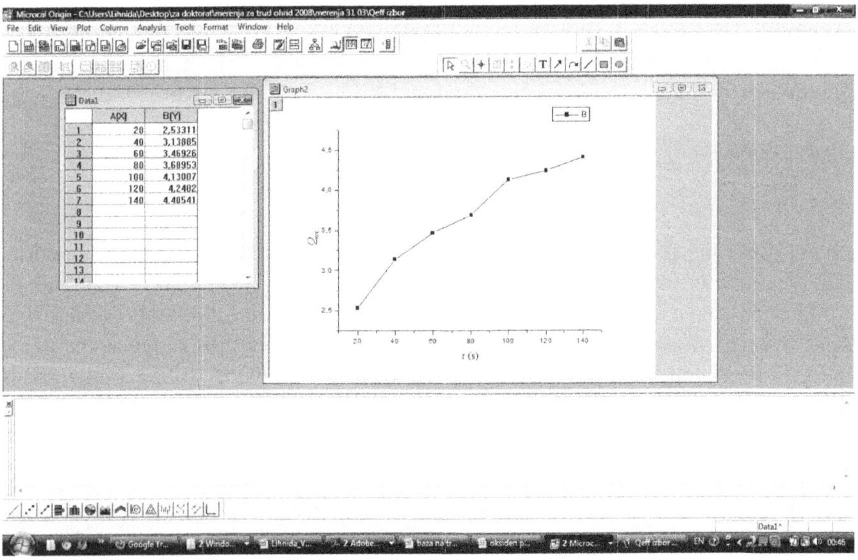

Fig. 6 Calculated values of oxide charges generated during CVS

4 Conclusions

The *C-V* characteristics of MOS capacitors contain a wealth of information about the semiconductor characteristics, which extend to MOSFETs. Here we have presented a case of experimental investigation of particular samples, supported by two comercial softwares, Agilent VEE and Origin Pro 8.1.

The powerful and user-friendly features of the VEE Pro Programming software, when teamed with a *C-V* analyzer, are enabling simultaneous and mesurements govern by the software, in cases when the direct contact with the samples is imposible or undesirable. The program can interact with many other programming languages including MATLab, which favors for calculations and further data treatment.

The use of Origin Pro 8.1 graphing software for graphical presentation of experimental data, its analysis, and high publication-quality graphing, is also very convinient, because it contains powerful tools for all analytic needs, including peak analysis, curve fitting, statistics, and signal processing and it also supports many common formats for importing data, and exporting results.

Acknowledgments. The authors are exceptionally grateful to Professor Elena Atanassova from the Institute of Solid State Physics, Bulgarian Academy of Sciences, for providing the samples of very high quality used in this study.

References

1. Chau, R.: Advanced Metal Gate/High-k Dielectric Stacks for High-Performance CMOS Transistors. In: American Vacuum Society 5th International Conference on Microelectronics and Interfaces, Santa Clara, California, pp. 1–3 (2004)
2. Choi, R., et al.: Fabrication of high quality ultra-thin HfO_2 gate dielectric MOSFETs using deuterium anneal International Electron Devices Meeting. In: IEDM 2002. Digest, pp. 613–616 (2002)
3. Lucovsky, G., Rayner, B., Zhang, Y., Whitten, J.: Experimental determination of band offset energies between Zr silicate alloy dielectrics and crystalline Si substrates by XAS, XPS and AES and ab initio theory: a new approach to the compositional dependence of direct tunneling currents. In: International Electron Devices Meeting, IEDM 2002. Digest, pp. 617–620 (2002)
4. Inumiya, S., et al.: Fabrication of HfSiON gate dielectrics by plasma oxidation and nitridation, optimized for 65 nm mode low power CMOS applications. In: Symposium on VLSI Technology, Digest of Technical Papers, pp. 17–18 (2003)
5. Atanassova, E., Paskaleva, A.: Challenges of Ta_2O_5 as high-k dielectric for nanoscale DRAMs. Microelectron Reliab. 47, 913–923 (2007)
6. Atanassova, E., Paskaleva, A.: The effect of the metal electrode on the electrical characteristics of Ta2O5 capacitors for DRAM applications. In: Duenas, S., Castan, H. (eds.) New Materials and Processes for Incoming Semiconductor Technologies, pp. 77–111. Transworld Research Network, Kerala (2006)
7. Novkovski, N.: Conduction and charge analysis of metal (Al, W and Au)-Ta_2O_5/SiO_2-Si structures. Semicond. Sci. Technol. 21, 945–951 (2006)
8. Novkovski, N.: The use of the visual programming language HP VEE in data acquisition and processing. In: SEFI WGP Seminar EMEPE 1998, Brno, Czech Republic, pp. 78–81 (1998)
9. VEE Pro User's guide – Agilent Technologies, `http://cp.literature.agilent.com/`

10. Srivastava, V.M., Singh, G., Yadav, K.S.: Application of VEE Pro Software for Measurement of MOS Device Prameters using *C-V* curve. International Journal of Computer Applications 1, 43–46 (2010)
11. Novkovski, N., Atanassova, E.: A comprehensive model for the *I-V* characteristics of metal Ta_2O_5/SiO_2-Si structures. Appl. Phys. A 83, 435–445 (2006)
12. Novkovski, N., Atanassova, E.: Origin of the stress-induced leakage currents in Al-Ta_2O_5/SiO_2-Si structures. Appl. Phys. Lett. 86, 152104 (2005)
13. Kar, S.: Extraction of the capacitance of ultrathin high-*k* gate dielectrics. IEEE Trans. Electron. Dev. 50, 2112–2119 (2003)
14. Stojanovska-Georgievska, L., Atanassova, E., Novkovski, N.: Alteration of *C-V* characteristics of metal-Ta_2O_5-Si capacitors by constant voltage stress. Physica Macedonica 58, 107–112 (2008)
15. Sze, S.M.: Physics of Semiconductor Devices, 2nd edn. Wiley, New York (1985)
16. Nicollian, E.H., Brews, J.R.: MOS Physics and Technology. Wiley, New York (1982)
17. Stojanovska-Georgievska, L., Novkovski, N., Atanassova, E.: Study on the frequency dependence of *C-V* measurements of unstressed HfO_2-Ta_2O_5 stacks. Physica Macedonica (2010) (in press)

Towards an Integrated Model of Optimizing the Efficiency of IT in Organizations

Fisnik Dalipi and Ajri Shej

Abstract. Nowadays, many organizations begin to implement internationally recognized frameworks or methodologies that are all aiming at optimizing the efficiency of IT: increasing speed, quality, benefits or reducing costs, they all pretend to be THE model to optimize the use of IT. But, which one to choose? Can an organization implement more than one framework/methodology? Are they complementary? What does the reality reflect? To answer those questions, we are going to briefly compare five of the most famous well-recognized frameworks/methodologies: IT Governance Framework; Capability Maturity Model (CMM); Control Objectives for Information and related Technology (COBIT); Information Technology Infrastructure Library (ITIL) and, finally the well known Total Quality Management (TQM). We will examine to what extent they are different, similar or complementary. As a conclusion, we will propose and design a model that could eventually be used to inderstand how these five frameworks complement or fit with each other.

Keywords: Information Technology, IT Governance Framework, Capability Maturity Model (CMM), Control Objectives for Information and related Technology (COBIT), Information Technology Infrastructure Library (ITIL), Total Quality Management.

1 Introduction

This paper, among others, prove that even though many people tend to think that the existing frameworks are mutually exclusive, they have elements in common

Fisnik Dalipi
Department of IT, Faculty of Math-Natural Sciences, Tetovo State University
e-mail: fisnik.dalipi@unite.edu.mk

Ajri Shej
South East European University
e-mail: ajri.shej@gmail.com

L. Kocarev (Ed.): ICT Innovations 2011, AISC 150, pp. 311–323.
springerlink.com

and can, sometimes, complement each other. It is undoubtedly evident that each of those frameworks is more appropriate for particular situations; it all depends on the initial problematic. Indeed, sometimes those models tend to be similar in some points and could, consequently, enforce each other; on some other points they are different and could complement each other.

2 TQM and the Four Areas of IT Governance Framework

The main idea behind IT Governance is the link between IT and business, more precisely, "the link of IT with the present and future business objectives" [1] in order to enhance the value of IT investments delivered to the business. The IT Governance Institute (2001) defines IT Governance as follows: "the responsibility of the Board of Directors and executive management. It (IT Governance) consists of the leadership and organizational structures and processes that ensure that the organization"s IT sustains and extends the organization"s strategy and objectives" [2]. Another important part of IT Governance is the attribution and definition of decision rights and accountability to ensure that the investments made in IT support the long term and short term business objectives: "IT Governance is not about what specific decisions are made. That is management. Rather, governance is about systematically determining who makes each type of decision, who has input to a decision and how these people are held accountable for their role" [3]. The starting point or driving force of the IT Governance Framework is the stakeholder value. The framework consists of 4 main areas [1]: Strategic alignment, Performance measurement, Value delivery and Risk management.

The first two are drivers whereas the other ones are outcomes. Another distinction needs to be made regarding those two last elements: value delivery focuses on the creation of business value whereas risk management concerns the preservation of business value [1].

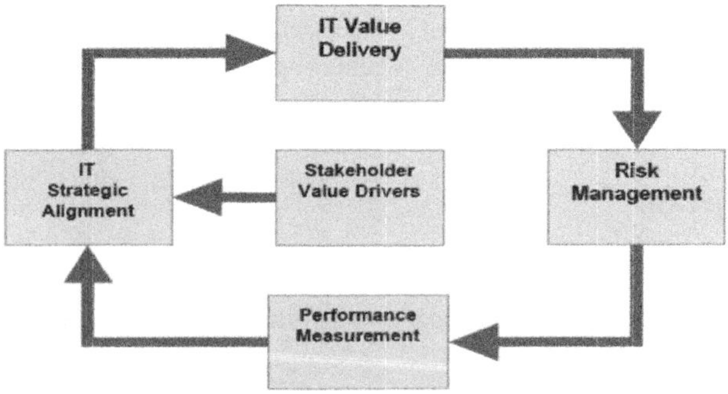

Fig. 1 IT Governance Framework (Van Grembergen W., De Haes S., Guldentops E. 2004)

As shown by the Figure 1, the IT strategic Alignment is a key driver to achieve business value (outcome). In its turn, performance measurement is a key driver to realize the strategic alignment [1].

In the recent years, managers have understood the increased importance of customer-defined quality in the production of a product or a service in order to meet customers" expectations. This gave birth to a new methodology: Total Quality Management or TQM [4]. TQM methodology is different in the sense that it is a proactive approach that builds quality into the process [5] and search the root causes of quality problems; „detection" has replaced „inspection" [6].

The importance of a quality product or service as defined by the Total Quality Management philosophy is that it has to satisfy customer"s demands and expectations, whether it is low price, prestige, usefulness or reliance of the product or service. As a result, there is not one universal definition of quality as it ultimately depends on what the customers mean by „quality". At the heart of TQM philosophy appear several elements, such as: employee empowerment, contonuous improvement, customer focus, etc.

According to Prof. Klefsjö (2003) "Six Sigma is a methodological way of working that supports" the TQM philosophy [7]. Indeed, Six Sigma is a methodology that uses statistical methods to assess the current quality process and, if need be, offers opportunity for improvements. Once the factors contributing the most to the variability of the process have been identified, there will be opportunities to detect the causes of this variability and to reduce it [8]. The result of this is a reduction in product defects [9], an increase in the quality of the product, a cost reduction, a better productivity, better employee morale and increased profits [8].

Finally, one of the big advantages of this methodology is that it allows to continuously keep a customer focus and, hence, to ensure a long term relationship between the firm and the customer [8].

In the frame of IT Governance Framework, Six Sigma in particular and TQM in general can play an important role in the alignment of business objectives with IT objectives. Indeed, as mentioned in the TQM philosophy, Six Sigma offers a "non subjective mechanism to communicate service quality improvement to business people, in business terms" [10]. As a result, it facilitates the **alignment** between business and IT goals. It also contributes to the mitigation of operational risks as it builds quality within the process. This refers to the **risk mitigation**"s part of the IT Governance Framework. Additionally, it offers the possibility to work with statistical tools and measure the **performance** of every process. Finally, it prioritizes the improvements initiatives of those processes that are critical to the success of the organization and, hence, has a positive impact on the **value delivered** to the business [10].

3 The Link between IT Governance Framework, COBIT and ITIL

3.1 IT Governance and COBIT

In its turn, COBIT is preoccupied with the alignment of IT with the business objectives. Indeed, it details 34 high-level objectives related to four domains which all aim at the satisfaction of one or many information criteria also called business needs. In other words, it helps to align the IT domain with the business domain which represents the so-called „strategic alignment" of the IT Governance Framework. It also includes some components of the „performance measurement" part. Indeed, COBIT advices to make use of Key Goals Indicators to measure to what extent the High-Level objectives have been achieved or the Key Performance Indicators to show „how well" the process is performing). COBIT also includes a „Risk Management" part as it, for example, provides audit guidelines to substantiate the risk of the minimum control objectives not being met or helps managers to balance risk and control of investment [11]. To summarize, COBIT is concerned with all the areas of the IT Governance Framework but provides more guidelines regarding the „how" to apply this four areas. Nevertheless, it remains quite general leaving a big part to the appreciation of the individuals. Finally, we have seen that COBIT ensures the alignment between IT and Business goals by first defining the business goals, then the IT goals, process goals, etc. and then ensuring that they all support each other. The definition of business goals mentioned in COBIT can be viewed as the „business plan" mentioned in the IT Governance framework.

3.2 IT Governance, COBIT and ITIL

The Information Technology Infrastructure Library or ITIL is seen as the de facto standard for IT Service Management (OGC, 2002, cited by [12]). IT Service Management is "the delivering and supporting of IT Services that are appropriate to the business requirements of the organization" (OGC, 2002, cited by [12]) or, similarly "the principles and practices of designing, delivering, and maintaining IT services to an agreed level of quality, in support of a customer activity" [10]. The goal of ITIL is not to impose business processes in an organization but rather to provide guidelines to plan, provide and support IT services [9].

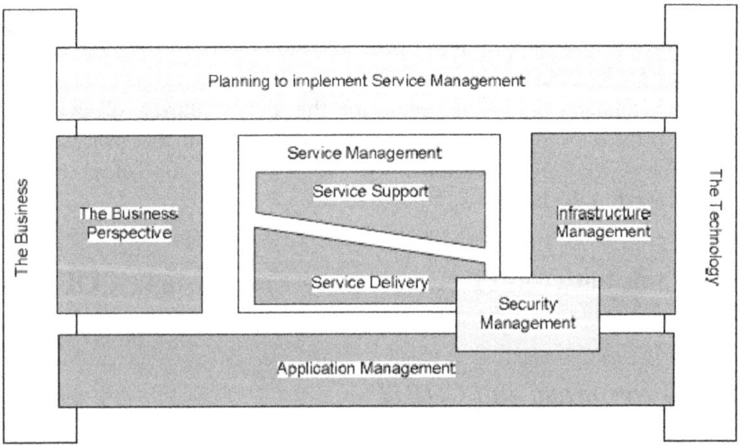

Fig. 2 The ITIL Framework (Sallé M. 2004)

As shown in Figure 2, the ITIL framework consists of consists of seven major areas [13]: Business perspective, Application Management, Service Delivery, Service Support, Infrastructure Management, Security Management, and Planning to Implement Service Management.

The Service Delivery and Service Support are the central elements of this framework. Service Delivery focuses on the delivery of services of quality by the IT to support business objectives and hence provides support to the business users; Service Support is concerned with the support of the users of IT services in day-to-day operations and ensures that those users have access to the right functions in order to support the business [14].

Regarding ITIL now, it is important to notice that, in opposition to what people usually think, COBIT and ITIL are not necessarily mutually exclusive. COBIT provides a more general framework that can suit any organization [15] and ITIL offers a detailed view, covering specific areas. As a consequence, ITIL could be "mapped to the COBIT Framework" [15] and, combined, they could "provide a powerful IT Governance, control and best-practice framework in IT service management" [15]. In fact, COBIT, although process oriented, tells you „what to do" but does not tell you precisely „how to do" it. On the contrary, ITIL focuses on the method and is far more prescriptive than COBIT. Indeed, ITIL defines "best practice processes for IT service management and support" [15]. Below (Figure 3) are illustrated the COBIT processes that are addressed by ITIL; we notice that most of them are part of the „Delivery and Support" domain, which is logical as we have seen that ITIL is mainly concerned with services support and delivery.

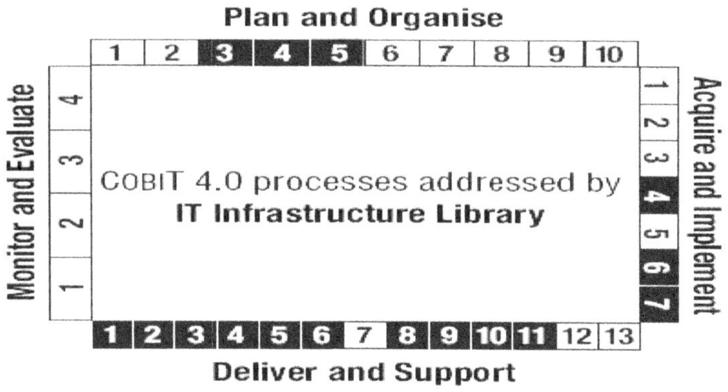

Fig. 3 COBIT processes addressed by IT infrastructure Library (IT Governance Institute 2006)

Finally, we can emphasize, from the definition of Service Support and Service Delivery, that ITIL is concerned with the alignment of IT services with the goals of the organization. Therefore, we can state that ITIL by its essence supports the IT Governance Framework. This supports what has been said above: ITIL is more focused than the COBIT framework but can be mapped to it (overall regarding the delivery and support domain) and they both support the IT Governance Framework.

4 The Link between CMM and Six Sigma

The Capability Maturity Model was developed by the Software Engineering Institute (SEI) at Carnegie Mellon University in around 1986 and illustrates the evolution of software processes through five levels of maturity; from immature levels to more mature ones. The CMM deals with the "planning, engineering and managing of software development and maintenance". As a result, if the key elements of each part are put in place, the CMM can help an organization to achieve their goals in a more effective and efficient way, meaning at lower costs, on schedule, with the right functionalities and the adequate quality. This model can be useful to benchmark a company against the industry in general as well as to address the planning of improvements of software process [16].

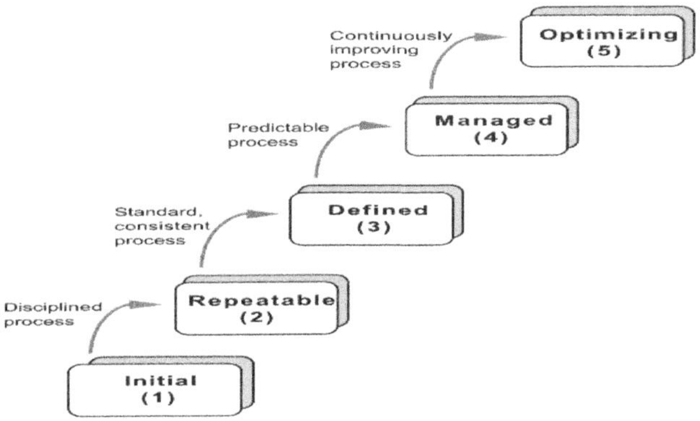

Fig. 4 Maturity Levels (Paulk M.C., Weber C.V., Garcia S.M., Chrissis M.B., Bush M. 1993)

At the first level, the environment proposed by the organization is not stable and does not permit to develop and maintain software. The organizations that stand at level 1 are incapable of predicting the software process capability: whether it will be delivered on time, within budget, with the right functionality and quality remains unpredictable and fundamentally depends on the ability and qualifications of individuals [16]. At the second leve, the environment becomes stable, allowing the planning and management of software developments to be based on former and successful experience but there is no evidence of formal training or communication of the standardized processes [16]. The third leve deals with the policies and procedures established within an organization to develop and manage software are well-defined, documented and communicated through formal training in order to ensure that every individual in the company is aware of what is expected from [16]. At the fourth level, quality goals are set by the company regarding software products and processes. In addition to that, the organization measures productivity and quality and set up a database to collect and analyze the available data from the defined software processes mentioned in Level 3. At the last level all processes and all individuals in the organization focus on continuous improvement. The objective is to underline weaknesses and analyze defects and to find, in a proactive manner, their root causes and correct those for improvement. In addition to this, the organization identifies best software engineering practices and propagates those into the entire organization.

4.1 Similarities

The Capability Maturity Models and Six Sigma have elements in common. Indeed, they are both striving for continuous improvement by trying to proactively eliminate the root causes of defect and quality problems. We had already clearly stated this for Six Sigma, but it is also true for CMM. Indeed, the organizations at higher maturity levels incur a great reduction in the time spent on rework and, consequently, free up time to be spent on other projects. According to [17], "IS organizations achieving at least a level 2 stage of maturity demonstrate a significant decrease in effort and numbers of problems reaching customers".

In addition to that, they both try to get processes under control and to minimize the variation in process performance in order to make the processes more reliable and increasingly efficient [18], What is more, they both rely on quantitative approaches to make decision regarding continuous improvement and errors detection [19]. In other words, CMM and Six Sigma both focus on "reducing variation, quantifying performance, and improving processes" [19].

4.2 Differences

In spite of those similarities, they differ on some points. The first difference finds its origin in the definition of defects. They both strive to reduce the number of errors, but Total Quality Management defines a defect as "any problem with service performance or a product fault that does not meet the customer"s specifications" [20]; the goal is to reduce the number of defects to 3,4 per million. On the other hand, CMM states that a defect is an action that has negative consequences or impact the process so that the work cannot be completed properly [20]. The goal of CMM is to improve the quality of the development process of a software product so that the quality of the latter progresses [20]. In other words, the difference lies in the fact that CMM is only designed to improve IT software processes whereas Six Sigma is implemented in order to improve all processes [19].

Another difference is the length of the Six Sigma project relative to the duration of CMM. Indeed, traditionally, a six sigma project lasts 3 to 6 months. On the contrary, to achieve the improvements that can be derived from the Capability Maturity Model, one must apply this model on the long term.

4.3 Complementarities

While adopting the Capability Maturity Model, it is difficult to focus the improvements of the process on customers" expectations [18-19]. As we have seen, one of the main reasons Six Sigma methodology"s success is that it forces you to stay customer focused. In that way, both models can be complementary and deliver greater improvements than if they were applied individually [18]. Indeed, if one applies Six Sigma to support the Capability Maturity Model, it may help to increase the performance of a process by using measures that are meaningful to, not only the customers, but also the organization. A difference between Six Sigma and CMM is that the latest, in opposition to Six Sigma, does not provide a methodology. Indeed CMM is a framework that specifies goals to achieve in order to progress in the frame of software process development [19] but it does not tell you how to do it. In other words, CMM tells you „how" to develop software, and Six Sigma specifies „what" to develop [19]; both could therefore be complementary.

5 The Link between ITIL and Six Sigma

Both ITIL and Six Sigma are designed to improve quality and enhance the satisfaction of customers [10]. They are often presented as mutually exclusive. However, there is some evidence that ITIL and Six Sigma are complementary and can be used together to facilitate the alignment between business and IT objectives and enhance the IT value delivered to the business [9].

ITIL tells the management „what to do" in the frame of Service Management but does not clarify the way to do it. The authors underline that "it is up to the IT staff to flesh out the details of process flow and create detailed work instructions, all in a way that makes sense for their organization" [10]. In this way, Six Sigma can be a useful agent to tell the organization „how" to improve quality. As a result, "together, they make a great combination for improving the quality of IT service delivery and support" [9].

Six Sigma offers statistical tools and techniques designed to support continuous improvements of quality services and products. It also allows the business to clearly state the success achieved with ITIL by means of metrics [9]. Hence, Six Sigma provides a methodology to measure the outcomes of the processes which can later be useful to improve any process [10]. In this way, Six Sigma is complementary with ITIL as this later framework does not provide such metrics. As stated by [9], the combination of the two methodologies helps "organizations to adopt best practices for service delivery and support [ITIL] by a quality process which ensure its success [Six Sigma]". Consequently, the impact of ITIL on the business will be greater when combined with Six Sigma and hence the benefits to the business will be higher.

6 Comparisons towards the Integrated Model

It seems important, at this point, to compare all the models seen until here. Beyond, you will find a summary table. The „plus" sign tells us the importance that the criterion holds in a given framework or methodology. One „plus" tells us that there is no real connection between this criteria and the corresponding framework, two „plus" indicates a relation and three „plus" illustrates that there is a real and strong relation and that this criteria is essential and a core element in the corresponding framework.

Table 1 Comparison between the methodologies

	Customer focus	IT oriented	What to do	How to do	Quantity	Business-IT alignment focused	Process griven
IT Governance	+	++	+++	+	++	+++	++
CMM	+	++	++	++	++	++	+++
ITIL	++	+++	++	+++	+	+++	+++
COBIT	+	++	+++	++	++	+++	+++
TQM	+++	+	++	+++	+++	+	+

+	No real connection
++	Related
+++	Strongly related

This table is a „model" that could eventually be used to understand „how" those five frameworks complement each other. The weaknesses of one framework are indeed susceptible of being fulfilled by other models" strengths. Of course, other criteria as well as additional frameworks could be added to this table. It is here important to notice that the scoring of this table is very subjective and can or is to be discussed and that a tool, to be professionally used, needs to be approved and validated, which is not the case here.

This diagram in Figure 5 has been designed in order to better understand how the five frameworks described above fit with each other. We will start with the bottom of the pyramid and then rise up to the top. We will first exclude TQM and Six sigma from the discussion and integrate them afterwards as they are more business oriented.

ITIL is at the bottom of the pyramidal construction; it is the more prescriptive and specific framework that we have seen. It focuses mainly on IT services and delivery and on the alignment between the IT services and the business goals of an organization. Support and Delivery are one of the four domains that are represented in COBIT. Consequently, it is in this area that ITIL complements COBIT: by being more specific and more prescriptive in the area of services support and delivery.

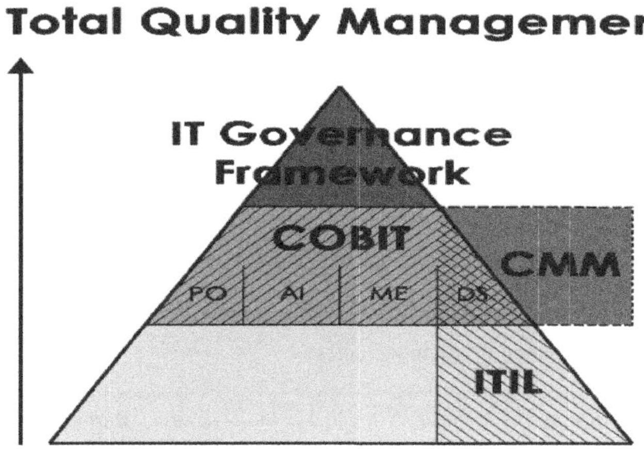

Fig. 5 Putting together all the methodologies

COBIT also consists of three other domains: Plan and Organize (PO), Acquire and Implement (AI) and Monitor and Evaluate (ME) which represent the second level of the pyramid. Related to those four domains are 34 high-level control objectives (related to 34 processes) that are all aiming at the satisfaction of business needs. Therefore, we can state that the COBIT framework aspires at improving the alignment between the IT goals and the business goals of an organization and hence improves the „strategic alignment" addressed by the **IT Governance Framework**. COBIT complements this lastly mentioned framework in other ways. In fact, it tells you into more details how to apply the main areas of the IT Governance Framework, that are Strategic Alignment (for the reason mentioned before), Risk Management (helping to mitigate the risk of the minimum control not being in place, advising managers to balance risk and control investments, or again mitigating the risk of a misalignment between business and IT goals), Performance Measurement (by using KPI and KGI to measure the performance of the processes in place for example).

In its turn, the **Capability Maturity Model** or CMM can be seen as the output of the COBIT Framework. It is in fact an evaluation of how COBIT is used. There exists one tailored CMM for each of the 34 high-level control objectives; it consists in scoring from 1 to 5 the way the different objectives and hence processes have been implemented; from ad hoc processes to documented, communicated, standardized and measured processes. As we have seen how ITIL addressed some of the COBIT processes, how ITIL but particularly COBIT focused on the strategic alignment (and also, for COBIT, on risk management and performance measurement), that way supporting the IT Governance Framework, and finally how the Capability Maturity Model was used in the frame of COBIT, we can at present move towards analyzing the position that Total Quality and Six Sigma occupy in this IT environment.

As mentioned before, **TQM** is more oriented towards business. However, it has been demonstrated that TQM and Six Sigma methodologies could, in some ways, complement the four frameworks that constitute the above pyramid. Six Sigma complements ITIL in the way that it guarantees the quality of the adopted practices in terms of service delivery and support; it complements the CMM in the way that, as the CMM it focuses on reducing variation, quantifying performance, and improving processes but it puts emphasis on the „critical-to-quality" factors meaning those factors that are critical to the success of the process concerned. Six Sigma could then allow an organization to move faster through the levels of the Capability Maturity Model. Finally, TQM complements the IT Governance Framework in the way that it introduces a common goal (the customer) in the organization making the alignment of efforts at all levels on the one hand and between IT and business on the other hand, much easier and more efficient. In other words, quality has become a key element in the choice of the customers. Indeed, they are more and more demanding, regarding the quality of a product or service as a competitive advantage. Having the factor „quality" defined as a critical success factor while adopting those frameworks could be an additional strength for the organization implementing them. On top of that, Six Sigma offers a non-subjective mechanism based on data and facts that facilitates the communication and hence the alignment between IT and Business goals. In addition to that, adopting those frameworks can bring multiple benefits to the organization implementing them. One of them, not the slightest, is that it creates standardized, well-documented processes based on best practices, efficient measurement and formal training. Being more and more advanced in the adoption of those frameworks is thus a step towards quality, hence towards Total Quality Management.

7 Conclusions

This paper, among others, prove that even though many people tend to think that the existing frameworks are mutually exclusive, they have elements in common and can, sometimes, complement each other. It is undoubtedly evident that each of those frameworks is more appropriate for particular situations; it all depends on the initial problematic.

Individually, all these methodologies have proved to bring multiple benefits to the organizations adopting them; from a reduction in costs to an increased quality of the product/service offered, but also offering a better communication with business partners, an increased customers" satisfaction, a better ability to monitor, and many others. In this paper, links between those frameworks have been identified; while some of them are more prescriptive, describing step by step „how to do" to increase the efficiency of IT, others are more general and focus on developing „what to do" without specifying exactly the way to achieve it. In that way, the methodologies reviewed in this paper could be complementary.

As one can notice, there are many links between those frameworks and, therefore, there are great opportunities to increase the benefits they offer individually by implementing them together. The importance is to find out how the weaknesses of one framework can be complemented by the strengths of another.

In conclusion, IT, although still a support function in some organizations or departments, is increasingly regarded as a strategic partner. This is overall the case in those organizations where IT is essential to the survival of the company or where the continuity of services is critical to the success of the company. For these organizations, IT is far more than a support function; it is a strategic partner that, if carefully and efficiently managed and aligned through the use of internationally recognized frameworks, can enhance the value delivered to the business and help maintaining the required level of quality.

References

1. Van Grembergen, W., De Haes, S.: IT Governance Structures, Processes and Relational Mechanisms: Achieving IT/Business Alignment in a Major Belgian Financial Group, University of Antwerp (2005)
2. De Haes, S., Van Grembergen, W.: IT Governance structures, processes and relational mechanisms: Achieving IT/business alignment in a major Belgian financiar grup. In: Proceedings of the 38th Hawaii International Conference on System Science. IEEE Computer Security (2005)
3. Brown, A.E., Grant, G.G.: Framing the Frameworks: A Review of IT Governance Research. In: Communication of the Association for Information Systems. vol. 15, Eric Sprott School of Business, Carleton University (2005)
4. Barnes, R., Pike, J.: TQM in Action: A Practical Approach to Continuous Performance Improvement. Chapman & Hall, London (1996)
5. Reid, R.D., Sanders, N.R.: Operations Management: An Integrated approach, 2nd edn. University of New Hampshire (2005)
6. Wilklinson, A., Redman, T., Snape, E., Marchington, M.: Managing with Total Quality Management: Theory and Practice. Macmillan Press Ldt (1998)
7. Lennartsson, M., Vanhatalo, E.: Evaluation of Possible Six Sigma Implementation Including a DMAIC project, Lulea University of Technology (2004)
8. Belkacem, A., Hadi, E.: Puissance Six Sigma. L'Usine nouvelle, Paris, Dunod (2005)
9. Aazadnia, M., Fasanghari, M.: Improving the Information Technology Service Management with Six Sigma. International Journal of Computer Science and Network Security 8(3), Iran Telecommunication Research Center (ITRC) (2008)
10. Den Boer, S.: Six Sigma for IT Management. ITSM Library (2006)
11. Belkacem, A., Hadi, E.: Puissance Six Sigma. L'Usine nouvelle, Paris, Dunod (2005)
12. Office of Government Commerce (2002), `http://www.ogc.gov.uk/`
13. Sallé, M.: IT Service Management and IT Governance: Review, Comparative Analysis and their Impact on Utility Computing, Trusted Systems Laboratory, HP Laboratories Palo Alto (June 2004)
14. Shaw, M.: IT Governance: IT Auditing Framework and Issues Dealing with Regulatory and Compliance Issues, Center for IT and e-Business Management, University of Illinois at Urbana-Champaign (2007)
15. OTGI & OCG (2005)

16. Paulk, M.C., Weber, C.V., Garcia, S.M., Chrissis, M.B., Bush, M.: Key practices of the Capability Maturity Model, Version 1.1, Software Engineering Institute, Carnegie Mellon University Pittsburgh, Pennsylvania (February 1993)
17. Gartner: Describing the Capability Maturity Model, Measure IT, Capability Maturity Model, Special edn. (2001), http://www.e-strategy.ubc.ca/__shared/assets/MeasureIT-GartnersCMMmodel1278.pdf
18. Murugappan, M., Keeni, G.: Blending CMM and Six Sigma to Meet Business Goals. IEEE Software 20(2), 42–48 (2003)
19. Tayntor, C.B.: Six Sigma Software Development. CRC Press (2002)
20. Le Vasseur, C.: Balancing Six Sigma and the Capability Maturity Model, Gartner Research (2008), http://www.gartner.com/4_decision_tools/measurement/measure_it_articles/2003_0424/bal_cmm.jsp

MRI Based Visualization and Evaluation of Carpal Tunnel Syndrome

Saso Koceski, Natasa Koceska, and Filip Shteriev

Abstract. This work presents a software tool for interactive visualization and enhanced evaluation of the carpal tunnel syndrome (CTS), based on MRI. CTS is one of the most common peripheral neuropathies due to the compression of the median nerve. The software provides a fully scalable real-time 3D reconstruction of the tendons and the median nerve, allowing multiplanar reformatting, realistic visualization of the integrity of myelin sheath, and color enhancement, without the use of contrast agents. Provided metrics and measurement tools for different parameters, including median nerve location, size, flattening, but also nerve adjacency and local deformation shape measures were also evaluated and the results are presented.

Keywords: MRI visualization, 3D volume rendering, carpal tunnel syndrome evaluation, metrics.

1 Introduction

The carpal tunnel (Figure 1) is a semi-rigid tunnel located at the base of the hand. The space is approximately 6cm in length from the wrist to the mid-palm. Tendons of the fingers and the median nerve course through the confines of the carpal tunnel and then diverge to the fingers [1].

The median nerve innervates the thumb, index, and long fingers as well as the radial side of the ring finger. The superficial tendons flex the proximal interphalangeal joint of each finger, and the deep tendons flex the distal interphalangeal

Saso Koceski · Natasa Koceska
Faculty of Computer Science, University "Goce Delcev"-Stip,
bul. Krste Misirkov bb. 2000 Stip, Macedonia
e-mail: saso.koceski@ugd.edu.mk, natasa.koceska@ugd.edu.mk

Filip Shteriev
SyNRG Software and IT Solutions, Skopje, Macedonia
e-mail: shteriev@gmail.com

L. Kocarev (Ed.): ICT Innovations 2011, AISC 150, pp. 325–334.
springerlink.com

joint [2]. Because of the rigidity of the boundaries of the carpal tunnel, any process that either increases the volume of the contents of the tunnel or decreases the size of the tunnel can result in symptoms.

Among the group of disorders, described as peripheral compression neuropathies, the carpal tunnel syndrome (CTS) is the most common one. These disorders occur because some peripheral nerves are compressed as a result of certain risk factors including superficial anatomic location, coursing through an area at higher risk of insult, or coursing along a narrow pathway through a bony canal [3]. Due to the anatomic position of the median nerve as it courses through the narrow carpal tunnel, it is at higher risk of insult than many other peripheral nerves.

CTS incidence is 1% in the general population and it is usually observed between 4th and 6th decades with a female to male ratio of 2-5:1. It is characterized by pain and discomfort in the wrist and the hand accompanied by paresthesia at the distribution of median nerve and atrophy of the thenar muscles.

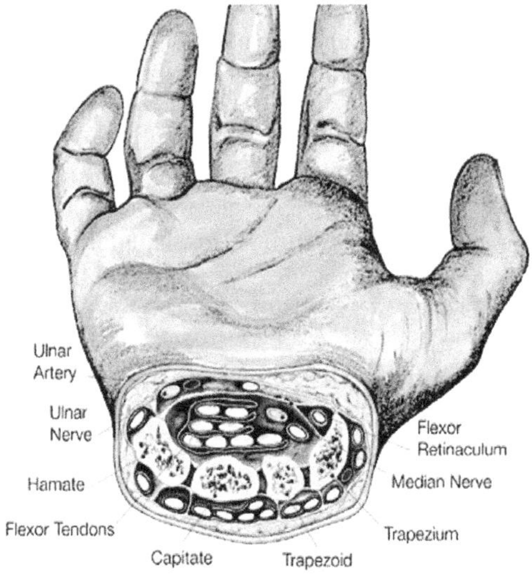

Fig. 1 Anatomy of the carpal tunnel [19].

Generally, symptoms in less severe cases are intermittent and often seem to intensify at night, which may be due to flexion of the wrist during sleeping. However, symptoms in severe cases are unrelenting, and thenar (thumb) functional muscle loss occurs [4]. Certain occupational groups have been identified as having an increased risk of CTS, including farmers, construction workers, factory workers, drivers, typists, cashiers, programmers [5]. The impact of CTS symptoms are felt not only at work, but also at home and are associated with activities involving strength, such as pushing open a window or pushing up from an armchair, and activities involving grip strength, such as writing with a pen or pouring from a container to a glass [6].

The economic burden on patients and their families and society in the form of direct and indirect costs, due to the CTS, is very big. The workers and their families must cope with inability to perform social activities and activities of daily living due to pain, clinical depression, and most significantly the long-term loss of earnings. Employers must deal with indirect costs of production interruption, accident investigation, and the recruiting and training of a new worker to replace the impaired worker [7]. According to the statistical data, in 1995, the economic cost of CTS per year in the United States was estimated to exceed $2 billion [8]. According to the US Bureau of Labor Statistics, there were 11,950 formally reported cases of CTS involving lost work days in 2008 among government workers [9]. All these factors have increased the demand for rapid and accurate detection and diagnosis of CTS.

American Academy of Orthopaedic Surgeons (AAOS) has published guidelines which state that a diagnosis of CTS should be made based on the history of the present illness, physical examination, and electrodiagnostic tests [10]. Recently, most studies have pointed that imaging techniques may provide important information in entrapment neuropathies, particularly in cases of equivocal electrophysiologic studies [11],[12]. Median nerve was firstly visualized on sonography [13]. Because the attenuation difference between the median nerve and other structures within the tunnel are very restricted, sonography is not sensitive enough to visualize the carpal tunnel. Computed tomography is also not sensitive enough for imaging of the carpal tunnel structures [14].

Because of its excellent resolution of soft tissues, MRI has often been used in studying the anatomy of the carpal tunnel to assess possible abnormalities associated with CTS and has been considered as the best imaging modality. Therefore, in this work we are presenting MRI software system that should help for CTS detection. To the best of our knowledge no prior software system that implements an integral set of tools and metrics, for CTS evaluation and detection, has been reported in the literature. The developed software system and its tools and metrics are evaluated and the results are presented.

2 Software Architecture Overview

The overall architecture of the developed software is presented in Figure 2.

Globally three layers can be identified in the proposed architecture:

Data logic layer: This layer implements software routines for reading DICOM files. It supports all kinds of DICOM image files: grayscale and color, single-frame and multi-frame, JPEG-compressed (lossy and lossless), JPEG-2000-compressed (reversible and irreversible), RLE-compressed and ZIP-compressed studies. Special routines for reading and writing proprietary project files as well as routines for report generation and multimedia files generation are also implemented at this level.

Program logic layer: 3D module implements routines for manipulation of three-dimensional volumetric data. To add robust interpretation of data, it provides volume reduction features that can be used to remove irrelevant parts of the

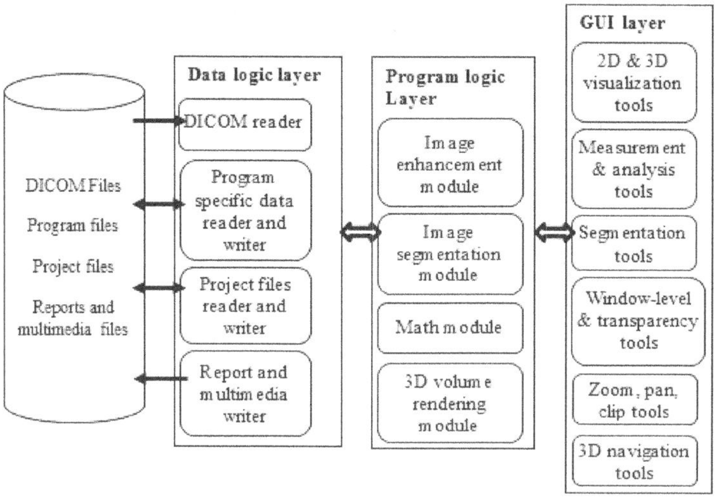

Fig. 2 Software architecture overview.

volume and highlight the relevant ones. It includes techniques like volume clipping planes, volume of interest definition, slabbing and masking. At this level image enhancement routines are also implemented. They can help to reduce some anomalies in the data due to the imperfection of the hardware. Due to the fact that location, size, and shape of the tendons and the median nerve have key role in CTS detection, image segmentation module is of particular importance.

GUI layer: This level provides the user with all necessary tools for visualization, inspection and analysis of the carpal tunnel content and CTS detection. The GUI is presented in Figure 3.

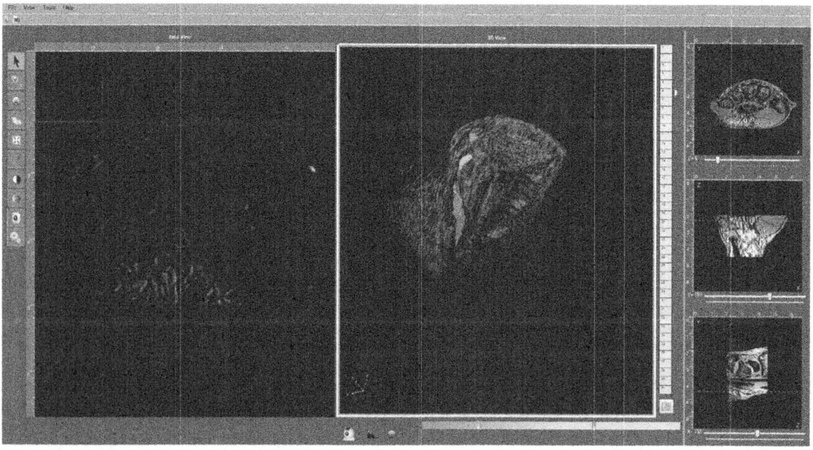

Fig. 3 GUI of the developed software [15]. It consists of axial view (where the segmentation is performed), 3D view, MPR views and navigation plains. Multi - plain volume clipping can help analyzing the wrist and CT structures

3 Using MRI for CTS Evaluation and Detection

Median Nerve Location and Carpal Tunnel Contents: As the nerve shifts to a position interposed between tendons, during wrist flexion and extension, it is probable that mechanical insult would occur during its extrusion between the loaded tendons [3]. So, it is very useful to know where the nerve is typically located in the carpal tunnel during various wrist positions. In order to avoid lacerating the nerve during surgery, it is also useful to determine whether bifurcation or trifurcation of the nerve exists within the tunnel. Therefore, the developed software platform provides realistic 3D visualization of the wrist from MRI, implements multiplanar reformatting and 3D navigation, which can be useful for identifying, visualizing the position and displacement of the nerve and other CT structures (Figure 4), as well as to track their propagation along the tunnel.

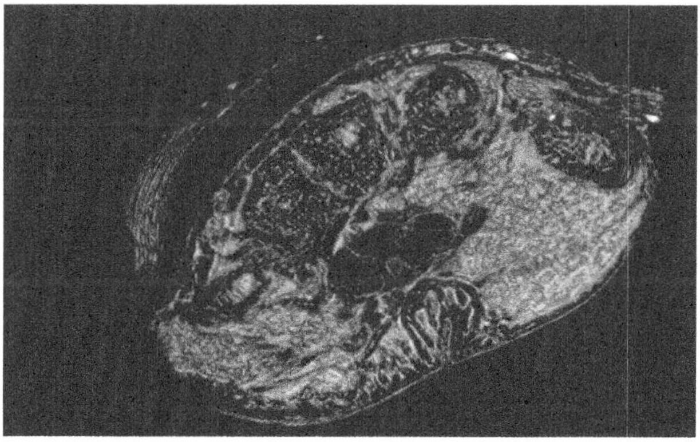

Fig. 4. Accurate volume rendering is very useful for CT structures identification.

In order to help the user to understand the volume data, to uncover important details in the wrist volume data, the software platform implements a real time visualization method based on GPU ray casting, which is capable of using multiple planes for convex volume clipping. Our approach allows the user to select arbitrary number of clipping planes. Since the ray direction is computed in the eye coordinate system and its bounding box is defined in the world coordinates, the ray direction must be transformed back to the world system and the intersections must be computed in the world coordinate system. So, from analytical geometry, the parameterized equation of the ray can be defined as:

$$P_w xyz = P_{eye} xyz + D_w * t \tag{1}$$

where P_w is the 4D homogeneous coordinates (x, y, z, 1), P_{eye} is the eye position (viewpoint) defined with its homogeneous coordinates $(x_{eye}, y_{eye}, z_{eye}, 1)$, t is the distance from the eye position, D_w is the normalized 3D vector of the ray direction (x_d, y_d, z_d). According to the analytic geometry a plain can be defined as:

$$Ax + By + Cz + D = S \cdot P_w = 0 \tag{2}$$

From Eq. 1 and Eq. 2 the intersection between the ray and the plain can be obtained as:

$$t = -\frac{Ax_{eye} + By_{eye} + Cz_{eye} + D}{Ax_d + By_d + Cz_d} = -\frac{S * P_{eye}}{Sxyz * D_w} \tag{3}$$

Since the clipping plane is dividing the convex volume in two parts, one of them that the normal of the plane points to is kept, whilst the other one is visually discarded. Several planes are often used for convex volume clipping and each ray may have several intersections with clipping planes. How to select two points as start and end points from these intersections is a key problem to correctly implement convex clipping. In the case in which a ray is parallel to one of the plains, the eye point P_{eye} will be in the visible part of the clipping plane S if $S*P_{eye}>0$. If a ray is not parallel to a plane, there must be an intersection. Ray-plane intersections are classified into two categories: intersections with angles less than 90 deg. with the plain normal and intersections with angles greater than 90 deg. with the plain normal. From these angle values the start end the end points are calculated, so each ray can be traced to compute volume integral from the start point to the end point on the GPU. Clipping planes can be moved in order to permit the user to analyze in 3D the whole wrist structure (Figure 3).

In addition to visualizing the location of the nerve, MRI is useful for studying the size and shape of the median nerve, carpal tunnel structures adjacency, carpal tunnel contents ratio, the local deformations of the median nerve surface tissue, and how these characteristics vary in normal subjects versus symptomatic patients. To determine these quantitative metrics, which can be used as CTS indicators, it is necessary to segment the carpal tunnel boundary, median nerve and tendons. Image segmentation is the problem of extracting (segmenting, cutting-out) foreground objects from background in an image. It is one of the most fundamental problems in computer vision. Fully automatic segmentation is still an open problem due wide variety of possible objects' combination, and so it seems that the use of human "hints" is inevitable. Recently, interactive image segmentation has become more and more popular in the field of medical image segmentation. The goal of interactive segmentation is to extract object(s) from background (or just split the image into continues classes) in an accurate way, using user knowledge in a way that requires minimal interaction and minimal response time.

In order to obtain shape and location information about the individual digital flexor tendons, the median nerve, and the carpal tunnel, interactive segmentation algorithm of the carpal tunnel structures in the axial plain, was implemented. The algorithm is modified version of the algorithm described in [22] and implemented using the dynamic programming methodology. This approach works well in the typical case where the foreground and background differ in appearance, as well as in challenging cases where the CT structure is clearly perceived, but the regions on both sides of the boundary are similar and cannot be easily discriminated. The final result is accurate, because it allows the user to enforce hard constraints on the boundary directly. The cross-sectional area of each structure inside the carpal tunnel was found by calculating the area inside each segmented structure boundary.

Traditional metrics: Numerous CTS studies have used MRI to quantify the size and shape of various structures within the carpal tunnel to identify pathological changes associated with CTS [16],[17]. Many studies have reported an enlargement and flattening of the median nerve as CTS indicators.. The cross-sectional area is traditionally computed as the area of an ellipse fit of the median nerve and the flattening ratio as the major axis length divided by the minor axis length, of the ellipse [18]. Some studies of normal carpal tunnels found the average cross-sectional area of the median nerve to be 7 mm^2 at the level of the pisiform and 8 mm^2 at the level of the hook of the hamate [16], and other studies reported 10 mm^2 at the pisiform and 9 mm^2 at the hook of the hamate [17]. On the other hand, several studies have found that patients with CTS typically have significantly flatter median nerves at the hook of the hamate than normal carpal tunnels, with ratios of 3.4 at the pisiform and 3.8 at the hook of the hamate [20]. However, another study found that there was no significant difference between the nerve flattening ratio of normal and CTS patient groups [21]. So, we argue that the flattening ratio can't be considered as a reliable metric, since two nerves may have completely different shapes visually, but have very similar flattening ratios. Instead, we propose a carpal tunnel contents ratio to be used as CTS evaluation metric. It can be defined as:

$$CR = \left(\sum_{i=1}^{n=9} St_i + Smn\right) / Sct \tag{4}$$

Where St_i is the cross-sectional area of each of the nine tendons, *Smn* of the median nerve and *Sct* of the whole carpal tunnel. A larger ratio would indicate either enlarged tunnel contents or a smaller tunnel, and thus there would be less space for the median nerve, which could lead to its damage.

New evaluation metrics: In addition to traditional size and shape measures, a new MRI evaluation metric, which calculates the distance of the closest adjacent structure (tendon or carpal ligament) to the nerve boundary, was developed. A given structure was considered to be adjacent to the nerve if it was within 1mm of the boundary of the nerve. To identify local deformations, the boundary of each segmented tissue structure was discretized into equal length line segments. Analysis of the internal angles between consecutive segments was considered as an indicator for impingement (angles > 240) or pinch (angles < 120) occurrence. To visualize the deformations, the location of local pinches and impingements were highlighted in red and then they were compared to the adjacency data to determine when an adjacent structure was the cause of the nerve deformation (Fig. 5).

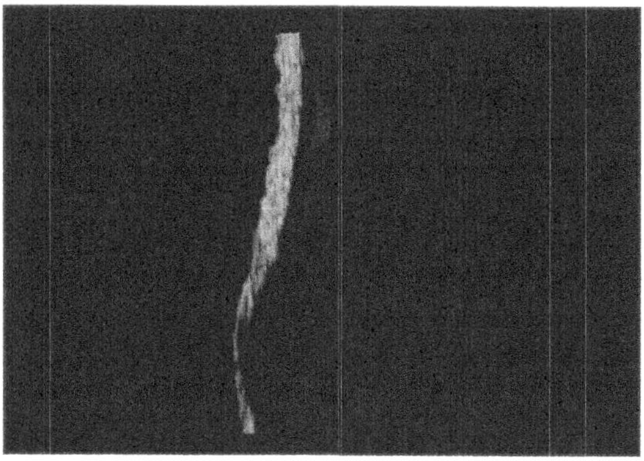

Fig. 5. Semi-transparent volume rendering.

For better visualization of local deformations semi-transparent volume rendering of separate carpal tunnel structures was implemented. The opacity in the image is derived by the compositing equation given by Eq. 5.

$$\alpha_{accum} = \alpha_0(1 - \alpha_{accum}) + \alpha_{accum} \quad (5)$$

where α_o is the opacity value. Each sample point contributes to the final image in different degrees and its contribution or visibility α_v is defined by $\alpha_o(1\text{-}\alpha_{accum})$. For each ray, we compute the voxel opacity α_o (importance) of every sample point on the ray and its visible opacity α_v.

4 Software Evaluation and Results

The software is implemented using native DirectX and C# .NET environment. Software evaluation was performed on a PC with windows7 OS, 4GB RAM, AMD Athlon™ II X4 620 Processor 2.6GHz, NVidia GeForce GTS 250 graphic card. The study for software and algorithm validation included 30 patients with CTS and 30 healthy volunteers (control group). MRI findings of 30 CTS wrist volumes and 30 normal wrist volumes including 40 T1 weighted axial plane images, were compared. The images were obtained by the wrist coil of a 1.5T MRI scanner. The slices were 3 mm thick with 1 mm gap intervals; matrix size was 256x256; FOV was 160 mm.

All examinations were evaluated by 30 radiologists (17 of them with more than 15 years of experience within the field). Radiologists were blinded to the clinical information of the patients or test persons. Segmenting the 9 digital flexor tendons, median nerve, and carpal tunnel boundary on images spanning the carpal tunnel took on average 1 minute per cross-sectional image, depending on the image quality.

To speed up the segmentation procedure and 3D reconstruction, the mask created on a single axial slice can be copied and positioned on the neighboring slices. Then if necessary, manual editing by the user was also enabled.

The opinions of users are rated through a 5-point Likert scale (5 for strongly agree; 1 for strongly disagree) for evaluation. It was reported that in all cases (even in most complicated) that the realistic 3D visualization was very helpful for CT contents analysis and can be used as an important tool in CTS detection and evaluation (4.80/5 with standard deviation 0.450). In particular, all radiologists reported that the multiplanar reformatting and clipping help them to identify and to follow the nerve and tendons propagation along the carpal tunnel (4.84/5 with standard deviation 0.640). Using the clipping feature, 100% of the radiologists successfully detected the bifurcation of the nerve, which was present in one of the patients. Using the nerve size metric tool, the radiologists found the cross-sectional area of the median nerve in normal patients to be 9-10 mm^2 at the pisiform and 10-11.2 mm^2 at the hook of the hamate, while in CTS patients the median nerve size at the hook of the hamate varies between 14-17mm^2.

Carpal tunnel contents ratios in the sections through the pisiform and the hook of the hamate were calculated as a validation measure using the cross - sectional areas of the structures on each section. This ratio ranged from 35-66% at the pisiform and 42-65% at the hamate. These values closely overlap the range typically reported (45-60%) at the narrowest part of the tunnel (the hamate), indicating the resulting models were realistic and consistent with prior work. The results also showed the CTS patient group had more local nerve deformations (pinch and impingement) than the normal subject group. This was especially evident at the hook of the hamate, compared to the pisiform, which corresponded to the largest change in the percent adjacency of the nerve. Fulfillment of the expectation was based on a significance level of 0.0005.

5 Conclusion

In conclusion we can underline the users' opinion that the developed software can be considered as a valuable and accurate platform for CTS evaluation and detection, but can also help in planning of surgery or revealing the CTS etiology.

References

1. Hochman, M.G., Zilberfarb, J.L.: Nerves in a pinch: Imaging of nerve compression syndromes. Radiol. Clin. North. Am. 42(1), 221–245 (2004)
2. Aroori, S., Spence, R.A.: Carpal tunnel syndrome. Ulster. Med. J. 77(1), 6–17 (2008)
3. Neal, S., Fields, K.B.: Peripheral nerve entrapment and injury in the upper extremity. Am. Fam. Physician 81(2), 147–155 (2010)
4. Jarvik, J.G., Yuen, E., Kliot, M.: Diagnosis of carpal tunnel syndrome: Electrodiagnostic and MR imaging evaluation. Neuroimaging Clin. N. Am. 14(1), 93–102 viii (2004)

5. Roquelaure, Y., Ha, C., et al.: Attributable risk of carpal tunnel syndrome according to industry and occupation ingeneral population. Arthritis Rheum 59(9), 1341–1348 (2008)
6. Keogh, J.P., Nuwayhid, I., Gordon, J.L., Gucer, P.W.: The impact of occupational injury on injured worker and family: Outcomes of upper extremity cumulative trauma disorders in maryland workers. Am. J. Ind. Med. 38(5), 498–506 (2000)
7. Foley, M., Silverstein, B., Polissar, N.: The economic burden of carpal tunnel syndrome: Long-term earnings of CTS claimants in washington state. Am. J. Ind. Med. 50(3), 155–172 (2007)
8. Aroori, S., Spence, R.A.: Carpal tunnel syndrome. Ulster Med. J. 77(1), 6–17 (2008)
9. BLS. Nonfatal occupational injuries and illnesses requiring days away from work for state government and local government workers, 2008. Bureau of Labor Statistics (2010)
10. AAOS. Clinical practice guideline on the diagnosis of carpal tunnel syndrome. American Academy of Orthopaedic Surgeons (2007)
11. Amrami, K., Felmlee, J.P., Spinner, R.J.: MRI of peripheral nerves. Neurosurg. Clin. N. Am. 19, 559–572 (2008)
12. Hof, J., Kliot, M., Slimp, J., Haynor, D.R.: What's new in MRI of peripheral nerve entrapment? Neurosurg. Clin. N. Am. 19, 583–595 (2008)
13. Fornage, B.D., Schernberg, F.L., Rifkin, M.D.: Ultrasound examination of the hand. Radiology 155, 785–788 (1985)
14. Jessurun, W., Hillen, B., et al.: Anatomical relations in the carpal tunnel: a computed tomographic study. J. Hand. Surg. 12, 64–67 (1987)
15. NerveVision (April 2011), `http://www.nervevision.com`
16. Middleton, W.D., Kneeland, J.B., Kellman, G.M., Cates, J.D., Sanger, J.R., Jesmanowicz, A., Froncisz, W., Hyde, J.S.: MR imaging of the carpal tunnel: Normal anatomy and preliminary findings in the carpal tunnel syndrome. AJR Am. J. Roentgenol. 148(2), 307–316 (1987)
17. Yao, L., Gai, N.: Median nerve cross-sectional area and MRI diffusion characteristics: Normative values at the carpal tunnel. Skeletal Radiol. 38(4), 355–361 (2009)
18. Mogk, J.P., Keir, P.J.: Evaluation of the carpal tunnel based on 3-D reconstruction from MRI. J. Biomech. 40(10), 2222–2229 (2007)
19. D'Arcy, C., McGee, S.: Does this patient have carpal tunnel syndrome? JAMA 283(23), 3110 (2000)
20. Uchiyama, S., Itsubo, T., Yasutomi, T., Nakagawa, H., Kamimura, M., Kato, H.: Quantitative MRI of the wrist and nerve conduction studies in patients with idiopathic carpal tunnel syndrome. J. Neurol. Neurosurg. Psychiatry 76(8), 1103–1108 (2005)
21. Monagle, K., Dai, G., Chu, A., Burnham, R.S., Snyder, R.E.: Quantitative MR imaging of carpal tunnel syndrome. AJR Am. J. Roentgenol. 172(6), 1581–1586 (1999)
22. Zadicario, E., Avidan, S., Shmueli, A., Cohen-Or, D.: Boundary snapping for robust image cutouts. In: IEEE Conference on Computer Vision and Pattern Recognition, CVPR 2008 (2008)

TMACS: A Tool for Modeling, Manipulation, and Analysis of Concurrent Systems

Jane Jovanovski, Maja Siljanoska, Vladimir Carevski, Dragan Sahpaski, Petar Gjorcevski, Metodi Micev, Bojan Ilijoski, and Vlado Georgiev

Abstract. This paper reports on an effort to build a tool for modeling, manipulation, and analysis of concurrent systems. The tool implements the CCS process language and can build labeled transition systems in Aldebaran format from CCS expressions. Furthermore, it can be used to reduce the state space of a labeled transition system and to check whether two labeled transition systems exhibit the same behaviour, using two behavioural equivalences: strong and weak bisimulation equivalence. The tool has the functionality needed to perform modeling, specification, and verification, illustrated on one classical example in the concurrency theory: the Alternating Bit Protocol.

1 Introduction

One of the process languages used to formally describe and analyse any collection of interacting processes is the Calculus of Communication Systems (CCS), process algebra introduced by Robin Milner and based on the message-passing paradigm [1, 2]. CCS, like any process algebra, can be exploited to describe both specifications of the expected behaviours of the processes and their implementations [3]. The standard semantic model for CCS and various other process languages is the notion of labeled transition systems, a model first used by Keller to study concurrent systems [4]. In order to analyze process behaviour and establish formally whether two processes offer the same behavior, different notions of behavioral equivalences over (states of) labeled transition systems have been proposed [5]. One of the key

Jane Jovanovski · Maja Siljanoska · Vladimir Carevski · Dragan Sahpaski · Petar Gjorcevski · Metodi Micev · Bojan Ilijoski · Vlado Georgiev
Institute of Informatics, Faculty of Natural Sciences and Mathematics, University "Ss. Cyril and Methodius", Skopje, Macedonia
e-mail: janejovanovski@gmail.com, maja.siljanoska@gmail.som, carevski@gmail.som, dragan.sahpaski@gmail.som, pgorcevski@t-home.mk, metodi.micev@gmail.som, b.ilijoski@gmail.som, vlatko.gmk@gmail.som

L. Kocarev (Ed.): ICT Innovations 2011, AISC 150, pp. 335–345.
springerlink.com

behavioral equivalences is the strong bisimilarity [6] which relates processes whose behaviours are indistinguishable from each other [7]. Weak bisimilarity [1, 8] is a looser equivalence that abstracts away from internal (non-observable) actions. These and other behavioral equivalences can be employed to reduce the size of the state space of a labeled transition system and to check the equivalence between the behaviours of two labeled transition systems.

This paper presents TMACS, a tool that can be used to automate the process of modeling, specification, and verification of concurrent systems described by the CCS process language. TMACS stands for Tool for Modeling, Manipulation, and Analysis of Concurrent Systems. It is given as a Java executable jar library with a very simple Graphical User Interface (GUI). TMACS can parse CCS expressions, generate labeled transition system from CCS in Aldebaran format [10], reduce a labeled transition system to its canonical form with respect to strong or weak bisimilarity and check whether the behavioral equivalence of two labeled transition systems modulo strong or weak bisimilarity.

1.1 Related Work

Different tools for modeling, specification and verification of concurrent reactive systems have been developed over the past two decades. Probably the most famous and most commonly used ones, especially in the academic environment, are the Edinburgh Concurrency Workbench (CWB) [11] and micro Common Representation Language 2 (mCRL2) [12, 13].

CWB is a tool for analysis of cuncurrent systems, which allows for equivalence, preorder and model checking using a variety of different process semantics. Although CWB covers much of the functionality of TMACS and more, it has a command interpreter interface that is more difficult to work with, unlike our tool that has a Graphical User Interface (GUI), and is very intuitive. As far as we know CWB does not have the functionality for exporting labeled transition system graphs in Aldebaran format, that TMACS has.

mCRL2, the successor of μCRL, is a formal specification language that can be used to specify and analyse the behaviour of distrubuted systems and protocols. However, to our knowledge, mCRL2 does not provide the possibility to define systems' behaviour in the CCS process language. Also, even though mCRL2 supports minimization modulo strong and weak bisimulation equivalence, it does not output the computed bisimulation, a feature that we have implemented in TMACS.

1.2 Outline

The contributions of this paper are organized as follows. Section 2 presents the implementation of the CCS process language and the generation of labeled transition systems as a semantic model of process expressions. Section 3 describes the minimization and the comparison of two labeled transition systems with respect to strong weak bisimilarity. It includes implementation details of two algorithms for computing strong bisimulation equivalence, the naive algorithm [5] and the advanced algorithm due to Fernandez [14]. It also entails a description of the saturation technique together with a respective algorithm for saturating a labeled transition system [5]. Next, in Section 4, we illustrate the application of TMACS for modeling, specification and verification on one classical example in the concurrency theory: the Alternating Bit Protocol [15, 16]. Finally, we give some conclusions and some directions for future development of the tool in Section 5.

2 CCS Parsing and Labeled Transition System Generation

Given a set of action names, the set of CCS processes is defined by the following BNF grammar:

$$P ::= \emptyset \mid a.P_1 \mid A \mid P_1 + P_2 \mid P_1|P_2 \mid P_1[b/a] \mid P_1\backslash a \tag{1}$$

Two choices were considered for the problem of building a parser in TMACS. The first choice was to build a new parser, which required much more resources and the second choice was to define a grammar and to use a parser generator (compiler-compiler, compiler generator) software to generate the parser source code. In formal language theory, a context-free grammar [18] is a grammar in which every production rule has the form $V \rightarrow w$, where V e is a nonterminal symbol, and w is a string of terminal and/or nonterminal symbols (w can be empty). Obviously, the defined BNF grammar for describing the CCS process is a CFG. Deterministic context-free grammars [18] are grammars that can be recognized by a deterministic pushdown automaton or equivalently, grammars that can be recognized by a LR (left to right) parser [19]. Deterministic context-free grammars are a proper subset of the context-free grammar family [18].

2.1 Generating the Parser

ANTLR (ANother Tool for Language Recognition) [20] is a parser generator that was used to generate the parser for the CCS grammar in TMACS. ANTLR uses LL(*) parsing and also allows generating parsers, lexers, tree parsers and combined lexer parsers. It also automatically generates abstract syntax trees [19], by providing operators and rewrite rules to guide the tree construction, which can be further processed with a tree parser, to build an abstract model of the AST using some programming language.

ANTLR was used to generate lexer, parser and a well defined abstract syntax tree which represents a tree representation of the abstract syntactic structure of the parsed sentence. The term abstract is used in a sense that the tree does not represent every single detail that exists in the language syntax, e.g., parentheses are implicit in the tree structure. From a development point of view it is much easier to work with trees than with a list of recognized tokens. One example of an abstract syntax tree is shown in Fig. 1 which is the result of parsing the expression:

$$A = (b.B \mid c.D + d.D)\backslash\{b, c\} \tag{2}$$

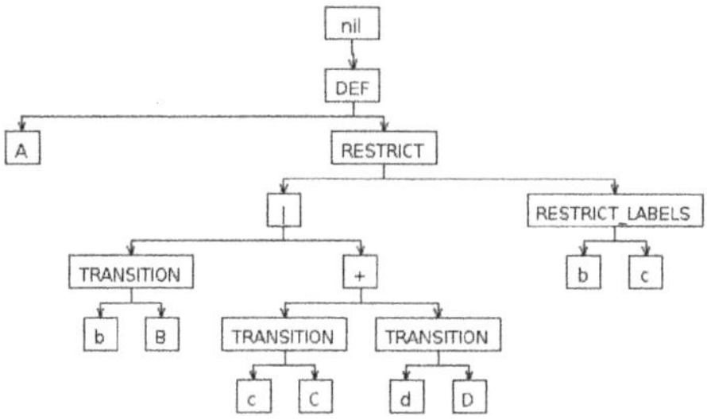

Fig. 1 Example of an abstract syntax tree obtained as a result of parsing the CCS expression in (2)

2.2 CCS Domain Model and Labeled Graph Generation

Although working directly with abstract syntax trees and performing all algorithms on them is possible, it causes a limitation in future changes, where even a small change in the grammar and/or in the structure of the generated abstract syntax trees causes a change in the implemented algorithms. Because of this, a specific domain model was built along with a domain builder algorithm, which has corresponding abstractions for all CCS operators, processes and actions. The input of the domain builder algorithm is an abstract syntax tree, and the output is a fully built domain model.

The algorithm for generation of labeled transition system implemented in TMACS is a recursive algorithm which traverses the tree structure of objects in the domain model and performs SOS rule every time it reaches an operation. In this fashion all SOS transformation are performed on the domain and as result a new graph structure is created which represents the labeled transition system which can be easily exported to a file in Aldebaran format.

3 Minimization and Comparison of Labeled Transition Systems

Bisimulation equivalence (bisimilarity) [6] is a binary relation between labeled transition systems which associates systems that can simulate each other's behaviour in a stepwise manner, enabling comparison of different transition systems [9]. An alternative perspective is to consider the bisimulation equivalence as a relation between states of a single labeled transition system, providing means for construction of smaller models of the system [9].

TMACS implements both options: reducing the size of the state space of a given labeled transition system and checking the equivalence of two labeled transition systems, using two behavioral equivalence relations: strong bisimilarity and observational equivalence (weak bisimilarity).

3.1 Minimization Modulo Strong Bisimilarity

The process of reducing the size of the state space of a labeled transition system $A = (S, Act, \rightarrow, s, F)$ in TMACS was implemented using an approach which consists of two steps:

1. Computing strong bisimulation equivalence (strong bisimilarity) for the labeled transition system;
2. Minimizing the labeled transition system to its canonical form using the strong bisimilarity obtained in the first step;

Two different methods were used for computing the strong bisimulation equivalence: the so called naive method and a more efficient method due to Fernandez, both of which afterwards serve as minimization procedures.

The naive algorithm for computing strong bisimulation over finite labeled transition systems stems directly from the theory of partially ordered sets and lattices [5] underlying Tarski's classic fixed point theorem [21]. This algorithm has time complexity of $O(mn)$ for a labeled transition system with m transitions and n states. Its implementation in TMACS takes as an input a labeled transition system in Aldebaran format, and generates the corresponding labeled graph as a list of nodes representing the states which use the following data structure:

- $S_p = \{(a,q) \mid p \xrightarrow{a} q,\ p,q \in S,\ a \in Act\}$ - set of pairs (a,q) for state p where a is an outgoing action for p and q is a state reachable from p with action a

The algorithm then computes the strong bisimulation equivalence and outputs it as pairs of bisimilar states $L = \{(p,q) \mid\ p \sim q,\ p,q \in S\}$.

The algorithm due to Fernandez exploits the idea of the relationship between strong bisimulation equivalence and the relational coarsest partition problem [22] solved by Paige and Tarjan in $O(m\log n)$ time [23]. Fernandez's adaptation of the algorithm of Paige-Tarjan has the same complexity as the original one, major difference being that a refinement step is made with only one element of *Act* in the original one. Our implementation of the algorithm takes as an input a labeled transition system in Aldebaran format, generates a labeled graph and then partitions the labeled graph into its coarsest blocks where each block represents a set of bisimilar states. To define graph states and transitions the following terminology is used represented by suitable data structures:

- $T_a[p] = \{q\}$ - an a-transition from state p to state q
- $T_a^{-1}[q] = \{p\}$ - an inverse a-transition from state q to state p
- $T_a^{-1}[B] = \cup\{T_a^{-1}[q], q \in B\}$ - inverse transition for block B and action a
- W - set of sets called splitters that are being used to split the partition
- $\text{infoB}(a,p)$ - info map for block B, state p and action a

The algorithm of Fernandez outputs the strong bisimulation equivalence relation over *Proc* as a partition $P = \{B_i \mid p \sim q,\ \forall p,q \in B_i, i = \overline{1,n}\}$ where B_i, $i = \overline{1,n}$, represent its equivalence classes.

Having computed the strong bisimulation equivalence, the next step in the reduction of the state space of the labeled transition system uses the bisimulation equivalence obtained in the first step in order to minimize the labeled graph. This reduction is implemented as follows:

1. All states in a bisimilar equivalence class B_i are merged into one single state $k = \bigcup p_j$, for $p_j \in B_i$;
2. All incoming transitions $r \xrightarrow{a} p_j$, for $p_j \in B_i$, are replaced by transitions $r \xrightarrow{a} k$;
3. All outgoing transitions $p_j \xrightarrow{a} t$, for $p_j \in B_i$, are replaced by transitions $k \xrightarrow{a} t$;
4. The duplicate transitions are not taken into consideration.

The procedure is repeated for all equivalence classes B_i, $i \in \overline{1,n}$.

3.2 Minimization Modulo Weak Bisimilarity

The minimization of a labeled transition system modulo weak bisimilarity is reduced to the problem of minimization modulo strong bisimilarity, using a technique called saturation. Saturation first precomputes the weak transition relation and then constructs a new pair of finite processes whose original transitions are replaced by the weak transitions [5].

If T is the set of all transitions of a labeled transition system with m transitions and n states, then its saturation T^* is constructed as using the following properties:

1. $(p,\tau,q) \wedge (q,\tau,r) \Rightarrow (p,\tau,r)$
2. $(p,a,q) \wedge (q,\tau,r) \Rightarrow (p,a,r)$
3. $(p,\tau,q) \wedge (q,a,r) \Rightarrow (p,a,r)$

Following the above properties, the algorithm for saturation in TMACS was implemented as follows:

1. First, the reflexive transitive closure of τ is computed as $\left\{(p,\tau,q) \mid p\left(\xrightarrow{\tau}\right)^* q\right\}$ and the following set is constructed $T_1 = T \cup \left\{(p,\tau,q) \mid p\left(\xrightarrow{\tau}\right)^* q\right\}$ for $p,q \in S$
2. Next, $T_2 = T_1 \cup \left\{(p,a,q) \mid (\exists q' \in S)\,(p,a,q') \in T \wedge q'\left(\xrightarrow{\tau}\right)^* q\right\}$ where $p,q \in S$ and $a \in Act$, is constructed
3. Finally, $T^* = T_2 \cup \left\{(p,a,q) \mid (\exists p' \in S)\,(p',a,q) \in T \wedge p\left(\xrightarrow{\tau}\right)^* p'\right\}$ for $p,q \in S$ and $a \in Act$, gives the saturation T^* of T

An illustration of a saturated labeled transition system before and after applying the saturation technique is given on Fig. 2.

Once the saturation algorithm is run and the original labeled transition system is saturated, the computation of weak bisimilarity amounts to computing strong bisimilarity over the saturated system. Afterwards, the process of minimization of the original system is the same as the process of minimization modulo strong bisimilarity applied on the saturated system.

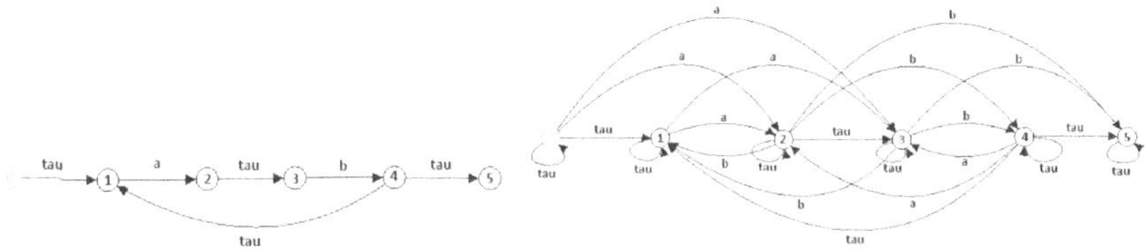

Fig. 2 Example of a labeled graph before and after applying the saturation technique

3.3 Comparison Modulo Strong Bisimilarity

The idea for the implementation of equivalence checking modulo strong bisimilarity was based on the following fact: Two labeled transition systems are (strongly) bisimilar iff their initial states are bisimilar [17]. This means that in order to check whether two labeled transition systems are bisimilar it is enough to check whether their initial states are bisimilar. This can be done using the following approach:

1. The two labeled transition systems are merged into a single transition system
2. Algorithm for computing the strong bisimilarity is applied to the merged system
3. A check is performed to see if the initial states belong to the same bisimulation equivalence class

3.4 Comparison Modulo Weak Bisimilarity

The comparison of two labeled transition systems modulo weak bisimilariy amounts to checking strong bisimilarity over the saturated labeled transition systems [5]. In other words, two labeled transition systems are weakly bisimilar iff their saturated systems are strongly bisimilar. Following this fact, we implemented the comparison of two labeled transition systems modulo weak bisimilarity by applying the saturation algorithm over the original labeled graphs in order to obtain their saturated labeled graphs, after which the process of comparison of the saturated labeled transition systems modulo strong bisimilarity was applied as described above.

4 Application

Here we demonstrate the process of formal modeling, specification and verification for one classical problem in the concurrency theory: the Alternating Bit Protocol.

4.1 Specification and Verification of the Alternating Bit Protocol

The Alternating Bit Protocol is a simple yet effective protocol, designed to ensure reliable communication through unreliable transmission mediums, and it is used for managing the retransmission of lost messages [5, 24].

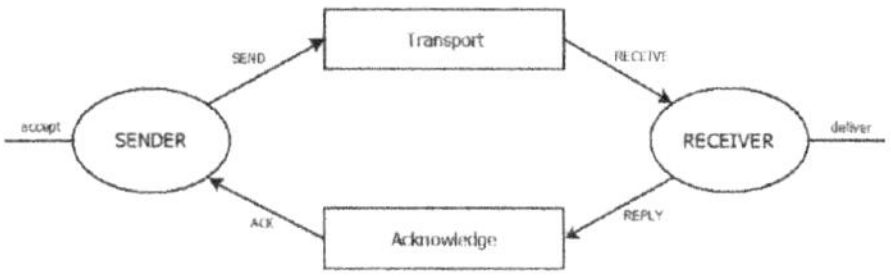

Fig. 3 Representation of the Alternating Bit Protocol

The representation of the Alternating Bit Protocol consists of a sender S, a receiver R and two channels - transport T and acknowledge A, as shown on Fig. 3.

The functioning of the alternating bit protocol can be described as follows [5]:

1. The sender S sends a message repeatedly until it receives an acknowledgment from the receiver R that contains the same protocol bit as the message being sent:

$$\begin{aligned} S &= \overline{send0}.S + ack0.accept.S_1 + ack1.S \\ S_1 &= \overline{send1}.S_1 + ack1.accept.S + ack0.S_1 \end{aligned} \tag{3}$$

The transport channel transmits the message to the receiver, but it may lose the message (lossy channel) or transmit it several times (chatty channel):

$$\begin{aligned} T &= send0.(T + T_1) + send1.(T + T_2) \\ T_1 &= \overline{receive0}.(T + T_1) \\ T_2 &= \overline{receive1}.(T + T_2) \end{aligned} \tag{4}$$

2. When the receiver R receives a message, it sends a reply to S which includes the protocol bit of the message received. If the message is received for the first time, the receiver will deliver it for processing, while subsequent messages with the same bit will be simply acknowledged:

$$\begin{aligned} R &= receive0.\overline{deliver}.R_1 + \overline{reply1}.R + receive1.R \\ R_1 &= receive1.\overline{deliver}.R + \overline{reply0}.R_1 + receive0.R_1 \end{aligned} \tag{5}$$

The acknowledgement channel sends an acknowledgment to S, and it can also acknowledge it several times or lose it on the way to the sender S:

$$\begin{aligned} A &= reply0.(A + A_1) + reply1.(A + A_2) \\ A_1 &= \overline{ack0}.(A + A_1) \\ A_2 &= \overline{ack1}.(A + A_2) \end{aligned} \tag{6}$$

3. When S receives an acknowledgment containing the same bit as the message it is currently transmitting, it stops transmitting that message, flips the protocol bit, and repeats the protocol for the next message.[24, 13]

Now, having described the behaviour of the alternating bit protocol components as above, the CCS process expression depicting the behaviour of the protocol as a whole can be obtained as a parallel composition of the processes describing the sender S, the transport channel T, the receiver R and the acknowledgement channel A, restricted on a set of actions L:

$$ABP \stackrel{def}{=} (S|T|R|A)\backslash L \tag{7}$$

where the set of restricted actions is

$$L = (send0, send1, receive0, receive1, reply0, reply1, ack0, ack1) \tag{8}$$

The above CCS expression represents the implementation of the protocol which details the proposed means for achieving the desired high-level behaviour the Alternating Bit Protocol should exhibit. This desired high-level behaviour consists of the fact that the alternating bit protocol should act as a simple buffer, therefore its CCS specification is defined as follows:

$$\begin{aligned} Buf &= accept.Buf' \\ Buf' &= \overline{deliver}.Buf \end{aligned} \tag{9}$$

In order to verify the alternating bit protocol, we need to prove that the implementation *ABP* meets the specification *Buf* with respect to some behavioural equivalence. Using TMACS we shall show that an observational equivalence between *Buf* and *ABP* can be found, i.e. that $ABP \approx Imp$. For that purpose, first we use TMACS to obtain the labeled graphs corresponding to the CCS representations of *Buf* and *ABP*, and afterwards we perform a comparison of the labeled transition systems modulo weak bisimilarity which yields a positive answer about the existance of weak bisimulation equivalence between *Buf* and *ABP*.

The weak bisimulation equivalence obtained by running any of the two bisimulation algorithms implemented in TMACS over the saturated labeled transition systems is given in Table 1.

4.2 Real-Life Application

It is worth to mention that TMACS is already being used on real-life problems by one software company in Macedonia specialised in building complete e-government solutions for improving business performance. In particular, TMACS was successfully used to analyse and verify the accuracy of the workflow of the Official Gazette of the Republic of Macedonia. This is an important indicator that the tool can be quite useful both for experimental as well as practical applications.

Table 1 Verification of the alternating bit protocol using weak bisimilarity

ABP implementation states	ABP specification states
$(S\|T\|R\|A)\backslash L$	*Buf*
$(S\|(T+T1)\|R\|A)\backslash L$	
$(S\|(T+T1)\|R\|(A+A2))\backslash L$	
$(S\|T\|R\|(A+A2))\backslash L$	
$(S\|(T+T1)\|\overline{deliver}.R1\|A)\backslash L$	
$(S\|(T+T1)\|\overline{deliver}.R1\|(A+A2))\backslash L$	
$(S1\|(T+T1)\|R1\|(A+A1))\backslash L$	
$(S1\|(T+T2)\|\overline{deliver}.R\|(A+A1))\backslash L$	
$(S1\|(T+T2)\|R1\|(A+A1))\backslash L$	
$(S\|(T+T2)\|R\|(A+A2))\backslash L$	
$(accept.S1\|(T+T1)\|R1\|(A+A1))\backslash L$	*Buf'*
$(S\|(T+T1)\|R1\|(A+A1))\backslash L$	
$(S\|(T+T1)\|R1\|(A+A2))\backslash L$	
$(S\|(T+T1)\|R1\|A)\backslash L$	
$(S1\|(T+T2)\|R\|(A+A2))\backslash L$	
$(accept.S\|(T+T2)\|R\|(A+A2))\backslash L$	
$(S1\|(T+T2)\|R\|(A+A1))\backslash L$	

5 Conclusions and Future Work

In this paper we presented TMACS, a tool for modeling, manipulation and analysis of concurrent systems. It includes recognizing CCS expressions, building labeled transition system graphs and checking equivalence between two labeled graphs with respect to strong and weak bisimilarity. The tool is simple, but yet functional enough to perform specification and verification of concurrent systems described as CCS expressions and/or labeled transition systems. We have successfully validated and tested the tool with large number of examples comparing the results obtained with mCRL2 and CWB, and the results obtained are encouraging and competitive.

Future work can include optimization of the tool's response times for very large labeled transition system graphs, as well as implementation of prunung strategy for infinite labeled transition system graphs. It would be also usefull if the labeled graph is better visualized where every SOS rule is shown, with its source and sink nodes showing the CCS expression for each of the nodes.

Furthermore, the minimization and comparison functionality for labeled transition systems which at the moment includes only strong and weak bisimilarity can be extended to include other behavioural equivalences and preorders as well. Another direction for future development can be to extend the tool to support not only CCS expressions and process semantics in general, but also other formal verification approaches as well.

References

1. Milner, R.: A Calculus of Communication Systems. LNCS, vol. 92. Springer (1980)
2. Milner, R.: Communication and Concurrency. Prentice-Hall (1989)
3. Bergstra, J.A., Ponse, A., Smolka, S.A.: Handbook of Process Algebra. Elsevier Science B.V. (2001)
4. Keller, R.M.: Formal Verification of Parallel Programs. Communications of the ACM 19(7), 371–384 (1976)
5. Aceto, L., Ingolfsdottir, A., Larsen, K.G., Srba, J.: Reactive Systems - Modeling, Specification and Verification. Cambridge University Press (2007)
6. Park, D.M.R.: Concurrency and Automata on Infinite Sequences. In: Deussen, P. (ed.) GI-TCS 1981. LNCS, vol. 104, pp. 167–183. Springer, Heidelberg (1981)
7. Roscoe, A.W.: Understanding Concurrent Systems. Texts in Computer Science. Springer, Heidelberg (2010)
8. Milner, R.: Operational and Algebraic Semantics of Concurrency Processes. In: van Leeuwen, J. (ed.) Handbook of Theoretical Computer Science, pp. 1201–1242. Elsevier and MIT Press (1990)
9. Baier, C., Katoen, J.-P.: Principles of Model Checking. The MIT Press (2008)
10. Fernandez, J.C.: Aldebaran: User's Manual, Technical Report, LGI-IMAG Grenoble (1988)
11. Moller, F.: Perdita Stevens, Edinburgh Concurrency Workbench User Manual (Version 7.1) (1999), `http://homepages.inf.ed.ac.uk/perdita/cwb/`
12. van Weerdenburg, M., et al.: mCRL2 User's Manual (2008), `http://www.mcrl2.org`
13. Alexander, M., Gardner, W.: Process Algebra for Parallel and Distributed Processing. Computational Science Series. Chapman&Hall/CRC Press (2009)
14. Fernandez, J.C.: An Implementation of an Efficient Algorithm for Bisimulation Equivalence. Science of Computer Programming, vol. 13, pp. 219–236 (1989/1990)
15. Lynch, W.C.: Computer systems: Reliable Full-duplex File Transmission over Half-duplex Telephone Line. Communications of the ACM 11(6), 407–410 (1968)
16. Bartlett, K.A., Scantlebury, R.A., Wilkinson, P.T.: A note on Reliable Full-duplex Transmission over Half-duplex Links. Communications of the ACM 12, 260–261 (1969)
17. Groote, J.F., Reniers, M.: Modelling and Analysis of Communicating Systems. Technical University of Eindhoven, rev. 1478 (2009)
18. Chomsky, N.: Three Models for the Description of Language. IEEE Transactions on Information Theory (1956)
19. Aho, A.V., Sethi, R., Ullman, J.D.: Compilers: Principles, Techniques, and Tools. Addison-Wesley (2006)
20. Parr, T.: The Definitive ANTLR Reference - Building Domain-Specific Languages. The Pragmatic Bookshelf (2007)
21. Tarski, A.: A Lattice-theoretical Fixpoint Theorem and its Applications. Pacific Journal of Mathematics 5(2), 285–309 (1955)
22. Kanellakis, P., Smolka, S.A.: CCS expressions, finite state processes and three problems of equivalence. In: Proceedings of the 2nd ACM Symposium on Principles of Distributed Computing, pp. 228–240. ACM Press (1983)
23. Paige, R., Tarjan, R.: Three Partition Refinement Algorithms. SIAM J. Comput. 16(6) (1987)
24. Kulick, S.: Process Algebra, CCS, and Bisimulation Decidability. University of Pennsylvania, pp. 8–10 (1994)

Apportionment Heuristics for Mapping Tasks in Heterogeneous Computing Systems

Igor Mishkovski, Sonja Filiposka, Dimitar Trajanov, and Ljupco Kocarev

Abstract. One of the biggest problems in heterogeneous computing is how tasks should be mapped in these kinds of environments. Because this problem of mapping tasks has been shown to be NP-complete, it requires heuristic techniques. Therefore, we present new schedulers based on the apportionment methods used in elections. In order to obtain the performances of these schedulers we compare them with other known and used heuristics in many different parameters. The presented heuristics can be used when the tasks are big and when they can be divided in smaller sub-tasks. The comparison in this paper shows that these apportionment methods can cope well with the other methods when the number of tasks in the system is no bigger than a certain level. The new apportionment scheduler, based on Hamilton's method, copes well with the existing ones and it outperforms the other schedulers when some conditions are met.

Keywords: Schedulers, Mapping Heuristics, Apportionment Methods.

1 Introduction

Distributed systems appear in the computer history as a result of many factors and influences. One of the most important factors was the price reduction of a computer system; this increased the number of computers that can be integrated in

Igor Mishkovski
Politecnico di Torino, Turin, Italy
e-mail: igor.mishkovski@polito.it

Sonja Filiposka · Dimitar Trajanov · Ljupco Kocarev
Faculty of Electrical Engineering and Information Technologies, Skopje, Macedonia
e-mail: {filipos,mite,lkocarev}@feit.ukim.edu.mk

Ljupco Kocarev
Macedonian Academy of Sciences and Arts, Skopje, Macedonia
e-mail: lkocarev@manu.edu.mk

L. Kocarev (Ed.): ICT Innovations 2011, AISC 150, pp. 347–356.
springerlink.com

distributed. Heterogeneous computing (HC), as a part of distributed system, uses different types of machines, networks and interfaces in order to maximize their combined performance and cost-effectiveness [1, 2, 3]. The machines in this type of systems are independent and for the end user of the system they look as if one machine. Heterogeneous computing is used for solving computationally intensive problems. Thus the most important parameter of one HC system is its speed. The speed of the HC system heavily depends on the schemes that assign tasks to machines (a.k.a. *matching*) and specify the order by which these tasks will execute on the proper machines (a.k.a. *scheduling*) [4]. This problem of matching and scheduling is known as mapping and it has been shown that is NP-complete [5, 6].

Given a set of independent tasks and a set of available resources, there are heuristics which try to minimize the total execution time of the set by finding an optimal mapping of tasks to machines. This process of mapping can be *dynamic* (map tasks as they arrive) or *static* (mapping is done prior to the execution of any of the tasks) [7, 8]. Static approach is adequate if the tasks to be mapped are known in advance, and if a good estimation exists for the resource's power and capacity. Dynamic mapping is used when if the tasks cannot be determined in advance and if the system performances are fluctuating.

The issue of mapping tasks in heterogeneous systems has been of major interest [9] and [11]-[19]. In [11] the scheduling is based on a predictor, which predicts the variance to make scheduling decisions and it fine-tunes the algorithm using a feedback mechanism. In [12] authors compare eleven static heuristics and they give a common ground for comparison and insights into circumstances where one technique outperforms another. Dynamic mapping heuristics are studied in [9]. In this paper two types of mapping heuristics are considered, immediate mode and batch mode heuristics. Furthermore, three new heuristics are introduced. Their simulation results reveal that the choice of which heuristic to use depends on: the structure of the heterogeneity among tasks and machines, and the arrival rate of the tasks. In [13] a compensation based algorithm is presented which uses a feedback mechanism to predict the execution time of the jobs. The authors in [14] describe Grid Computing systems that have a game theoretic approach in processor scheduling. They also provide an analysis and comparison of these systems. In [15] low-complexity efficient heuristics for task scheduling among heterogeneous processors are presented. The comparison study showed that their algorithms outperform the previous approaches in terms of performance and cost. In [16] the authors propose a metascheduler which uses various preemptive methods to minimize job execution time, such as stopping a bigger job so that a smaller one can proceed first. In [17] different scheduling schemes are characterized with respect to varying arrival times and burstiness in the job arrival rate. Using the insight, authors propose approaches to improve the strategies regarding turn-around time. In [18] authors propose, evaluate, and compare eight dynamic mapping heuristics. The tasks in the study have priorities and multiple soft deadlines. They calculate the value of the task using the priority of the task and the completion time of the task with respect to its deadlines.

In the simulations in this paper, it is assumed that no communication exists between the tasks, the tasks are very big (i.e. the tasks have many instructions) and that they can be additionally divided into smaller sub-task, each machine in the system executes one task at a time in the order in which tasks are assigned. The dynamic mapping applied in this paper is non-preemptive and it works in *batch mode* [9]. In the batch mode the scheduler does not map the tasks as they arrive, but instead it maps the tasks in prescheduled times, also called *mapping events*. The key feature of the scheduler is to finish the execution of the tasks as soon as possible, i.e. to reduce the *makespan* [10].

The major contribution of this work is introducing new heuristic which is based on apportionment methods. Another gain of this paper is that the comparison between the new heuristic and the existing ones is done using many different parameters, such as: the system load; the average number of busy machines; the average turnaround time of the executed tasks; the average response time of the executed tasks; how many tasks were finished; the number of deployments of the tasks that were involved and the load balance among the machines. Another important parameter that we consider is the *heuristic execution times*. This parameter tells how much time is needed for a certain heuristic to map the arrived tasks on the available machines. .

The remainder of the paper is organized as follows. Description of the apportionment methods used in elections and summary of their characteristics is found in Section 2. Section 3 describes how apportionment methods can be adjusted for the purpose of mapping tasks in heterogeneous computing environments. In Section 4 the simulation scenario is presented and this section examines the obtained results from the simulation study. In Section 5 the conclusion and future work are presented.

2 Methods of Apportionment

The problem of how many representatives should be allotted to some important institution (i.e. US Congress) exists since the beginning of democracy and elections. One might come up with an idea of one man, one vote, i.e. pure proportional method. But candidates are human beings and they cannot be divided. Thus, because this apportionment cannot be done perfectly it must be done in a manner as near perfection as can be. These methods of apportionment should fulfill certain rules in order to be objective [20], as example we refer to the US Congress:

1. No state's number of representatives should decrease, if the total number of representatives increases.
2. Every state should have within one (exclusive) of their quotient. For example if a state should receive 3.4 representatives it can receive 3 or 4. If the state should receive exactly 3 representatives, it should receive 3, but not 2 or 4. In future, we will call having this property -satisfying Quota.
3. All states abide by the same formula for representation.
4. Methods do not artificially favor large states at the expense of the smaller ones and vice versa.

2.1 Hamilton's Method (HM)

The algorithm of this method is as follows:

1. Calculate the Standard Divisor (SD).
2. Calculate each state's Standard Quota (SQ).
3. Initially assign each state its Lower Quota (LQ).
4. If there are surplus seats, give them, one at a time, to states in descending order of the fractional parts of their Standard Quota.

The standard divisor is the average number of people per seat over the entire population and it can be calculated as $SD=TP / p$, where, TP is the total population and p is the number of seats in the congress.

The standard quota is the fraction of the total number of seats a state would be entitled if the seats were not indivisible and it is calculated as $SQ=SP / SD$, where, SP is the state population.

The lower quota (LQ) is the standard quota rounded down.

But this method violates the rule number one: an increase in the total number of seats to be apportioned causes a state to lose a seat (a.k.a. Alabama paradox).

2.2 Jefferson's Method (JM)

Jefferson came up with what is known as the method of greatest divisors. Suppose we are given state populations p_1, p_2, .., p_N and representative apportionment a_1, a_2, .., a_N. We can calculate a divisor $L(s) = a(s)+1$ for each state s. Then states can be ranked using the $p(s)/L(s)$ ratios. The higher this ratio, the more deserving this state is to get another representative. In this method everybody starts with zero representatives. The representatives are always assigned to the state with the current highest ratio (rank-index). The first N representatives are assigned one to each state. This is naturally enforcing US Constitution rule about each state having a minimum of one representative. The divisor choice $L(s) = a(s)+1$ is natural, because it ranks how much better off the state will be if it was given one more representative. The divisor choice Jefferson's Method uses is arbitrary, since other methods use divisors such as $L(S) = a(s)+1/2$ - the Webster's method (WM), or $L(S) = \sqrt{a(s)*(a(s)+1)}$ - the Huntington-Hill method (HHM). The Jefferson's method favors large states and it does not satisfy lower quota. Webster's method does not satisfy upper or lower quota, but it does not favor large or small states and the Huntington-Hill method favors small states.

3 Using Methods of Apportionment as Heuristics for Mapping Tasks

The difference between mapping tasks in HC environments and allocation of places in some institution is that in HC environment machines can be divided among multiple tasks. Thus, the pure proportional model (PPM) can take part in the process of mapping tasks. With this proportional method each task gets a portion of the heterogeneous system as a whole. The portion that task i gets for the system is equal to:

$$port_i = \frac{size_i}{\sum_j size_j} \tag{1}$$

where, $size_i$ is the size of the task i and the sum in the nominator is the sum of the instructions of the arrived tasks. It is obvious that with this method one machine can be shared among many tasks and the tasks are executed in Round-Robin fashion. This additionally requires time for context switch which certainly reduces the overall performance of the system. Having this fact in mind in this part we will briefly present how these apportionment methods used in the US Congress can be adjust for mapping tasks in HC environment. The analogy is that instead the US Congress is now the HC environment and the available seats are the machines in the HC environment. The states are replaced by the tasks and the total population of a state is an equivalent to the number of instructions that a certain task has.

For example we will use the Hamilton's method. TP is the sum of the number of instructions of the available tasks and p is the number of available processors. Then the standard quota is calculated as the fraction of the total number of processors a task would be entitled to if the processors were not indivisible and for tasks I, i.e. $SQ_i=TS_i / SD$. The lower quota (LQ) is the standard quota rounded down.

Table 1 Hamilton's method example

Task	# of million instructions	SQ	LQ	Leftover	Surplus	Final mapping
1	6	0.02	0	0.02		0
2	271	0.68	0	0.68		1
3	85	0.21	0	0.21		0
4	153	0.39	0	0.39		0
5	200	0.50	0	0.50		0
6	406	1.02	1	0.02		1
7	161	0.41	0	0.41		0
8	242	0.61	0	0.61	1(3)	1
9	217	0.55	0	0.55	1(4)	1
10	243	0.61	0	0.61	1(2)	1
Totals	**1831**	**5**	**1**	**4**	**4**	**5**

For example, in Table 1 the task mapping is shown for 5 processors and 10 users in the system and where every user has a different task to be executed on the system.

As we can see, only tasks 2, 8, 9 and 10 would be executed in the system for that certain round, while the other tasks have to wait for another round. Intuitively, this heuristics gives excellent performance when the arrival of the tasks is not so dense and there is a big number of free machines. In this case one bigger task can be divided to several smaller ones and they can be executed in parallel on the machines in the HC environment. On the other hand, there will be some delay because of the task partitioning. If this delay is not so big then this is a perfect case when this apportionment heuristics can be used. But, when the number of tasks is very big and the HC environment is overloaded then these heuristics act as Biggest Job First heuristic. Thus, in their nature these heuristics change their attitude dynamically, i.e. in an overloaded cluster they use one-to-one mapping, and many-to-one mapping when the number of free machines is bigger. The Jefferson's approach will favour bigger tasks.

4 Simulation Model and Results

In order to analyze how the proposed heuristics cope with other heuristics used for mapping tasks in HC environments we have implemented the following ones: **OLB** (*Opportunistic Load Balancing*), **MET** (*Minimum Execution Time*), **MCT** (*Minimum Completion Time*), **Min-min**, **Max-min**, **SA** (*Simulated Annealing*).

All of these heuristics do not consider the case when tasks are very big and they can be divided in a certain number of sub-tasks. In that event, one task can be executed on a certain number of processors – space sharing. This can be done in computing systems where communication delays can be neglected. If this is a case one of the proposed apportionment mappings (in Section II and III) can be used for mapping tasks in HC environments.

All of the schedulers have been simulated in Matlab. Firstly, we create a virtual system with p heterogeneous machines and u users of the systems. We differentiate between small and big users of the system. The big users (called predators) generate big tasks with greater probability than the small users (called victims). The simulation is discrete, i.e. we use time steps in our simulation and every time step can last a certain number of seconds. The arrival time of the tasks is simulated using a uniform distribution. Each task is characterized by its size expressed in instructions and machines are characterized by how many instructions in second they can execute. In every time step the simulator schedules the arrived task to a certain machine depending on the chosen scheduler.

We analyze the schedulers using these eight parameters:

1. The system load;
2. The average number of busy machines;
3. The average turnaround time of the executed tasks;
4. The average response time of the executed tasks;
5. The number of finished tasks;
6. Provided that all tasks were finished, the time when they finished;
7. The number of deployments of the tasks that were involved
8. The load balance among the machines.

The simulation lasts 1000 rounds, the number of users in the system is 25 and every user generates 4 tasks. The number of the machines present in the system is 10. There are 20 predators and 5 victims. There are 20 users that generate big tasks (from 100 to 1000 million instructions) with probability of 0.9 and small tasks (from 1 to 100 million instructions) with 0.1. There are 5 users that generate small tasks with probability 0.9 and big ones with 0.1. The CPU power of each machine is obtained from the normal distribution where the mean is 10 million instructions in second and the variance is 3 million instructions.

First, we investigate the percentage of the simulation when the system was busy. In fig. 1 we can see that with the proportional scheduler the system is kept busy near 100% of the time, while with the Hamilton's method (HM) this percentage is lower (about 50%), the other schedulers kept the system busy for almost all of the time. This means that in this scenario HM leaves space (i.e. computational power) for other tasks and it used only half of the time steps to execute the given set of tasks.

Another parameter is the average number of machines that were kept busy during the simulation. In fig 2 we can see that the PPM scheduler keeps most of the machines busy, the HM keeps busy about 4 machines, while the MET scheduler keeps busy only the best machine for the whole simulation.

Fig.3 shows how many tasks were finished during the simulation. We can see that the worst performance is given by the MET scheduler, while only HM and PPM succeed to execute all the given tasks.

Another important parameter is the average turnaround time achieved by a certain scheduler (fig. 4). As expected, the best performance is given by the HM scheduler, and the worst performance is obtained using MET and then by Min-min. The HM is better than the rest because it may map one task to a number of machines and it is better from PPM because no time is lost for context switch which is needed for the Round-Robin used in the PPM. However, PPM and HM must divide the task and this means that we assume that tasks can be parallelized.

Fig.5 shows that the HM finishes all the tasks earlier than the other schedulers that succeed to finish all the tasks before the last round. The MET heuristic does not to finish all the tasks till the end of the simulation.

The average rounds that the tasks spend in the queue (i.e. the time between its deployment and arrival) is shown in fig.6. We can see that the HM copes well with the other heuristics. The best performances are given by Max-Min and the worst by Min-Min.

Depending on the type of the scheduler, tasks can be deployed to one or to many machines. Here the PPM and the HM scheduler do not cope with the rest of the schedulers (fig. 7). These results are obvious and mean that if all of these schedulers were implemented, the most time will be lost for deploying the tasks using the PPM and HM method. From the next figure we can see that the HM and PPM schedulers give good load balance among the machines, expectedly the worst load balance is given by the MET scheduler. We can see that good balance is also shown by the following: Min-Min, Max-Min, Simulated Annealing and MCT favors machines 3, 4 and 8 because these machines work on higher frequencies than others.

In fig. 9 the heuristics execution time is shown. The graph was made relative to Min-Min heuristics because this heuristic shows worse results than the other heuristics. It is obvious that the MCT heuristic loses the smallest amount of time for scheduling and it is also easy to implement. It is worth mentioning that the Simulated Annealing is very dependable on its scheduling parameters (i.e. heating rate and number of rounds). In this simulation the HM heuristic shows similar results as Simulated Annealing.

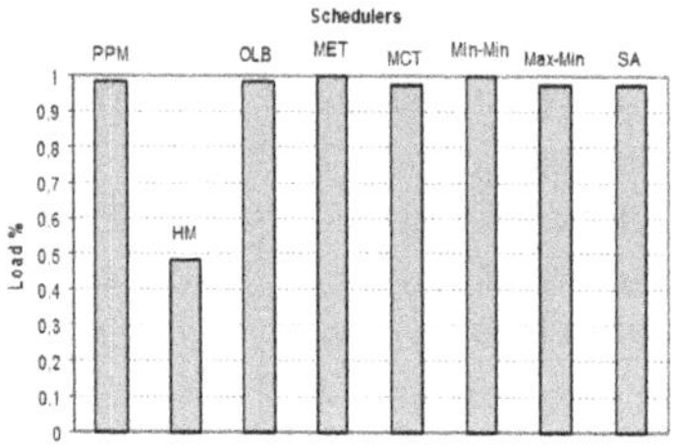

Fig. 1 The load of the system

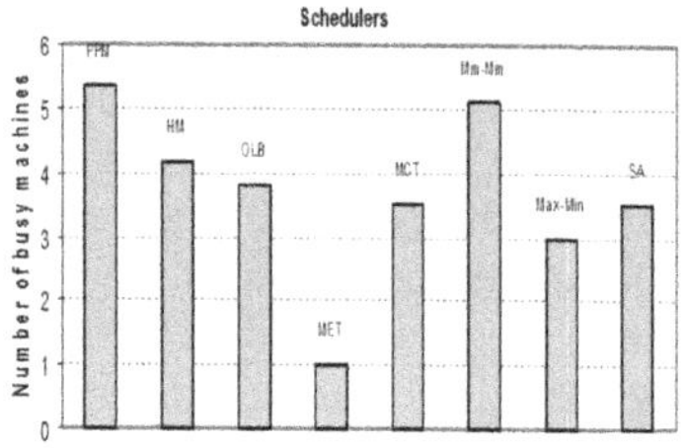

Fig. 2 Average number of busy machines

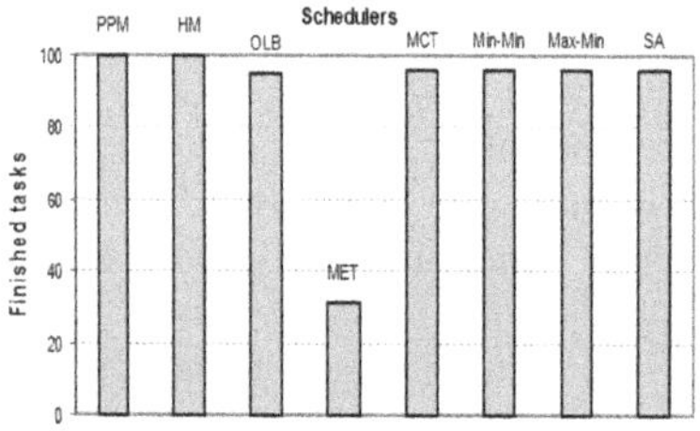

Fig. 3 Number of finished jobs

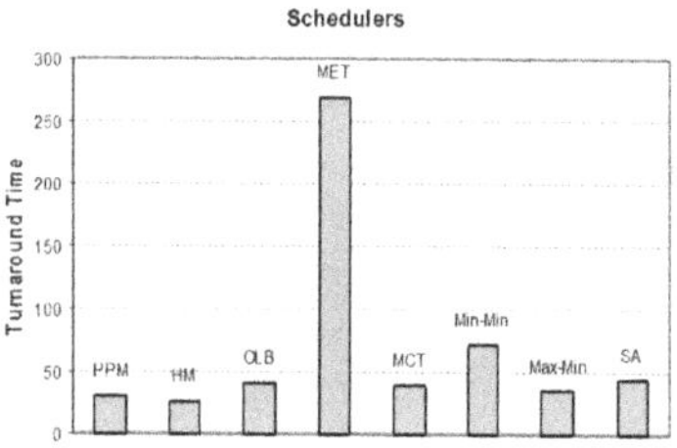

Fig. 4 The average turnaround time

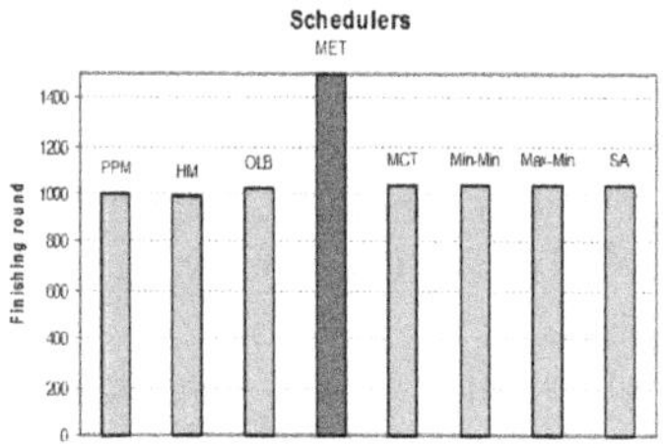

Fig. 5 Finishing round

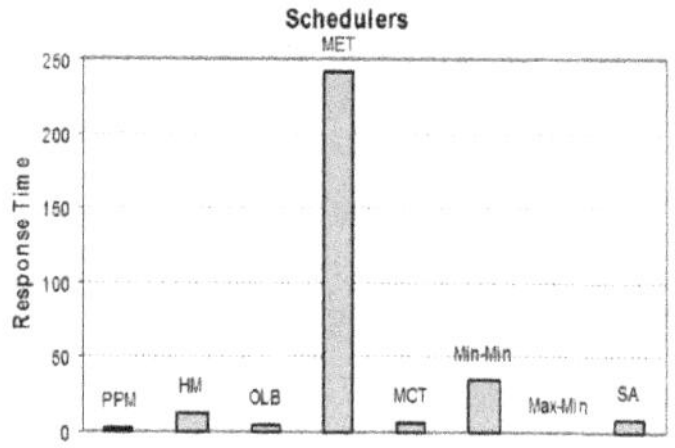

Fig. 6 Average response time

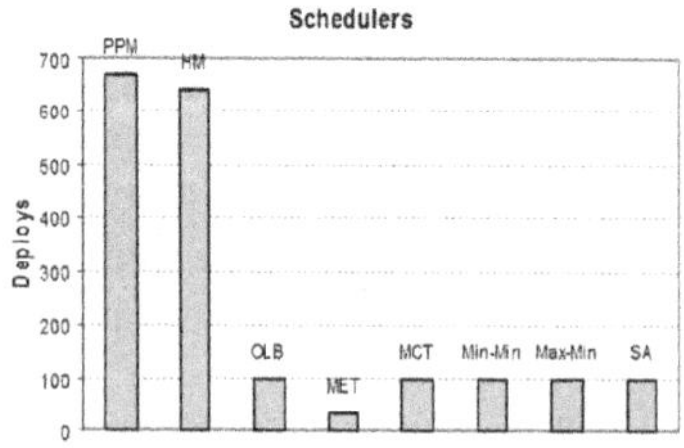

Fig. 7 Number of deploys

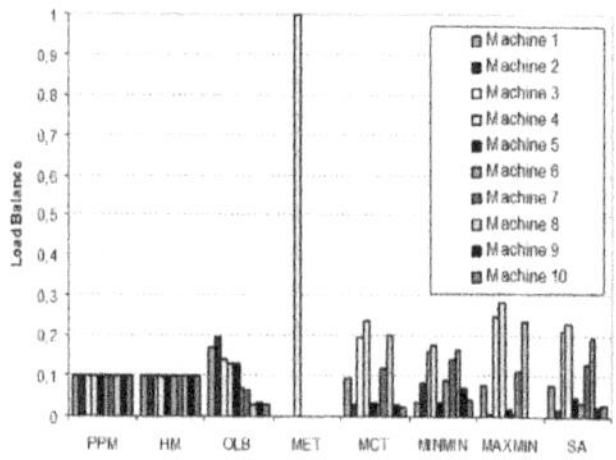

Fig. 8 Load balance among the machines

For broader analysis we reduced the number of machines present in the system from 10 to 5. This means that the system was more overloaded than in the first scenario and that the HM scheduler will dynamically change the mapping from one task-to-many machines to one task-to-one machine and that it is similar to

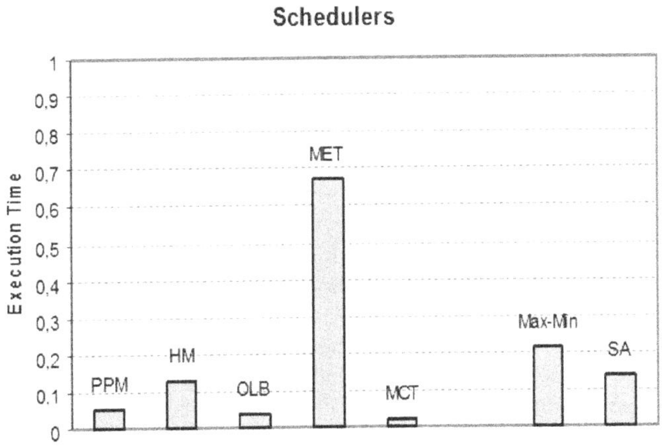

Fig. 9 Execution time for different heuristics

bigger tasks first kind of scheduler and certainly its performances will degrade. Results showed that Max-Min outperforms others in most of the performances and that HM copes well with the others in load balance. HM also finished the jobs earlier that the other schedulers.

5 Conclusion

In this paper we have implemented new heuristic based on the apportionment method used in elections. We also implemented other schedulers which are known and used in HC environments. The mappings were dynamic in batch mode. The implemented method is called the Hamilton's method. We also suggested a way how and when these apportionment methods can be used in mapping tasks in HC environments.

Additionally, we made a comparison of these heuristics in different parameters: the system load; the average number of busy machines; the average turnaround time of the executed tasks; the average response time of the executed task; the number of finished task; provided that all tasks were finished, the time when they finished; the number of deployments of the tasks that were involved and the load balance among the machines. The comparison is made using simulation in matlab and the results show that the Hamilton's method copes well with the other schedulers and it has good load balance. Moreover it finishes all the tasks earlier than any other scheduler.

This comparison shows under which circumstances one should choose the right scheduler in Heterogeneous Computer Systems.

References

1. Eshaghian, M.M. (ed.): Heterogeneous Computing. Artech House, Norwood (1996)
2. Freund, R.F., Siegel, H.J.: Heterogeneous processing. IEEE Comput. 26(6) (June 1993)

3. Maheswaran, M., Braun, T.D., Siegel, H.J.: Heterogeneous distributed computing. In: Webster, J.G. (ed.) Encyclopedia of Electrical and Electronics Engineering, vol. 8, pp. 679–690. Wiley, New York (1999)
4. Braun, T.D., Siegel, H.J., Beck, N., Bölöni, L., Maheswaran, M., Reuther, A.I., Robertson, J.P., Theys, M.D., Yao, B.: A taxonomy for describing matching and scheduling heuristics for mixed-machine heterogeneous computing systems. In: 1998 IEEE SRDS, pp. 330–335 (1998)
5. Fernandez-Baca, D.: Allocating modules to processors in a distributed system. IEEE Trans. Software Engrg. 15(11), 1427–1436 (1989)
6. Ibarra, O.H., Kim, C.E.: Heuristic algorithms for scheduling independent tasks on non-identical processors. J. ACM 24(2), 280–289 (1977)
7. Brawn, T.D., Siegel, H.J., Beck, N., Bölöni, L., Maheswaran, M., Reuther, A.I., Robertson, J.P., Theys, M.D., Yao, B., Freund, R.F., Hensgen, D.: A comparison study of static mapping heuristics for a class of meta-tasks on heterogeneous computing systems. In: 8th IEEE HCW 1999, pp. 15–29 (1999)
8. Wang, L., Siegel, H.J., Roychowdhury, V.P., Maciejewski, A.A.: Task matching and scheduling in heterogeneous computing environments using a genetic-algorithm-based approach. J. Parallel Distrib. Comput. 47(1), 8–22 (1997)
9. Maheswaran, M., Ali, S., Siegel, H.J., Hensgen, D., Freund, R.F.: Dynamic Mapping of a Class of Independent Tasks onto Heterogeneous Computing Systems. J. of Par. and Dist. Comp. 59(2), 107–131 (1999)
10. Pinedo, M.: Scheduling: Theory, Algorithms, and Systems. Prentice Hall, Englewood Cliffs (1995)
11. Yang, L., Schopf, J., Foster, I.: Conservative Scheduling: Using Predicted Variance to improve Scheduling Decisions in Dynamic Environments. In: ACM/IEEE SC 2003 Conference (SC 2003) (2003)
12. Braun, T., et al.: A comparison of eleven static heuristics for mapping a class of independent tasks onto heterogeneous distributed computing systems. J. of Par. and Dist. Comp. 61(6), 810–837 (2001)
13. Tao, Y., Wang, X., Gozali, J.: A Compensation-based Scheduling Scheme for Grid Computing. In: 7th ICHPCG in Asia Pacific Region (2004)
14. Forghanizadeh, S.: Grid Processor Scheduling based on Game Theoretic Approach, CPSC532A Final Project (2005)
15. Hariri, S., Topcuoglu, H., Wu, M.-Y.: Task scheduling algorithms for heterogeneous processors. In: 8th IEEE HCW 1999, pp. 3–14 (April 1999)
16. Vadhiyar, S., Dongarra, J.: A Metascheduler for the Grid. In: 11th IEEE International Symposium on High Performance Distributed Computing, HPDC 2002 (2002)
17. Holenarsipur, P., Yarmolenko, V., Duato, J., Panda, D.K., Sadayappan, P.: Characterization and enhancement of Static Mapping Heuristics for Heterogeneous Systems. In: Proc. of the 7th Int. Conf. HPC, December 17-20, pp. 37–48 (2000)
18. Kim, J.-K., et al.: Dynamic Mapping in a Heterogeneous Environment with Tasks Having Priorities and Multiple Deadlines. In: Proc. of the 17th Int. Symp. on PDP, April 22-26, p. 98.1 (2003)
19. `http://www.ctl.ua.edu/math103/apportionment/appmeth.htm`
20. Shapiro, R.: Methods of Apportionment, Apportionment of Representatives in the United States Congress House of Representatives and avoiding the Alabama Paradox, Engineering, University of Bridgeport, USA (December 5-13, 2008)

Semantic Stored Procedures Programming Environment and Performance Analysis

Marjan Efremov, Vladimir Zdraveski, Petar Ristoski and Dimitar Trajanov

Abstract. Supporting the idea of the semantic web, we developed a Semantic Stored Procedures (SSP) programming environment for the Oracle11g database, similar to the existing ones, available for the stored procedures in the relational databases. This show-case SSP-environment supports the basic test/store/execute/remove scenarios for sem_match-queries, but is also a quite extensive system. An eclipse plug-in that provides a GUI for these operations and java API which helps programmers to use the SSP-call functionality were released in the package with the Oracle11g SSP-extension. The performances of our system are comparable (much better in most of the cases) with the existing adapters, as shown with the measurements in comparison with the well-known Jena adapter. Thus, a complete development environment for SSP-programming is now available and the semantic query management, as a simple concatenation of strings inside the code, seems to becoming just a forgotten nightmare of our semantic web programmer's life.

1 Introduction

The technologies of the Semantic Web allowed development of novel approaches to data storage and retrieval. A semantic search system is essentially an information retrieval system which employs semantic technologies in order to enhance different parts of the information retrieval process. This is achieved by semantic

Marjan Efremov
Open Mind Solutions – Skopje, bul. Kliment Ohridski 20a/1-3, 1000 Skopje, Republic of Macedonia
e-mail: marjan.efremov@oms.com.mk

Vladimir Zdraveski · Petar Ristoski · Dimitar Trajanov
Faculty of Electrical Engineering and Information Technologies – Skopje, Rugjer Boskovic bb, 1000 Skopje, Republic of Macedonia
e-mail: {vladimir.zdraveski,petar.ristoski, dimitar.trajanov}@feit.ukim.edu.mk

L. Kocarev (Ed.): ICT Innovations 2011, AISC 150, pp. 357–366.
springerlink.com

indexing and annotation of content, query expansion, filtering and ranking the retrieved information [1]. The semantic search also introduces additional possibilities, such as search for an online ontology [2], search for online (distributed) knowledgebase, retrieval of facts from the ontology and knowledgebase and question answering.

In terms of providing better database support for the Semantic Web based applications, Oracle invented the Oracle 11g semantic database [2], which provides environment for storage and retrieval of semantic data, represented as RDF triples [3]. Together with the database, a Jena adapter for Oracle Database was written [4]. By use of the adapter, any Jena-based semantic application can be migrated on the Oracle 11g database. The performance and opportunities obtained by the Oracle 11g are comparable with the basic Jena (file based) solution. Retrieval is made by use of Jena Graph API [5] or with direct use of SPARQL [6]. With the use of SPARQL appears the problem, which is the impossibility of storing the queries inside the database. In the current environment, a query is managed as a simple string that is quite difficult scenario for the programmer.

Our solution of the problem is based on the existing stored procedures philosophy. A stored procedure (sometimes called a proc, sproc, StoPro, StoredProc, or SP), a subroutine available to applications accessing a relational database system, is a very essential part of the software, when any kind of a database storage is used.

We succeeded to implement a semantic stored procedures programming environment for sem_match queries. Sem_match is a query format, used by the Jena adapter for Oracle 11g (to query the database, programmer's SPARQL is converted to sem_match and then sent to the database), and is comparable with the SPARQL in terms of power, but more powerful in terms of response time. We designed a simple eclipse plug-in [7] as a user frontend for query testing, storing and removing scenarios.

2 Related Work

The Oracle 11g is the latest database release by Oracle, leading and only commercial database with native RDF/OWL data management. It is open, standards-based, scalable, secure, reliable and performing RDF management platform. Based on a graph data model, RDF data (triples) are persisted, indexed and queried, like other object-relational data types. Application developers use the power of the Oracle Database 11g to design and develop a wide range of semantic-enhanced business applications. [8] Stores rich, real-world relationships in the data beyond columns, table joins and Boolean to obtain more semantically complete query results. Enables machine-driven discovery of new relationships using the native Oracle Database inference engine, ontologies, and RDFS /SKOS/OWL semantics and user defined rules. Inferred data is persisted in the database for faster querying. [9] Oracle Database Strengths are Scalability (Trillions of triples), Availability (tens of thousands of users), Security (protect sensitive business data), Performance (timely load, query & inference), Accessibility (to enterprise applications) and Manageability (leverage IT resources) [10].

On the other hand, there is the well-known Jena. It is a Java framework for building Semantic Web applications and provides an API to extract data from and write to the RDF graphs. The graphs are represented as an abstract "model". A model can be sourced with data from files, databases, URLs or a combination of these. A Model can also be queried through SPARQL and updated through SPARUL [11]. Jena also provides support for OWL [12].

Sesame [13] is another existing standard framework for processing RDF data. It can be deployed on top of a variety of storage systems, relational databases, in-memory, file systems, keyword indexers, and offers a large scale of tools to developers to leverage the power of RDF and related standards. Sesame fully supports the SPARQL query language for expressive querying and offers transparent access to remote RDF repositories using the exact same API [14] as for local access.

Adapters that enable semantic frameworks to be deployed over Oracle Database 11g were also released. The Jena Adapter for Oracle Database provides a Java-based interface to Oracle Database Semantic Technologies by implementing the well-known Jena Graph and Model APIs. Similarly, The Sesame Adapter for Oracle Database [15] integrates the popular Sesame Java APIs with Oracle Semantic Technologies support.

All these tools are quite complete and easy to use, except that you have to build your queries as a concatenation of strings and send the whole query every time you need to execute it. Thus, the management of the queries becomes quite chaotic in large applications and sometimes it is almost impossible to code and/or make changes. Our main problem is that there isn't (wasn't) an environment similar to the stored procedures framework or of any other kind that provides those functionalities for the Semantic Web querying process.

3 Solution Description

In the context of solving the described problem we developed a simple show-case Semantic Stored Procedures Framework and implemented the most required operations and tools.

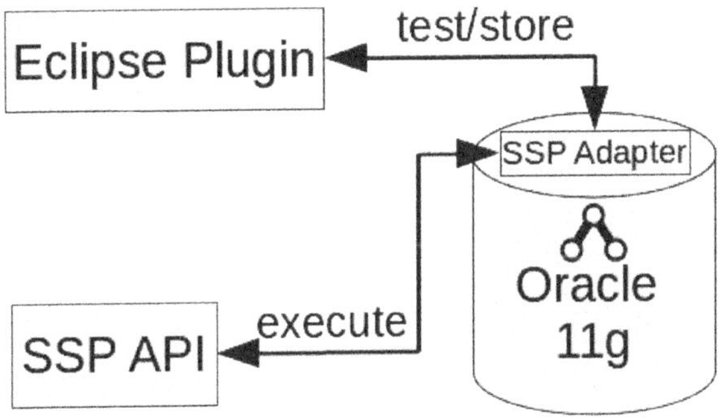

Fig. 1 The System Architecture

The system architecture, shown in Fig. 1, consists of the main block, the Semantic Stored Procedures (SSP) Adapter placed inside the Oracle 11g and provides the whole functionality related with the semantic database. Further, the eclipse plug-in was designed to provide interaction with the adapter during the development phase and make it easier. And at the end, classes and methods (the SSP API) for a Semantic Stored Procedures call were written.

3.1 *Client Side (The Eclipse Plug-In)*

The user of the plug-in is able to create a new semantic stored procedure via the New File GUI Wizard. After completing his sem_match query in the editor (shown in Fig. 2) he could test it, store it and remove it from the database directly via the GUI. These commands are shown in the context menu in Fig. 2. The server address and other server access parameters are red from an xml configuration file, placed inside the client project.

If the query contains variables, then the "Test Query"-command opens a dialog (shown in Fig. 3) that requires values for the variables to be entered, so the query can be tested properly.

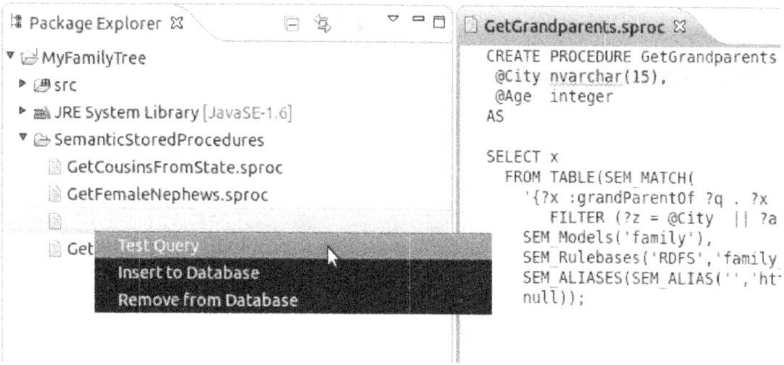

Fig. 2 The Eclipse plug-in

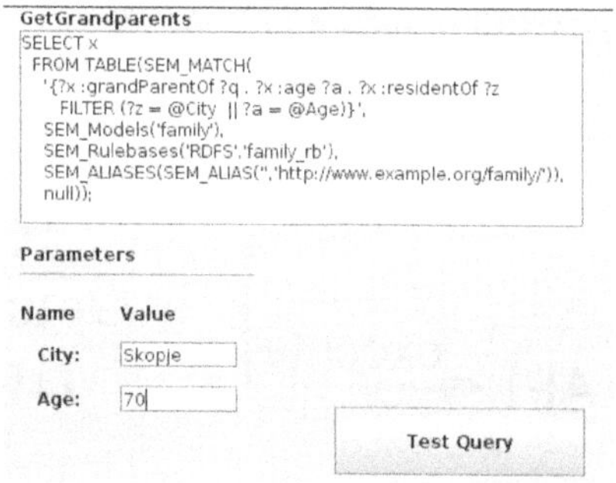

Fig. 3 The variable parameters

3.2 Server Side

First of all, we must set the semantic database, create a semantic model, create a table which will hold the semantic data, create a semantic entailment, some additional custom rule bases etc. [16] After the prerequisite step the database is ready to be queried.

On the server side four functionalities are implemented. They allow testing any SEM_MATCH query, creating and storing a procedure, execute previous created procedure (it can have variables in sem_match query and in runtime will be replaced with concrete values) and delete a previous saved procedure from database. All of this is programmed in stored procedures and kept in the database, on the server machine. The goal is all of the requested queries to be executed on the server side, the client just send the demand throw the plug in.

RDF/OWL data can be queried using SQL. The Oracle SEM_MATCH table function, which can be embedded in a SQL query, has the ability to search for an arbitrary pattern against the RDF/OWL models, and optionally, data inferred using RDFS, OWL, and user-defined rules. The SEM_MATCH function meets most of the requirements identified by W3C SPARQL standard for graph queries. [17] This function has the following attributes:

SEM_MATCH(query VARCHAR2,

> models SEM_MODELS, rulebases SEM_RULEBASES, aliases SEM_ALIASES, filter VARCHAR2, index_status VARCHAR2, options VARCHAR2)

RETURN ANYDATASET;

Only the query attribute is required. The other attributes are optional (that is, each can be a null value) [18]. The query attribute is a string literal (or concatenation of string literals) with one or more triple patterns, usually containing variables. (The query attribute cannot be a bind variable or an expression involving a bind variable).

Our goal is to skip any adapter and just do a direct connection between the client and the server, in our case with thin client from the Eclipse. With this, we achieve better performance, by avoidance of the transformation from one type to another type of query and converting the result data among different types of queries. In our case, the data grid from the execution of the query is converted to an xml file, with concrete defined format. Although this server-side formatting iterations slightly slows down the system, our framework allows us to define the xml format of the response, unlike the others, that return a standard grid-like result.

3.3 Advantages and System Usability

A simple preview of the available semantic architectures is shown in Fig.4. Here are presented the two Jena architectures compared to our solution. It is

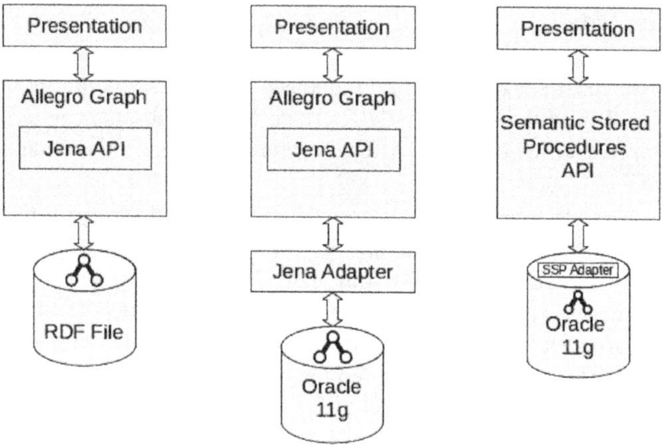

Fig. 4 Standard Jena from file (left), Jena adapter for Oracle 11g (middle), Semantic Stored Procedures Environment (right)

obvious that our solution places the SSP Adapter inside the Oracle 11g database (on the Server side) and that is the key advantage of the SSP solution.

Further, the SSP solution can generate customized format of the response data directly at the server-side, thus minimizing the parsing time on the client-side. The query is stored in the database once and then only called by its name. That makes the system more user-friendly and saves on network traffic and reduces the response time of the query, as will be shown with the measurements bellow.

4 Performance and Comparison with the Jena SPARQL

The thin lines on the graphs from Fig. 5 represent execution time of a single query and the thick lines represent the average execution time, averaged from 50 query executions. We tested with 7 different queries with increasing complexity. Testing was done on a local machine, so there is not network delay influence.

The SSP solution (darker lines) is almost twice faster than the one based on the Jena adapter for Oracle 11g, in case of small data sets. Performance decreases with the size of the dataset, especially because the size of the response dataset. That is a consequence of the server-side parsing of the result. The result of the execution of the sem_match query is a multi line (multi row) data set, which must be concatenated and formatted in one big set, to be easy operable and dynamically transformed. When executing sem_match query or storing sem_match query in a procedure, first of all, the requested columns in the "SELECT" of the query are parsed, spaces are removed and the final string is replaced in exact place in the dynamically built xml file. Aliases of the columns in the select statement are not allowed, it handles only column names.

Whatever the case is, first the requested sem_match query is executed and the result data set is treated like CLOB and formatted using Oracle XML Functions [19]. That is a reason why our framework has worse times than SPARQL over Jena.

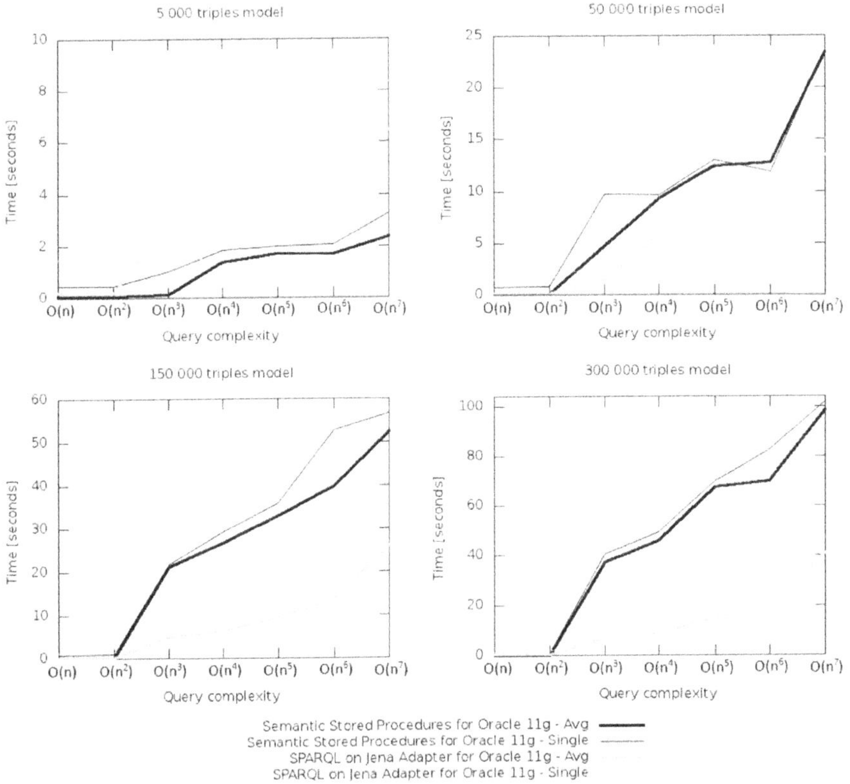

Fig. 5 The Performance Measurements

When the result is not formatted, the SSP solution response time is about a half of the Jena Adapter for Oracle 11g solution, in both of the tests, the single query and the average of many executions of it, shown in Fig. 6. In this measurement, Jena is also called without parsing the result, thus only the effective database retrieval time is measured in the both systems.

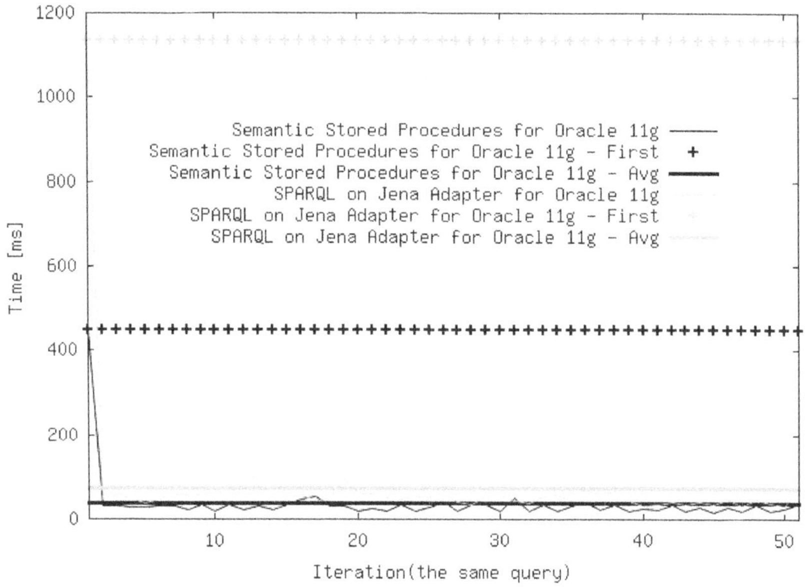

Fig. 6 Response time without server-side parsing

5 SEM_MATCH vs. SPARQL Power

There are two ways to query a semantic data stored in the Oracle Database: the SEM_MATCH-based SQL statements and the SPARQL queries through the Jena Adapter. The queries using each approach are similar in appearance, but there are important behavioral differences [20]. Further details about a comparison between the sem_match and the SPARQL and how the different parts are positioned, when a query is built can be found in [21].

SPARQL queries involving DISTINCT, OPTIONAL, FILTER, UNION, ORDER BY, and LIMIT are converted to a single Oracle SEM_MATCH table function (this is supported in the new Oracle release 11.2) [22]. If a query cannot be converted directly to SEM_MATCH because it uses SPARQL features not supported by SEM_MATCH (for example, CONSTRUCT), the Jena Adapter employs a hybrid approach and tries to execute the largest portion of the query using a single SEM_MATCH function while executing the rest using the Jena ARQ query engine. In this case, the Jena Adapter converts the inner UNION query into a single SEM_MATCH table function, and then passes on the result set to the Jena ARQ query engine for further evaluation [23].

Thus, there is a 1:1 mapping between the SPARQL and the sem_match, so you can do the same things with both of them, except that SPARQL is a bit more human-understandable and more widespread.

6 Conclusion and Future Work

As we mentioned previously, the SSP solution only implements the pattern provided for the SQL stored procedures in scope of the Semantic Web. The need of the stored procedures scenario is already proven, thus the need for a Semantic Stored Procedure is obviously clear. The measurements only confirmed our logical expectations. Thus, the show-case scenario was successfully implemented and the SSP environment is worth to be finalized and spread among the Semantic Web programmers.

The SSP eclipse plug-in can be improved in various aspects. The UI can be redesigned to a more user friendly one. The semantic stored procedure editor will offer syntax highlighting and code completion, so the user can easily construct complex queries. The plug-in will also offer synchronization with the database, so the user will be aware of the current state of the database and can take actions over the database via the plug-in GUI.

The user will be offered to write SPARQL queries instead of SEM_MATCH queries, so it will be much easier to use the plug-in. In order to work with SPARQL queries, a SPARQL to SEM_MATCH query converter will be developed (or used an existing), so every SPARQL query will be translated into one or more SEM_MATCH queries, which will be stored in database as a semantic stored procedure.

The query result is a XML formatted string, thus further extensions with the XSLT can take place and enable the user to provide a XML Schema string as a variable, within the query call and receive an answer formatted according to that XML Schema.

References

1. Strasunskas, D., Tomassen, S.L.: On Variety of Semantic Search Systems and Their Evaluation Methods. In: Proceedings of International Conference on Information Management and Evaluation, University of Cape Town, South Africa, pp. 380–387. Academic Conferences Publishing (2010)
2. Pan, J.Z., Thomas, E., Sleeman, D.: Ontosearch2: Searching and querying web ontologies. In: Proc. of the IADIS International Conference, Oracle 11g Database, pp. 211–218 (2006), `http://www.oracle.com/technetwork/database/enterprise-edition/overview/index.html` (accessed March 24, 2011)
3. RDF, Resource Description Framework (2010), `http://www.w3.org/RDF/` (accessed March 27, 2011)
4. Jena Adapter for Oracle Database, `http://download.oracle.com/docs/cd/E18283_01/appdev.112/e11828/sem_jena.htm`
5. Jena – A semantic web, java framework, Official API documentation and examples for Jena libraries (2010), `http://jena.sourceforge.net/` (accessed March 20, 2011)
6. SPARQL, `http://www.w3.org/TR/rdf-sparql-query/` (accessed March 20, 2011)

7. Eclipse - plug-in-based editor, http://eclipse.org/ (accessed March 10, 2011)
8. Semantic Technologies Center – Oracle, http://www.oracle.com/technetwork/database/options/semantic-tech/whatsnew/index-088828.html (accessed March 5, 2011)
9. Oracle Database 11g Semantic Features, http://www.oracle.com/technetwork/database/options/semantic-tech/whatsnew/semtech-makes-ed-sm-195114.html (accessed March 5, 2011)
10. Oracle Database 11g Semantics Technical Talk, http://www.oracle.com/technetwork/database/options/semantic-tech/whatsnew/oracle-33.pdf?ssSourceSiteId=otnjp (accessed March 5, 2011)
11. SPARUL, SPARQL update, http://www.w3.org/Submission/SPARQL-Update/ (accessed April 10, 2011)
12. OWL, Web Ontology Language, http://www.w3.org/TR/owl-features/ (accessed)
13. Sesame, A semantic web, java framework, http://www.openrdf.org/about.jsp (accessed April 10, 2011)
14. API, application programming interface, http://en.wikipedia.org/wiki/API
15. Sesame Adapter for Oracle Database, http://download.oracle.com/docs/cd/E18283_01/appdev.112/e11828/sem_sesame.htm (accessed April 10, 2011)
16. Murray, C.: Oracle Database Semantic Technologies Developer's Guide 11g Release 2 (11.2) (2010), http://download.oracle.com/docs/cd/E18283_01/appdev.112/e11828/sdo_rdf_concepts.htm#CIHHEDAC
17. Oracle Feature Overview, Oracle Database Semantic Technologies (2009), http://www.oracle.com/technetwork/database/options/semantic-tech/semtech11gr2-featover-131765.pdf
18. Oracle Semantic Technologies Overview, http://download.oracle.com/docs/cd/B28359_01/appdev.111/b28397/sdo_rdf_concepts.htm (accessed March 10, 2011)
19. Generating XML Data from the Database, http://download.oracle.com/docs/cd/B19306_01/appdev.102/b14259/xdb13gen.htm (accessed March 30, 2011)
20. SEM_MATCH and Jena Adapter Queries Compared, http://download.oracle.com/docs/cd/E18283_01/appdev.112/e11828/sem_jena.htm (accessed March 20, 2011)
21. Wu, Z., Perry, M., Kolovski, V.: Oracle Database Semantic Technologies: Understanding How to Install, Load, Query and Inference (2010)
22. Compilation of SPARQL queries to a single SEM_MATCH Call, http://www.filibeto.org/sun/lib/nonsun/oracle/11.2.0.1.0/E11882_01/appdev.112/e11828/sem_jena.htm#sthref298 (accessed March 20, 2011)
23. Murray, C.: Oracle Database Semantic Technologies Developer's Guide 11g Release 2 (11.2) (2010), http://download.oracle.com/docs/cd/E18283_01/appdev.112/e11828/sem_jena.htm

Performance Impact Correlation of Message Size vs. Concurrent Users Implementing Web Service Security on Linux Platform

Sasko and Aristotel Tentov

Abstract. The implementation of web service security standards provides end-to-end security attaching signature and encryption headers to SOAP messages, but it outcomes with message overhead, as well as requires complex cryptographic operations for each message or some parts of it. This leads to extra CPU time and memory for processing information related to WS-Security, as well as elapsed time for parsing the increased XML message. This paper describes a series of experiments focused on understanding the performance impact of different security-level mechanisms based on WS-Security. We create a baseline of performance data that can be used to explore if the parameter "message size" or "number of concurrent messages" impacts more to the server performance for different message types with or without different WS-security mechanisms implemented on web service hosted on web server on Linux platform. Also, we analyze the correlation between these two parameters.

Keywords: web service security, performance, response time, throughput, SOAP message size, Linux Operating System.

1 Introduction

Web service technologies become a baseline for business-to-business transactions today. More and more confidential and even highly confidential data is sent through

---

Sasko
Ss. Cyril and Methodius University / Faculty of Computer Science and Engineering,
Rugjer Boshkovik 16, 1000 Skopje, Macedonia
e-mail: sashko.ristov@finki.ukim.mk

Aristotel Tentov
Ss. Cyril and Methodius University / Faculty of Electrical Engineering and Information Technologies, Rugjer Boshkovik bb, PO Box 574,
1000 Skopje, Macedonia
e-mail: toto@feit.ukim.edu.mk

L. Kocarev (Ed.): ICT Innovations 2011, AISC 150, pp. 367–377.
springerlink.com

web services, both in internal and external company business services and environment. Thus, web services must provide strong security guarantees in those open dynamic environments with heterogeneous platforms, to ensure message confidentiality, data integrity, authentication, authorization, non-repudiation, service availability, as well as web service identification to the clients.

Therefore, web service security standards were developed as a part of a comprehensive framework [12], such as WS-Security [11], which can be considered as a de facto standard to provide a SOAP message security model. It is based on open W3C-approved standards, XML Encryption [10] and XML Signature [9] and is intended to provide a model for securing the messages in end-to-end communication and the standardized mechanisms to attach digital signature and encryption in the SOAP message headers.

The implementation of these security standards outcomes with the message overhead, as well as requires complex cryptographic operations for each message or some parts of it. This leads to extra CPU time and memory for processing information related to WS-Security, as well as elapsed time for parsing the increased size of XML message. On the other hand, increasing the number of concurrent requests also impacts negative to the server performance.

In this paper, we introduce experimental approach and methodology for testing the influence of implementing various WS-security technologies to the web server performance parameters, those are "response time" and "throughput", with various message size and different number of concurrent requests on web service hosted on Linux server operating system, as one of the best and the most used operating system platform as web server. We developed clients' and servers' web services using (1) no security, (2) XML Signature, and (3) both XML Signature and XML Encryption.

At the beginning, we analyze the message overhead generated to the SOAP messages using XML Signature and XML Encryption. In Section 2 we present the methodology used in experiments. Section 3 presents the results of the experiments. It also analyzes the results and compares the performance impact of the message size and number of messages. Section 4 summarizes the results from the experiments and in the Section 5 we present the plans for future work.

1.1 Related Work

We find many articles comparing the performance impact of different web service security techniques. In [1] authors analyze two most commonly used distributed models in Java: Web services and RMI. Performance comparison of WS Security Kerberos Token profile and X.509 Token Profile is analyzed in [3]. In [5] authors analyze communication and encryption delays, as well as message overhead for different security mechanisms, such as SSL/TLS and XML Encryption and XML Signature.

Other related work is shown in [4] where authors evaluate existing and new techniques for WS-Security signature performance optimizations to speed up end-to-end message integrity assurance and authentication. Also, [7] identifies opportunities for optimizing WS-Security by examining the overhead of WS-Security operations.

Other related articles compare the performance impact of the different security-level mechanisms based on WS-Security. In [6], the authors compare the performance of a Web service without security with other Web services using the WS-Security to add encryption and digital signatures in SOAP messages. Impact of adding WS-Security to SOAP-based web services is analyzed in [2]. In [8], the authors compare the performance of WS-Security operations and choice of signature and encryption algorithms to non-secure messages using various message sizes and complexities. WS-Security optimizations are not considered.

In [16], we analyzed the agent-based web service, its absolute and relative response time overhead implementing security and increasing number and size of messages.

None of the related work neither experiment nor analyzed the performance impact of the different security-level mechanisms based on WS-Security, using various message size and concurrent requests.

This paper describes a series of experiments focused on understanding the performance impact of different security-level mechanisms based on WS-Security. We create a baseline of performance data that can be used to explore if the parameter "message size" or "number of messages" impacts more to the server performance for different message types with or without different WS-security mechanisms.

2 Experiments

The experiments were run on client-server architecture. Web services were run on Dell PowerEdge 1950 server, with 4 x 1.995 GHz CPU cores Intel Xeon E5405 and 4GB RAM. Using VMware ESX Server 3i [14], Linux Ubuntu 10.10 was installed with 1GB RAM memory and 2 CPU cores. Jetty [13] is used as a web and application server. As a client platform, we used notebook with 64-bit Windows 7 operating system, with 4GB RAM, CPU Intel Core 2 Duo P8600 2.4GHz, and 1Gbps Network card. Client and server were in the same LAN segment to exclude the network impact shown in [1]. We used SOAPUI [15] to create various server load tests.

During the test experiments, response time and throughput is measured for various number and size of concurrent request for different message types.

The test experiment basic goal is to measure the performance cost of implementing signature and encryption to SOAP messages. For that purpose, as a test data we used web service which accepts two string parameters and returns string which is concatenation of the two input strings. The first part of the experiment consists of series of test cases to examine the impact of increasing the message size to the server response time. The second part consists of series of test cases to examine the impact of increasing the number of the concurrent messages to the server response time.

All the test cases are done using (1) regular SOAP messages, (2) signed messages and (3) both signed and encrypted messages. In (1), the web server receives regular message, parses it and returns the concatenated string into SOAP message as a response. In (2), the web server receives signed SOAP message, verifies it, and if so, parses it and returns back the result parameter string r_1 into signed message as a response. In (3), the web server receives encrypted and signed message and then it decrypts and verifies it, and if so, the server returns the encrypted and signed message with parameter r_1 as a response.

Every test case is run in a time of 60 seconds, *N* messages are sent with *M* bytes each, with variance 0.5s. Parameters *N* and *M* are changed for each test case.

This paper interests server response time neither in burst nor in overload mode, but only in regular mode. The idea is to establish the server performance baseline for different WS-Security techniques in normal server workload mode.

SOAPUI is installed at the client side for simulating different test cases. Client is in the same LAN as the web server, with network response time (ping) smaller than 1ms, and during the tests, does not exceed 10ms and there isn't any lost packet. Thus, we can assume that the response time measured with SOAPUI is the same as the server response time. The sizes in kilobytes of the original (regular, unsecured) SOAP messages sent in test cases are approximately: 0.2, 2, 10, 20, 60, 70, 100, and 1000. The server load is done with various numbers of the messages (requests) in order to retain normal server workload mode. That is, from minimum 1 to maximum 1000 requests per second.

We expected to observe that response time will be increased while increasing the number and the size of the messages. The main question is: Which parameter impacts more to the server performance? Is it the number of concurrent users, or message size for different security-level mechanisms based on WS-Security?

Before each test, monitors are checked. After each test, all the server performance parameters are examined if they are returned into nominal state. If not, the server is restarted and returned into its nominal state. During the tests, network latency is measured to ensure that it does not affect to the response time results.

3 Results and Analysis

This section describes the results of the performed tests to measure the implication of the security mechanisms. We also analyze the results to understand the impact of implementing different security-level mechanisms on the Linux server performance.

3.1 Message Overhead Introducing XML Security

First, we measured the message overhead when creating (1) unsecured XML message, (2) signed XML message, and (3) signed XML message and then encrypted. To compute the overhead, we extracted the message sizes for each security mechanism.

As shown in Fig. 1, security mechanisms that do not use encryption, such as the regular SOAP message and Signed message add a constant amount of bytes to the regular parameters. The mechanisms with encryption included add a linear amount of bytes comparing to parameters size.

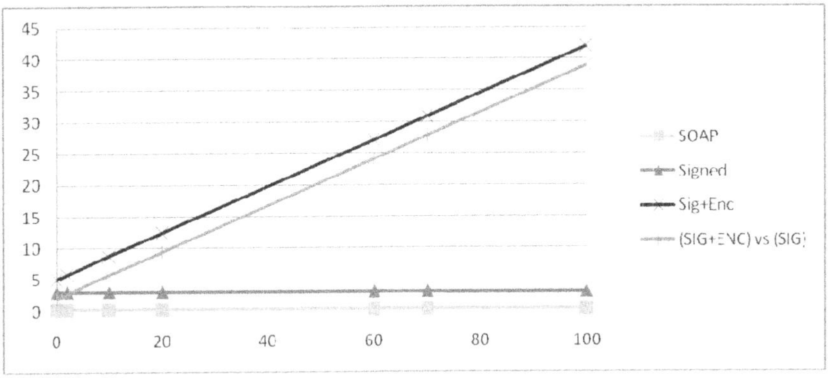

Fig. 1 Message overhead in kilobytes per various security mechanisms.

In [2] authors give the similar analysis, but we improved it with the fact that signed and then encrypted messages add linear overhead not only to original SOAP message, but also to signed messages, what means that adding the encryption to regular or signed message adds a linear overhead.

Using simple linear regression, for the signed and encrypted message overhead over the parameters, we get the equation that best fit the points.

$$y = 0.3684 * x + 5093.7995. \tag{1}$$

where y is the signed and encrypted message overhead, and x is the sum of sizes of the original input parameters s_1 and s_2, all in kilobytes.

3.2 Message Size Performance Impact

Next, we measure the impact of the message size for a constant number (1, 10, and 100) of concurrent messages per second. The results are shown in Fig. 2, Fig. 3 and Fig. 4.

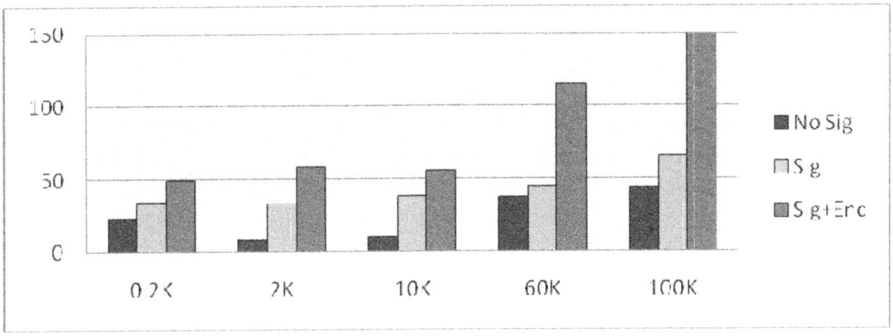

Fig. 2 Response time in ms per various security mechanisms for 1 message / sec

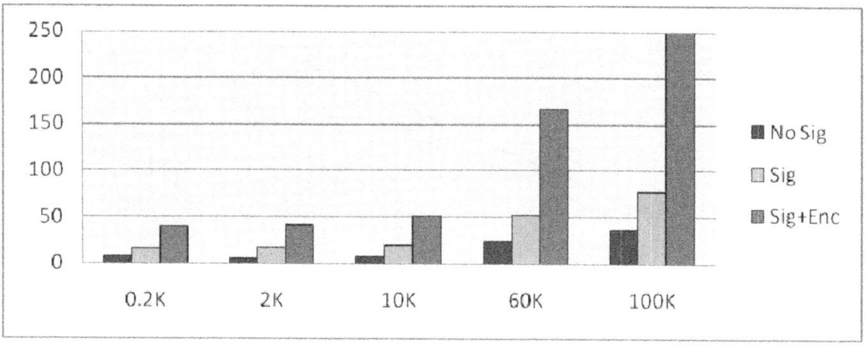

Fig. 3 Response time in ms per various security mechanisms for 10 messages / sec

As we can see, common for all figures is that increasing the message size, response time is also increasing for all three message types, which decreases the server performance. Also, increasing the message size have more influence to response time of both encrypted and signed messages, than only signed, especially with larger payload of 100 messages in second.

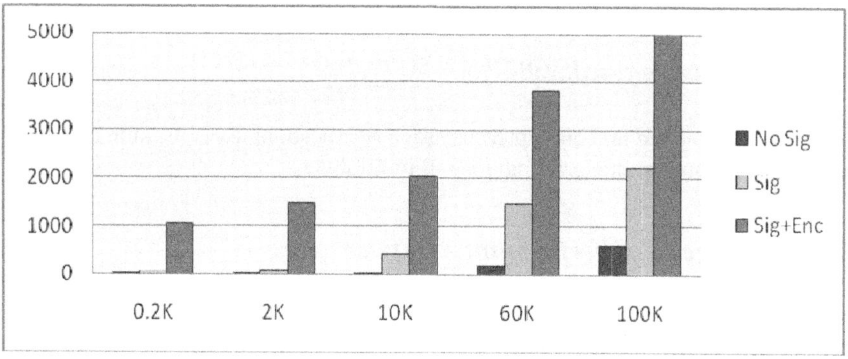

Fig. 4 Response time in ms per various security mechanisms for 100 messages / sec.

3.3 Concurrent Messages Performance Impact

Next, we measure the impact of the number of concurrent messages per second for a constant message size (2, 60 and 100K). The results are shown in Fig. 5, Fig. 6 and Fig. 7. As we can see, common for all figures is that increasing the number of concurrent messages per second, response time is also increasing for all three message types, which decreases the server performance. Also, increasing the number of concurrent messages have more impact to response time of both encrypted and signed messages, than only signed, especially with payload of larger messages.

3.4 Which Parameter Impacts More?

The main question here is: What is worse payload for the server: is it large number of small messages, or small number of large messages? Or maybe in the middle, that is medium number of medium sized messages? And what is the answer for all these questions for the three message types.

We found the answers in the next performance parameter, that is, throughput, measured in bytes per second. We analyzed the real vs theoretical throughput for each test, and found the intersection of each two curves, as shown on Fig 8. The X axis is the number of concurrent messages, and Y axis is throughput sized in MB/min (per test). Right from the intersection, that is, for larger number of concurrent messages, the real throughput is smaller than in the left, which means that we can use the intersection dot of each test as maximum throughput. This feature characterizes all the test cases we measured.

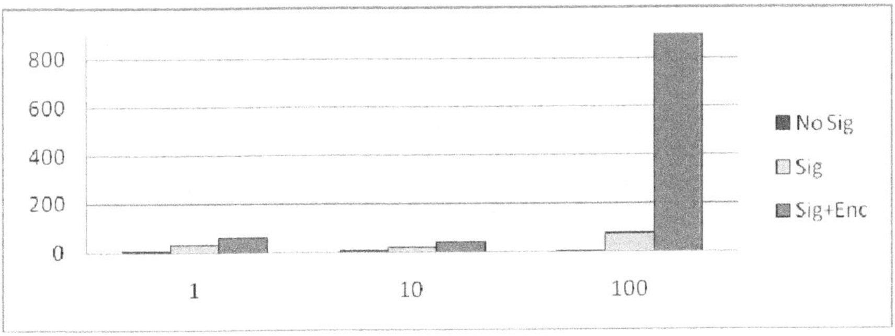

Fig. 5 Response time in ms per various security mechanisms for message size of 2K

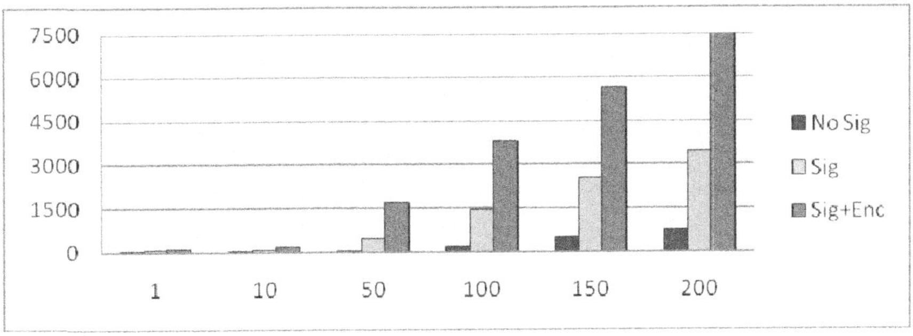

Fig. 6. Response time in ms per various security mechanisms for message size of 60K

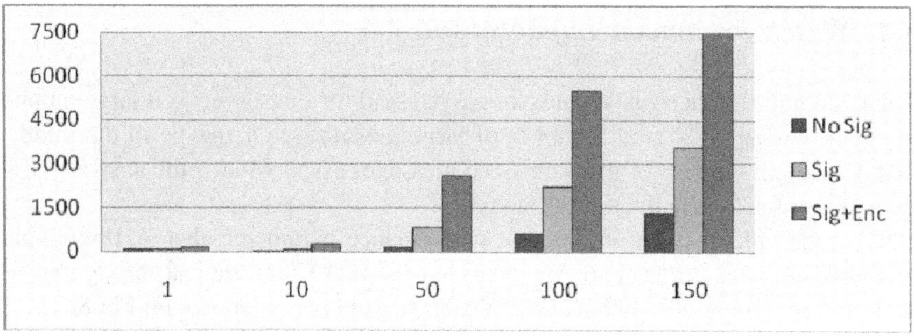

Fig. 7 Response time in ms per various security mechanisms for message size of 100K

From Table 1 we conclude that server reaches maximum throughput in the range of 200-500 messages (without security) with size of 10K-60K. The maximum throughput we measured is 480.5MB per minute, or 8MBps. Thus, the best payload for the server is with medium sized messages and medium number of messages.

For signed messages, the results of the tests are shown in Table 2. We conclude that server reaches maximum throughput in the range of 10-20 signed messages with size of 60K-100K. The maximum throughput we measured is 84MB per minute, or 1.4MBps. Thus, again, the best payload for the server is with medium number of messages and medium sized messages, but smaller to messages without security.

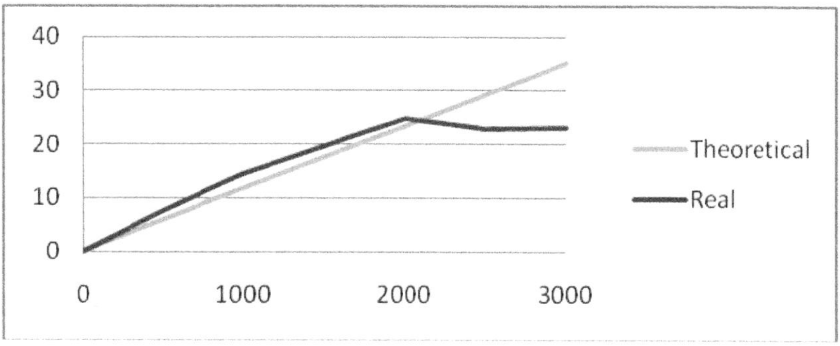

Fig. 8 Response time in ms without security mechanisms for message size of 0.2K

For signed and then encrypted messages, the results are shown on Table 3. We conclude that server reaches maximum throughput in the range near of 10 signed and then encrypted messages in a second with size near to 60K. The maximum throughput we measured is 55.81MB per minute, or 930KBps. Thus, again, the best payload for the server is with medium number of messages and medium sized messages, but smaller to messages without security or signed only.

Table 1 Maximums for different message size without security.

Concurrent Messages	Message Size (KB)	Max. Throughput (MB/min)
2085	195	24.39
1344	2189	176.52
621	10189	379.98
397	20189	480.56
109	60141	393.45
84	72189	362.22
59	102189	363.74
1	1002189	37.44

Table 2 Maximums for digital signed messages with different size.

Concurrent Messages	Message Size (KB)	Max. Throughput (MB/min)
126	195	13.97
121	2189	27.91
73	10189	51.67
18	60141	69.39
18	72189	79.65
14	102189	84.05
1	1002189	37.80

In Fig. 9 is shown the comparison of maximum throughputs for the three message types for a different message size.

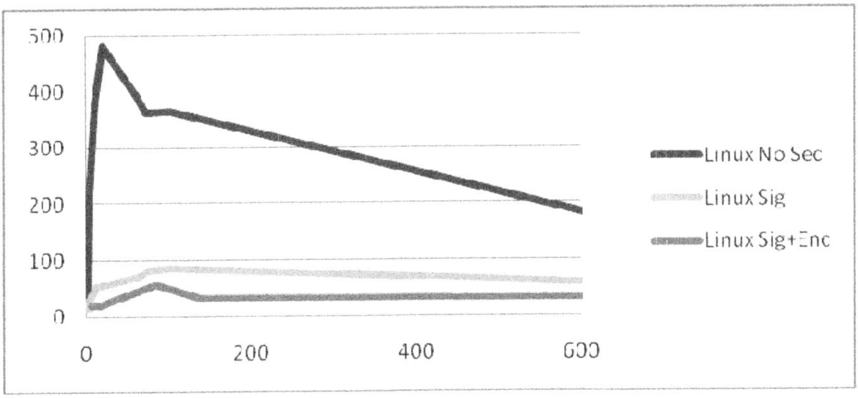

Fig. 9 Curve of maximums throughput for a various message size in Kilobytes

Table 3 Maximums for digital signed and then encrypted messages with different size.

Concurrent Messages	Message Size (KB)	Max. Throughput (MB/min)
51	195	20.24
19	2189	19.72
15	10189	27.56
11	60141	55.81
4	102189	30.57
1	1002189	28.40

4 Conclusion

In this paper we have done performance analysis and comparison on two web server performance parameters, "response time" and "throughput", via web services with two main input factors "message size" and "number of messages" implementing different security-level mechanisms based on WS-Security. Experimentally we establish performance baseline for introducing signing and encryption of the SOAP messages. We measured linear equation introducing encryption of the both signed and unsigned messages, and a constant overhead in signed messages.

We defined quantitative indicators to determine risk of introducing signing and encryption to the SOAP messages, for various message size and number of concurrent messages. At the end, we defined the range of best server performance for various message size and number of messages for signed and encrypted messages. The best payload for server is with medium number of messages and medium sized messages.

With these results and analysis, we believe that IT quality managers can predict server response time and throughput for a given message type, number and size of concurrent messages, in order to prevent server overhead and customer dissatisfaction.

5 Future Work

During the experiments, we used a constant Linux operating system platform and Java web services. Also, all the experiments were done on the same web server hardware and software. Also, the request and response messages were equal.

Our intention is to define the equation which will have as input parameters: RAM, number of CPU cores, operating system, number of messages, message size, type of messages and different sized requests and responses. As an output, we intend to get server response time and throughput.

References

1. Juric, M.B., Rozman, I., Brumen, B., Colnaric, M., Hericko, M.: Comparison of performance of Web services, WS-Security, RMI, and RMI–SSL. Journal of Systems and Software - Special issue: Quality software 79(5), 689–700 (2006)

2. Novakouski, M., Simanta, S., Peterson, G., Morris, E., Lewis, G.: Performance Analysis of WS-Security Mechanisms in SOAP-Based Web Services. Technical Report CMU/SEI-2010-TR-023, USA (2010)
3. Moralis, A., Pouli, V., Grammatikou, M., Papavassiliou, S., Maglaris, V.: Performance Comparison of Web Services Security: Kerberos Token Profile Against X.509 Token Profile. IEEE Computer Society, Washington, DC, USA (2007)
4. Van Engelen, R.A., Zhang, W.: An Overview and Evaluation of Web Services Security Performance Optimizations. IEEE Computer Society, Washington, DC, USA (2008)
5. Shopov, M., Kakanakov, N.: Evaluation of a single WS-Security implementation. In: Proceedings of the International Conference for Automatics and Informatics 2007, Sofia, Bulgaria, pp. VI-39–VI-42 (2007)
6. Rodrigues, D., Estrella, J., Branco, K.: Analysis of Security and Performance Aspects in Service-Oriented Architectures. International Journal of Security and Its Applications 5(1), 13–30 (2011)
7. Engelen, R., Zhang, W.: Identifying Opportunities for Web Services Security Performance Optimizations. In: IEEE Congress on Services - SERVICES, pp. 209–210 (2008)
8. Liu, H., Pallickara, S., Fox, G.: Performance of Web Services Security. In: Proceedings of the 13th Annual Mardi Gras Conference, Los Angeles, pp. 72–78 (2005)
9. Eastlake, D., Reagle, Solo, J.D. (eds.) XML-Signature Syntax and Processing W3C Recommendation (2002), http://www.w3.org/TR/xmldsig-core/
10. Eastlake, D., Reagle, J. (eds.) XML Encryption Syntax and Processing. W3C Recommendation (2002), http://www.w3.org/TR/xmlenc-core/
11. Nadalin, A., Kaler, C., Hallam-Baker, P., Monzillo, R. (eds.): Web Services Security: SOAP Message Security 1.1 (WS-Security, 2006), (OASIS Standard) (March 2004), http://docs.oasis-open.org/wss/v1.1/wss-v1.1-spec-os-SOAPMessageSecurity.pdf
12. Security in a Web Services World: A Proposed Architecture and Roadmap A joint security white paper from IBM Corporation and Microsoft Corporation, Version 1.0 (2002), http://download.boulder.ibm.com/ibmdl/pub/software/dw/library/ws-secmap.pdf
13. Jetty Web server and javax.servlet container, http://www.eclipse.org/jetty/
14. VMware ESX, http://www.vmware.com/products/vsphere/esxi-and-esx/index.html
15. SOAPUI testing tool, http://www.soapui.org/
16. Ristov, S., Tentov, A.: Security Based Performance Issues in Agent-based Web Services Integrating Legacy Information Systems. In: Proceedings of the WASA 2011. CEUR Workshop Proceedings, vol. 752, pp. 45–51 (2011) ISSN 1613-0073

The Comparison of Virtual Machine Migration Performance between XEN-HVM, XEN-PV and Open-VZ

Igli Tafa, Ermal Beqiri, Elinda Kajo, Hakik Paci, and Aleksander Xhuvani

Abstract. The aim of this paper is to compare the performance between three hypervisors: XEN-PV, XEN-HVM and Open-VZ. We have simulated the migration of a virtual machine by using a warning failure approach. Based on some experiments we have compared CPU Consumption, Memory Utilization, Total Migration Time and Downtime. We have also tested the hypervisor's performance by changing the packet's size from 1500 byte to 64 byte. From these tests we have concluded that Open-VZ has a bigger CPU Consumption than XEN-PV, but the Total Migration time is smaller than in XEN-PV. XEN-HVM has a worse performance than XEN-PV, especially regarding to downtime parameter.

Keywords: Hypervisor, XEN-PV, XEN-HVM, Open-VZ, CPU Consumption, Memory Utilization, Downtime.

1 Introduction

One of the most interesting technologies in the field of information technology nowadays is Virtualization. This technology gives some advantages regarding costs, sources and energy consumption, tolerance to failures, isolation to different

Igli Tafa · Elinda Kajo · Hakik Paci · Aleksander Xhuvani
Polytechnic University of Tirana, Faculty of Information Technology,
Computer Engineering Department, Tirana, Albania
e-mail: itafaj@gmail.com
e_kajo@yahoo.com
hakikpaci@gmail.com
axhuvani@yahoo.com

Ermal Beqiri
Tirana University, Tirana University, Mathematics & Statistics & Applied
Informatics Department
e-mail: ermalfr@yahoo.fr

L. Kocarev (Ed.): ICT Innovations 2011, AISC 150, pp. 379–394.
springerlink.com

attacks etc. Anyway, in this technology there are some black holes which have to do with the performance of the applications related to communication speed, sources or their energy consumption.

To realize a virtualization, it's needed to establish a hypervisor. The hypervisor is the administrator and the manager of the sources used by the virtual machines. The hypervisor can be established above the hardware and this is called full virtualization, or it can be established above the operating system and this is called OS virtualization. The full virtualization has got the advantage to integrate physical machines with different characteristics for example Intel x86 with AMD without doing any modification in Operating System's kernel. The performance offered by this type of virtualization is not high (i.e the communication with I/O devices is slow etc). To increase the performance is used the para-virtualization approach. This approach requires the modification of the kernel of Guest Operating Systems. It is also required that the processors have the same characteristics. In this way, the communication between applications in virtual machines and I/O devices will be realized through virtual I/O drivers which are built over the hypervisor. It will offer an increase of communication speed between applications and I/O devices.

One of the most important points in the technology of virtualization is live migration. This means that if a virtual machine which is running an application undergoes a discontinuity or it's CPU is loaded heavily, then the application can be migrated from this virtual machine to another. The virtual machine migration includes the transfer of page memories that are working; the transfer of the sources that are participating in this application (i.e network card, disc etc) and CPU's status. Each virtual machine has got it's own CPU, it's own physical memory (which is shared between different virtual machines), it's own I/O etc. Memory migration is the most crucial point of virtual machine's migration. There are some methods of it's migration, but the most used is the iteration method with Pre- Copy approach. In this method, initially transfers the modified pages which are recorded in a memory table which manages by XEN Hypervisor. This table traced by a bitmap method which is mark with '1' when a memory page is written which is called "downtime", is calculated in the order of milliseconds, despite of migration capacity, this is because the migration time is not high. In our article we have studied three types of hypervisors and we have measured their performance regarding the file's time transfer, CPU efficiency and memory utilization from a virtual machine to another in the same physical host or between virtual machines in different physical hosts. Hypervisors are Xen-PV, Xen-FV and OpenVZ,. Xen hypervisor is above the bare hardware, whereas OpenVZ is a hypervisor which is above the Host Operating System. Xen Hypervisor can transform from Para-Virtualization level to Full-Virtualization level if we set up **Xen/Qemu** emulator. So we can setup the emulation of I/O drivers in user's space. To achieve in this result it is necessary the hardware to support this technology. Intel and AMD processors recently support virtualization (Intel VT dhe AMD- V). Using these processors we can setup the Full Virtualization technique. This means that a Guest-OS in our Virtual Machine could be an Windows XP OS. These processors use VTX/SVM instructions. Generally Full Virtualization decreases the communication's performance with I/O disks. This technique often is called Xen-HVM. Usually Host-OS is called Dom0 and Guest-OS is called DomU. Unlike Xen-HVM where an application uses two system calls to manage a driver, Xen-PV uses special calls which offer the possibility to access virtual drivers which are managed by Dom0 and can be connect directly with the hardware. Referring to Xen-PV, the hypervisor lies in ring 0, whereas Guest-OS lies in ring 1. The applications are in the third ring, the second ring is not used, like it's shown in Figure 1. In Xen-HVm the ring 0 is reserved for GuestOS and the virtualized hardware lies in ring 1.

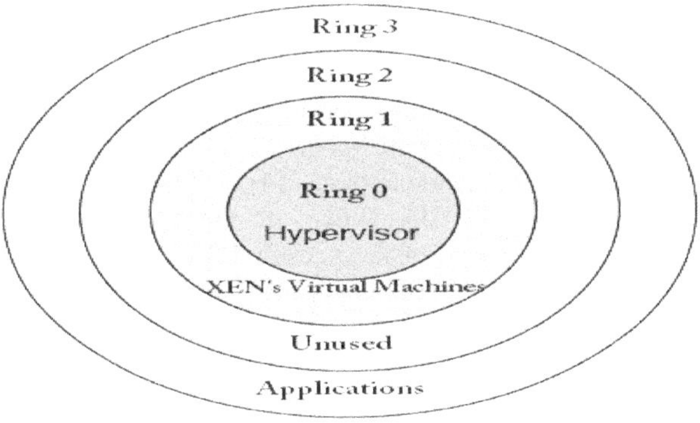

Fig. 1 The rings of x86 architecture and the way they are used by XEN

Open-VZ is an OS Virtualization. Guest-OS is called container or Virtual Private Server. Unlike Xen where each Guest has got it's own kernel, in OpenVZ all the containers share the same kernel with Host-OS. Anyway every Guest-OS has it's own IP, it's own I/O and it's own memory. Since every Guest-OS is a process in Open-VZ hypervisor, this method offers a better possibility than Xen in scalability but weaker in isolation. Open-VZ can modify the Linux's kernel giving to every unmodified Linux-based OS the possibility to be executed as a process in Linux.

All the above hypervisors support the SMP (Symmetric Multi Processor) technique. This means that some Guest-OS can use some host processors at the same time.

2 Related Works

In reference [15] is shown the difference between Open-VZ and Xen and is analyzed their performance. From those experiments, Open-VZ has got a higher speed then Xen. The speed in reading process is almost the same for both Hypervisors. Performance of Open-VZ is better than Xen because it introduces a smaller complexity then XEN. Guest-OS in Open-VZ seems like process and share the same kernel with the Host OS.

In [3] are used different measurements using different tools like ''**pktgen**'', a module that incorporates in Linux's kernel and it serves in traffic packet generation (packets with different sizes) from one host to another. Another benchmark called '' **stress tool**'' is used to measure CPU consumption and memory utilization. It is noticed that when packet's size decreases from 1500B to 64B, it is not utilized the whole bandwidth in the hypervisors, anyway Open-VZ has the best performance. In [3] is evaluated the situation when there are used different streams, which distributed packets with different sizes for both hypervisors from a virtual machine to another in a computer network connected with a gigabit switch. In all experiments, Open-VZ has the best performance.

In [2] is tested the performance between Xen and Open-VZ. In this system is built for the first time the multilayer approach where Web Server lies in a layer, DB in another layer and the PHP interface in the last one. This multi- layer approach, increases the scalability of the system's performance, the isolation and the speed is higher than in analogue cases [4]. The tool used to measure their performance is called **RUBIS**. When the number of the applications increase, the average response time of the packets RTT in Open-VZ is four times smaller than in the first case. This occurs because Xen has a bigger overhead than OpenVZ. Based on [2] there are a lot of miss cache for instruction in L2 cache. Anyway, from [2] is seen that Open-VZ consumes more CPU because of the share kernel of host and guests and the CPU fair sharing between containers. To measure the overhead here is used the tool **''Oprofile''**. This tool generates data anytime there is a Hardware event, i.e anytime it causes a miss cache. This tool [5] is adapted to the XEN performance and is called **''Xenoprof''**. To measure the CPU consumption in XEN is used the tool '' **xentop –b**'' which gives detailed information about the CPU consumption of every Guest-OS. There isn't any specific tool to measure the CPU consumption of the containers in Open-VZ, anyway here the data are measured from the generation of the time report which gives the time CPU consumes in every container in /proc/vz/vstat. To measure the performance for each hypervisor it is used the **RUBIS** benchmark which can increase the CPU load and the number of threads which are generated by a script in C. It is evaluated the throughput, the response time of the packets and CPU consumption. In all cases is concluded that Open-VZ has the best performance.

In [6] is shown that the creation of a multilayered disc increases the migration performance of virtual machines. Xen has lower scalability than Open-VZ, because in Open-VZ the GuestOSs are processes, so they consume a lot of memory and processing [4]. In reference [8] is compared the performance of CPU consumption for the same applications between XEN-PV, Xen-FV. It is seen that XEN-PV consumes less CPU. To measure the CPU performance here is used the tool **''SAR''**. XEN-PV has the highest speed of writing SAS disc.

3 Background

In this article we want to test three parameters: CPU Consumption, Memory Utilization and the Migration time of the virtual machine due to a warning failure.

To create a warning failure it is used a tool of Cent-OS 5.5 called **''Heartbeat''**. By using this tool we can get information if any machine has ''dead'' status. So the Hypervisor will migrate the applications (actually not only the applications) that were running in the ''dead'' machine to another one in order to continuing the process without any delay. In our article we will realize a script in C which will simulate the stop of the virtual machine in the physical host, regardless that in the reality it hasn't stopped yet. So, the virtual machine in the first physical host will finish its execution only at the moment when this machine will continues the execution in the second physical host by migration process. Thus the performance will be better than in the case when ''Heartbeat'' acts as usual [9]. Study of an uncontrolled failure will be an object in the future. However, we will examine the case when the virtual machine is relocated in the same physical host. The virtual machine's migration realized some steps:

a. The migration of memory pages that are in RAM which belong to the application that was being executed in the virtual machine.
b. The migration of the drivers of I/O devices.
c. The migration of virtual I/O discs as part of the activity of the virtual machine.
d. The migration of CPU-status.

This method is called pre-copy [10]. The purpose is that during the migration of the applications, we can reduce the time of the down-time as much as possible. In pre-copy approach, the down-time is lower, but there is a problem with the total time of the migration as a result of the iteration of dirty pages, which are saved in a bitmap table in RAM. As we know all the virtual machines have the possibility to share the common memory, I/O discs, CPU etc and all these processes are managed by the hypervisor. The application that is going to be tested is a 180MB application (online game). We will examine the parameters which we mentioned above by using the hypervisors: XEN-PV, Xen-FV, Open-VZ

We have used a computer and have exploited it in all the possible cases. The parameters of the computer we have used are:

Intel Core i7 920, Quad Core +, L2 4x256 KB, L3 = 8 MB, Asus, Three Channel DDR3 1600 Mhz, RAM 3x2GB, 64 bit processor, Hyperthread Technology, Freq 3.2 Ghz, VT Support, Turbo Boost Support.

We will start the experiment with XEN-PV and then with the two other hypervisors. The purpose is to find the hypervisor with the better performance during the migration of a virtual machine.

4 The Experimental Phase

4.1 The Simulation of Warning Failure in x0 Virtual Machine

Referred to figure 2, initially we will prepare a warning failure of x0 virtual machine. It means that x0 virtual machine in reality is under executing, but the hypervisor and the other virtual machine which is built over the hypervisor are informed from heartbeat tool (which is included in CentOS 5.5), that it is stopped as we explained in section 3. To simulate a warning failure we should create a script in C programming language and we call it ***heartcare***. This script is located in /proc and sends a message to heartbeat every time we want to execute it. At this moment heartbeat is informed for the virtual machine which should get interrupted (in our example it is x0 virtual machine), and at the same time heartbeat informs x1 virtual machine and Xen Hypervisor for this situation. Thus the hypervisor begins to migrate x0 virtual machine to x1 virtual machine based on pre-copy approach, which is explained in section 3.

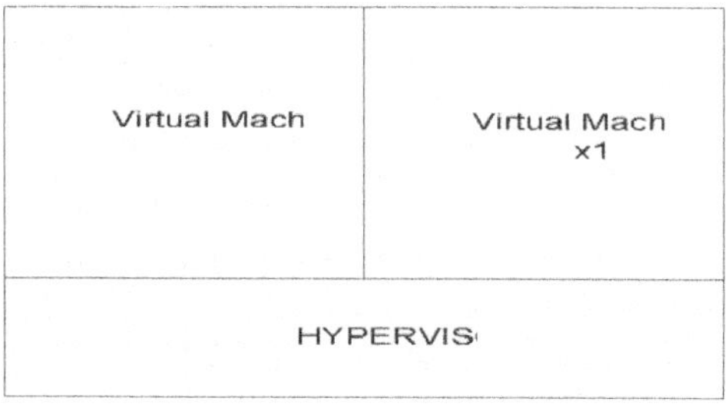

Fig. 2 Two Virtual machines that Lay above the Hypervisor

4.2 Xen-PV

As we explained in section 3 initially we have installed Xen as hypervisor, above it is installed Dom0 with Cent-OS 5.5 version and 2 virtual machines DomU (GuestOS) each has Ubuntu 10.04 Server installed. In x0 virtual machine is executed a 180 MB application.

4.2.1 CPU Consumption in Xen Hypervisor before and after the Migration of x0

At first we will evaluate the CPU consumption of Xen Hypervisor before migrating the x0 virtual machine. The migration occurs at the moment when a warning failure signal from ***heartcare*** script is sent to "Heartbeat" tool. To evaluate the CPU consumption in Xen, first we have been located in /proc directory and typed the command xentop–b. The output results of this command are saved in a matrix form in a script called ***XenCProc*** which is located in /proc/xen. This script presents the CPU consumption for every 5 sec. At the moment when we push s key in the keyboard it will give us the average of CPU consumption up to this moment. The value is **2,23%**. This is because the resources, memory consumption, virtual disks I/O, virtual network etc are not being used heavily.

After x0 virtual machine is migrated, the CPU consumption at the first moment increases slightly, then it is increased up to **9,63 %**, in 1,65 sec; this is the peak of consuming, because of page faults. When the page faults increases, the CPU consumption increases too. This result depends from the iteration of dirty pages which are maintained by bitmap table in "Grant Shared Table" located in

RAM and managed by the Hypervisor. As it looks in table 1 after 2,54 sec CPU consumption decreases to **3,11%**. After 3,66 sec CPU consumption is decreased to **2,11%**. This is the stabilized value. If we compare both cases before and after migration, the CPU consumption after the stabilization phase in the second case decreases up to 0,12%. The reason is the reduction of resources which were implemented to x0 virtual machine.

Table 1 CPU consumption in Xen Hypervisor after x0 virtual machine is migrated to x1.

CPU	Time (sec)
2,23 %	0
9,63%	1,65
3,11%	2,54
2,11%	3,66

4.2.2 Memory Utilization in Xen Hypervisor before and after x0 Migration

To evaluate the memory utilization in Xen before migration we will use the tool named "MemAccess" located in /etc [11]. Initially the memory utilization is 10,6 %. After the migration of x0 virtual machine, memory utilization increases 10,7% for 1,55sec. This is the peak of memory utilization value. After that value the memory utilization will be stabilized at 10,5% at 2,04 sec (see table 2). If we compare the memory utilization inertia with CPU consumption, it is clear that the memory has more stability because of it`s native nature. During the migration in memory are just added some extra code (pages migrated from x0 virtual machine). This extra code is replaced in dirty bit map table located in Grant Shared Table.

Table 2 Memory Utilization in Xen Hypervisor after x0 virtual machine is migrated to x1.

Memory Utilization	Time
10,6%	0
10,7%	1,55
10,5%	2,04

As it look from table 2 the stability of memory utilization happens after 2,04 sec from the migration process. If we compare table 1 and table 2 again, the peak of memory utilization happens after 1,55 sec while the peak of CPU consumption after 1,65 sec. This is because the iteration process does not affect directly to memory but it affects CPU consumtion. Also the CPU should adapt some additional parameters during the migration such as memory management, I/O disk refresh etc.

4.2.3 Average Total Time Migration of x0 Virtual Machine to x1

Initially we should clarify that the migration has occured in the same physical host. At the moment when ***heartcare*** script send a message to heartbeat tool to crash x0 virtual machine, a counter is programmed to start and it is implemented into that script. This counter will evaluate the total transferring time. At the end of migration another message is sent to heartcare script. This message is sent from ***XenCProc*** script because the last of phase of pre-copy migration is dedicated to CPU status of x0 virtual machine [10]. The CPU status can be identified using ***XenCProc*** script because the CPU status is the first argument saved in stack [12]. The ID of CPU status is in the end of the transfer. At the final transfer, ***XenCProc*** sends a message to ***heartcare*** script. The total time is shown in display. The average total transferring time in our test is evaluated **2,66 sec**. This is a very effective time, because the application installed is 180 MB (of course just a little size of this application is being transfered, because most of this is located in the hypervisor which is similar to a SAN device between two virtual machines, this is not part of our study).

4.2.4 Downtime during the Migration of x0 Virtual Machine to x1 Virtual Machine

This is a very critical case, because live migration phase depends from this parameter. To evaluate the downtime we will refer to XenCProc. Based on [10] downtime is evaluated as the transfer time of CPU status. Thus we should evaluate the total transferring time of Program Counter Register (also the same thing will be done with the execution instructions at the moment when the warning failure occurs in x0 virtual machine) of x0 virtual machine to x1 virtual machine. PC register is encapsulated in the shared memory of the Hypervisor. So in the same manner, with total transferring time transferring we should identify the last process. As we know when an interrupt occurs, CPU saves its status and PC counter. So we should identify the ID of the first process. This ID is recorded in ***XenCProc*** at the moment when heartbeat stops x0 virtual machine, then it passes to /proc/xentop file. The downtime algorithm is:

1. Heartcare sends a message to XenCProc
2. XenCProc saves the ID of the first process
3. Then we type xentop command
4. ID process is transported to xentop file
5. CPU status iss transferred, it send automatically a sys call to the hypervisor
6. Xen look the xentop file and starts the CPU status in x1 virtual machine
7. The downtime is saved at XenCProc
8. It is ahowed in display

The downtime is evaluated **4 ms**. It is a small value. There are some reasons:

1. We are doing a migration inside a physical host
2. CPU is very fast, see section 3.
3. There are some extra parameters such as Turbo BOOS
4. The application is not big (It can be considered small, only 180 MB)
5. There are no data dependency [13] etc.

Now we will repeat from a-d the experiments by changing the MTU (Message Transfer Unit). By changing the MTU value, the packet size will change automatically. It will affect the transferring time, downtime, memory utilization and CPU consumption too. The data packets are transfered from network virtual driver of x0 virtual machine to x1. Both virtual drivers form a team and are connected by a bridge soft which is managed by Xen. To change the packet data size we can change MTU from 1500B, which is standard of Ethernet Network Adapter, to 500 B and 64 B. For each VM we type the command:

Ifconfig eth0 mtu 500

This is a temporary value and we suppose that packet data size is 500 B. We should clarify that the results taken till now belong to the case when the packet data size is 1500 B.

MTU = 500 B

Table 3 CPU consumption in Xen Hypervisor after x0 virtual machine is migrated to x1.

CPU rate consumption (%)	Time (sec)
2,26 %	0
11,47%	1,87
3,87%	2,96
2,24%	3,91

Table 4 Memory Utilization in Xen Hypervisor after x0 virtual machine is migrated to x1.

Memory Utilization	Time
11,3%	0
11,9%	2,30
11,2%	2,41

MTU=64 B

Table 5 CPU consumption in Xen Hypervisor after x0 virtual machine is migrated to x1.

CPU rate consumption (%)	Time (sec)
9,24 %	0
21,13%	2,78
16,32%	4,50
8,03%	5,21

Table 6 Memory Utilization in Xen Hypervisor after x0 virtual machine is migrated to x1.

Memory Utilization	Time
11,9%	0
12,6%	3,42
11,6%	4,05

If we compare the tables 1-6, we see that the CPU consumption increases when packet data size decreases. The same thing happens with the memory utilization. The reason is the increasion of the overhead because small packets have more context switch and more overhead [14]

Table 7 The Average Total Migration time and Downtime for different MTU sizes

Packet data size	Average Total Migration time of x0_VM	Downtime
500 B	3,37 sek	6ms
64 B	5,12 sek	9ms

4.3 Xen-FV

If we want to use Xen as a Full virtual machine we should have a hardware that supports it. As we see in section 3 the parameters of our computer match with our requirements [15]. Also we should built QEMU on Xen, thus we should emulate the hardware in user space [16,17]. The Full virtualization in Xen has the same characteristics as VMWare which means that we can built OS with different native nature and different architecture, such as Windows in DomU. Also in Full virtualization it is not necessary to modify kernel OS Host or Guest. Nevertheless Full Virtualization has some disadvantages such as the increase of access time in I/O disks, because there are 2 trap instruction to access a disk[18]. The Full virtualization includes an additive complex layer presented by QEMU emulation software. In order to emulate network drivers in both GuestOS we should install e1000 emulator in /root directory.

MTU 1500 B

Table 8 CPU consumption in Xen-HVM after x0 virtual machine migration in x1.

CPU rate consuming (%)	Time (sec)
2,66	0
10,52	2,14
4,08	3,25
3,16	4,31

Table 9 Memory Utilization in Xen-HVM after x0 virtual machine is migrated to x1.

Memory Utilization	Time
11,5	0
11,9	2,26
11,5	2,84

MTU = 500 B

Table 10 CPU consumption in Xen-HVM after x0 virtual machine is migrated to x1.

CPU rate consuming (%)CPUCPUCPU rate consumption (%)	Time (sec)
3,72	0
14,59	2,51
6,11	4,33
3,6	5,70

Table 11 Memory Utilization in Xen-HVM after x0 virtual machine is migrated to x1.

Memory Utilization	Time
12,4	0
13,2	2,92
12,4	3,54

MTU=64 B

Table 12 CPU consumption in Xen-HVM after x0 virtual machine is migrated to x1.

CPU rate consumption (%)	Time (sec)
11,8	0
24,5	3,91
19,1	6,84
11,7	8,16

Table 13 Memory Utilization in Xen-HVM after x0 virtual machine is migrated to x1.

Memory Utilization	Time
13,02	0
13,7	5,7
12,9	6,16

If we compare the tables 8-13, the CPU consumption and memory utilization in Xen-HVM are bigger than in Xen- PV.

Table 14 Average Total Migration time and Downtime for different MTU size

Packet data size	Average Total time migration of x0_VM	Downtime
1500 B	4,10 sec	8 ms
500 B	5,24 sec	11 ms
64 B	6,08 sec	16 ms

As it look from the table 14, downtime and Average Total time are increase when the number of packet size is decrease.

4.4 OpenVZ

4.4.1 The Evaluation of CPU Consumption in OpenVZ

To evaluate the CPU consumption in OpenVZ we don't have any specific tool nevertheless we can measure the CPU wasted time in /proc/vz/vstat. To evaluate the CPU consumption we create a script in C which is called **traceproc.** It traces the active and idle processes in hypervisor by scanning the status of each process in vstat file. Each process has a wake bit in Process Status Register, if it is 1 this process is active and if it is 0 the process is idle. In **Traceproc** script located in /proc/vz we have implemented a formula:

The availability of the process= (Time for each active process)/(Total CPU time) x100% (1)

The sum of the availability active processes = CPU Availability (2)

In reality this formula doesn't calculate the CPU availability, because when the processes are idle they still spend CPU time, consequently their consume CPU. Thus for the idle process we should build a semaphore variable [7] in order to make them sleep. In this way they will not consume CPU. Semaphore variables are built in a script in C called **semaphore**, which records the ID of all idle processes. This information is taken from **Traceproc** script. For each passive process we generate a thread which sends a signal to these processes. In this manner, the passive processes are transformed in sleep processes. At the moment when CPU sends an interrupt message for one of the sleeping processes, the semaphore script is the first that takes this signal. This script reads the ID of calling processes, records it in a specific address into a specific register and then calls the specific thread. The thread wakes up the sleeping process. Thus the process can take the interrupt launched from CPU. This is a very dangerous approach because the script is implemented in user space, it means that after the interrupt request from CPU, the generated thread can't wake the process up. So the process is going to sleep forever. Nevertheless after this modification to **traceproc** script we will evaluate the CPU consumption by using the formula:

CPU consumption= Sum of active processes/ Total nr of process (3)

We should emphasize that this script gives us an approximately value of CPU consumption in OpenVZ hypervisor.

4.4.2 The Evaluation of Memory Utilization in OpenVZ

We have to use a tool named **stream_tool** [8] to evaluate the memory utilization. There is one problem, this tool cannot evaluate the dynamic changing i.e the iteration of dirty pages while the x0 machine migrates to x1. So we should build a script that finds the number of page faults and multiples them with the page size. Nevertheless we cannot find the appropriate number of transferred pages in a unit of time in case a page miss occurs. So we should implement another tool called **Bonnie ++**, which calculates the bandwidth transfer for 2 disks. We take RAM_VM1 as first disk and RAM_VM2 as second disk and we can calculate the total number of transferred pages for each iteration by using the formula:

The nr of transferred pages= Total size transferred (B)/ Page size (4)

The calculated from **stream** benchmark at 0 time:

Total memory utilization= (Time before a page fault occurs) + (Nr of transferred page while a page fault occurs) x (Nr of page faults) x (Page size) (5)

All these formulas are implemented in **Mem0 script**, written in C language.

4.4.3 The Evaluation of Transferring Time and Downtime

To evaluate the transferring time we can use the same script we did in previous cases, but this script is located in /proc/vz.

MTU 1500 B

Table 15 CPU consumption in OpenVZ Hypervisor after x0 virtual machine is migrated to x1.

CPU rate consumption (%)	Time (sec)
2,24 %	0
9,67%	1,52
3,20%	2,24
2,18%	3,04

Table 16 Memory Utilization in OpenVZ Hypervisor after x0 virtual machine is migrated to x1.

Memory Utilization	Time
10,8%	0
10,9%	1,42
10,8%	1,57

MTU = 500 B

Table 17 CPU consumption in OpenVZ Hypervisor after x0 virtual machine is migrated to x1.

CPU rate consuming (%)	Time (sec)
2,28 %	0
11,48%	1,77
3,88%	2,91
2,26%	3,84

Table 18 Memory Utilization in OpenVZ Hypervisor after x0 virtual machine is migrated to x1.

Memory Utilization	Time
11,2%	0
11,7%	1,59
11,2%	2,22

MTU=64 B

Table 19 CPU consumption in Xen Hypervisor after x0 virtual machine is migrated to x1.

CPU rate consuming (%)	Time (sec)
9,76 %	0
22,53%	2,1
17,11%	3,29
8,62%	4,27

Table 20 Memory Utilization in Xen Hypervisor after x0 virtual machine is migrated to x1.

Memory Utilization	Time
11,9%	0
12,8%	3,11
11,6%	3,58

If we compare tables 15-20 in OpenVZ, the CPU consumption and memory utilization is just a little bit more then the parameters in Xen, the reason is that all the Containers and Hosts share the same fair resources such as CPU, but the transition time in OpenVZ is smaller then that on Xen-PV, because in OpenVZ each container is considered a process.

Table 21 Average Migration Total time and Downtime with different MTU sizes

Packet data size	Average Total Migration time of x0_VM	Downtime
1500 B	2,06 sec	3 ms
500 B	2,57 sec	4 ms
64 B	3,72 sec	5 ms

In Table 21 is presented the Average Total Migration time of x0_VM is smaller than that of XEN. The same thing happens with downtime. The reason is the low complexity of OpenVZ, Overhead and Context Switch; because each container is considered a process.

5 Conclusions

From the above experiments we conclude the following results:

1. CPU Consumption and Memory Utilization in XEN-PV are lower than in Open-VZ because Open-VZ uses the same kernel for Host OS and Guests OS by fair sharing the CPU. XEN has got a better sharing of the CPU between Host OS and Guests OS.
2. XEN-HVM consumes more CPU because of the emulator's complexity (QEMU).
3. All the parameters for the three hypervisors increase with the decrease of packet's size. This causes a slower performance.
4. The Total Migration Time and Downtime are smaller in Open-VZ than in XEN because in OPEN-VZ the overhead is smaller (every OS works as a process).

6 Future Work

As a future intention we would want to test and compare the performance for five hypervisors XEN-HVM, XEN- PV, Open-VZ, KVM-FV and KVM-PV in a WAN. Also we will test these hypervisors not using a warning failure, but simulating an un-warning failure.

References

1. Che, J., He, Q., Ye, K., Huang, D.: Performance Combinative Evaluation of Typical Virtual Machine Monitor (2010)
2. Schlosser, D., Duelli, M., Goll, S.: Performance Comparison of Hardware Virtualization Platforms (2010)
3. Padala, P., Zhu, X., Wang, Z., Singhal, S., Shin, K.G.: Performance Evaluation of Virtualization Technologies for Server Consolidation (2007)
4. Clark, C., Fraser, K., Hand, S., Hanseny, J.G., July, E., Limpach, C., Pratt, I., Warfield, A.: Live Migration of Virtual Machines (2009)
5. Menon, A., Santos, J.R., Turner, Y., Janakiraman, G.J., Zwaenepoel, W.: Diagnosing performance overheads in the Xen virtual machine environment. In: Proceedings of the First ACM/USENIX International Conference on Virtual Execution Environments (VEE), pp. 13–23 (2005)
6. Haselhorst, K., Schmidt, M., Schwarzkopf, R., Fallenbeck, N., Freisleben, B.: Efficient Storage Synchronization for Live Migration in Cloud Infrastructures (2010)
7. Tanenbaum, A.: Modern Operating System, ch. 2, 4th edn. Semaphores (2009)
8. Nussbaum, L., Anhalt, F., Mornard, O., Gelas, J.-P.: Linux-based virtualization for HPC clusters (2009)
9. Braastad, E.: Management of high availability services using virtualization (2006)
10. Hines, M.R., Deshpande, U., Gopalan, K.: Live migration with post copy (2007)
11. Heo, J., Zhu, X., Padala, P., Arbor, A., Wang, Z.: Memory Overbooking and Dynamic Control of Xen Virtual Machines in Consolidated Environments (2009)
12. Tanenbaum, A.: Modern Operating System, ch. 2, 4th edn., p. 85, Processes (2009)
13. Henessy, J., Peterson, D.: Computer Organization and Design, ch. 4, 4th edn., p. 330, Pipeline Data Path and Control, (2010)
14. Tanenbaum, A.: Modern Operating System, ch. 3, 4th edn. Size of Page Memory (2009)
15. Regola, N., Ducom, J.-C.: Recommendations for Virtualization Technologies in High Performance Computing (2010)
16. Tang, C.: FVD: a High-Performance Virtual Machine Image Format for Cloud Computing (2008)
17. Berrangé, D.P.: Taking full advantage of QEMU in the Xen userspace (2007)
18. Tanenbaum, A.: Modern Operating System, ch. 1, 4th edn. Virtual machines (2009)

Modifications and Improvements on CEN/BII Profiles

Kiril Kiroski, Marjan Gusev, Magdalena Kostoska, and Sasko

Abstract. This paper introduces a new approach to modeling of e-Ordering and e-Invoicing solutions. Today, electronic means to conduct purchases and sales are essential activities for companies operations, and they need to decide how to implement e-Ordering and e-Invoicing, what software solutions to use, and how to integrate them successfully with their current software. The nature and application field of electronic services such as e-Ordering and e-Invoicing dictates increased interoperability between different platforms. EU-introduced Pan-European Public Procurement Online project, presents an effort for unifying the large variety of procurement solutions from various EU and non-EU countries, and uses CEN/BII profiles to introduce interoperability between them. This paper presents the summary of our analysis of the CEN/BII profiles, and introduces modifications and improvements to this model, to help the flexibility and ease of further modification of the model.

Keywords: e-Invoicing, e-Ordering, CEN/BII profiles.

1 Introduction

Business deals of each company always concern a sort of purchase or selling. Computer applications and Internet introduce automation of processes including invoices and purchase orders. There are a lot of applications supporting creation, modification and archiving of these documents, and although they have common characteristics, they have a lot of differences, mostly in solving interoperability issues and possibility to integrate, or work as part of cloud computing paradigm [1].

During our previous work, we analyzed a number of different software platforms working on the problem of electronic invoices and orders [1], and we introduced a

Kiril Kiroski · Marjan Gusev · Magdalena Kostoska · Sasko
University Ss Cyril and Methodius, Faculty of Computer Sciences and Engineering, Rugjer Boshkovikj 16, Skopje, Macedonia
e-mail: kiril.kjiroski@finki.ukim.mk
marjangusev@gmail.com
magdalena.kostoska@finki.ukim.mk
sashko.ristov}@finki.ukim.mk

L. Kocarev (Ed.): ICT Innovations 2011, AISC 150, pp. 395–404.
springerlink.com

new approach to benchmarking their sophistication [2]. The next logical step was to use the results and knowledge derived from our research. We established that the solutions achieving highest score on our benchmark "...support not only functionalities we found interesting for our work, but also features like benchmark KPIs, Business analysis, expend and asset tools, which also makes them much expensive solution for companies or organizations who need nothing more than sophisticated e-Ordering and e-Invoicing software, and we believe that a smaller, much compact solution can be designed to use this difference in target field."

This conclusion to our work led us to projects that turn much more attention to the essence of exchanging electronic documents (i.e. availability and accessibility, SaaS approach, and interoperability), and disregarding the appearance and appeal of the solution. Thus, PEPPOL [3] became the main target of our interest, so we started researching its characteristics, and we acquainted with the idea of introducing software platform for public procurement interoperable with most of its EU counterparts.

During our research, we also studied different electronic purchase solutions, such as EBiz4All [4], BBG's e-Shop [5], IBX [6]. These software platforms tend to satisfy interoperability goals of the European Union, by closely following respectable countries' laws and regulations, and insisting on interoperability between different countries and platforms, thus respecting the goal of achieving single pan-European market. Our research showed that achieving this level of interoperability involves following common standards - Business Interoperability Interfaces for Public procurement in Europe (CENBII), proposed by the European Committee for Standardization.

What also became evident during our research was that, although these profiles elaborately define all stages involved in electronic procurement process, they also do not provide for unanticipated events, including canceling an order or revoking an invoice, they do not allow minor modifications of the process – which is a question addressed by organizational interoperability requirements, and that CEN/BII profiles assume future modifications or additions to profiles will include reworking of all profiles containing modified feature.

Thus, we will first briefly introduce CEN/BII profiles and included interoperability requirements, and then we will give our view on modifying the profiles to allow them faster and easier adoption of future changes.

2 CEN/BII Profiles Overview

European Committee for Standardization introduced CEN/BII profiles [7] as means to "...facilitate the internal market and to achieve the Lisbon Objective of making Europe the most knowledge competitive society by 2010, electronic working, based on the development of interoperable public electronic procurement and business (eProcurement and eBusiness)..." The purpose of these profiles is to ensure interoperability between the two parties on multiple levels:

- Legal interoperability: With intention to work between different countries, one of the main issues is to ensure that obligations established between the parties retain legal consequences on both supplier and customer. Legal interoperability is also concerned with the subject of digital signatures and their validity in cross-border exchange.

- Organizational interoperability: Cross-border procurement has to be processed by following agreed steps, in their entirety, so as to ensure that all parties follow respective procedures, and completely fulfill their responsibilities. This is accomplished by following the profile to the letter, since adopting a profile implies supporting all the rules defined for that profile.

- Semantic interoperability: All interactions and relied messages between the parties must be distinctive and unambiguous. An electronic document received by one of the parties should be "...should be automatically elaborated and loaded into an IT system." The system must be able to parse and validate the document in a standard way and using standard methodologies.

- Technical interoperability: Exchanging messages between parties has to be carried out following a number of mandatory rules, and the supplied transport infrastructure is utilized via so called Access Points, which can be operated by senders and receivers of the messages, or by independent third parties [8]. Major technical requirements that must be supplied are [7]:

• Support for message level receipts.
• The e-business transport infrastructure must compensate for unreliable transport layer protocols by resending business documents following a recognized pattern.
• Support for authentication of the sender and the receiver of a business document.
• Exchanging business document in confidentiality.
• Support for the integrity of business document.
• Both parties in a business transaction must be able to prove that the transaction has taken place. Support for non-repudiation of business transactions.
• Support for reliable exchange of business documents.

The CEN Business Interoperability Interfaces (BII) specification is a meant to facilitate effective public procurement solution with focus on cross-border interoperability and closing the gap between systems built on UN/CEFACT and OASIS/UBL 2.0. CEN/BII profiles can be used to create interoperability between users of the two standards. Models built using profiles provided in CEN/BII specification tend to achieve development of standardized solutions which will enable borderless public e-procurement process in Europe. The whole e-procurement process is divided in five groups (Publication, Tendering, Sourcing, Ordering and billing, and Support), each of them containing several profiles, where each profile involves one or more business processes. These processes are shown on Fig. 1. Ordering (e-Ordering) and Billing (e-Invoicing) are topic of our research in this paper.

3 Model Modifications and Improvements

We previously established the major characteristics of the existing software solutions, measured their success in implementation of the major software goals, and evaluated the level of completion of the objectives we established the system should possess. Our goal is to follow international standards regarding ordering and invoicing software, and thus ensure interoperability with most European platforms. According to these objectives, we will describe several modifications and improvements over the CEN/BII model, and elaborate the choice of using SaaS deployment model over more traditional application service provider (ASP) model.

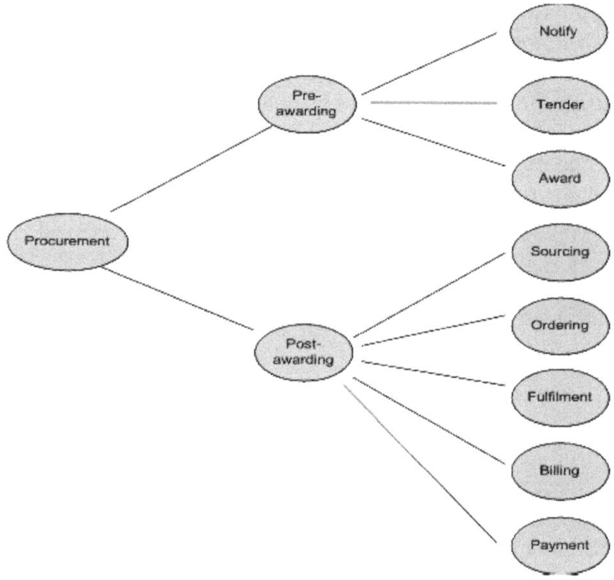

Fig. 1 Business processes used to describe CEN/BII profiles.

During our research, we noticed that CEN/BII model, although robust and encompassing, has a lot of redundant business profile's content. This is not a big problem when designing a solution according to specific requirements while using specific business profile. But, there are situations when lack of segmentation can seriously increase effort required to modify both the model and solution, such as:

- New features are to be added to the solution, requiring to completely replace business profile,
- Modifying the solution in such a manner, so it no longer corresponds to the business profile it uses, and requires adding or removing fields from the data model,
- Introducing completely new, or replacing existing CEN/BII profiles,
- Lack of certain features, such as completely cancelation of an order, revoking an already created invoice, and making orders using supplier's catalogue,
- Decreased flexibility of the model, due to non-existence of opportunity to combine profiles and thus create an advanced profile from basic ones.

Driven by these observations, we turned our efforts on designing modified and improved profiles, which will provide the needed flexibility and modular nature, and provide the developers with easier and simpler tools for building the application.

3.1 Developing a Cloud Computing Solution Model

Cloud computing represents today's most modern and attractive model for developing solutions that require highest levels of availability and accessibility, and through use of CEN/BII profiles and XML for exchanging information and data across internet, fulfill the requirement for semantic interoperability between end parties.

Cloud computing represents three different service levels:

- Infrastructure as a Service (IaaS) – consumer is provided with processing, storage, networks and other computing resources, and he is able to deploy and run arbitrary software. Root of the IaaS is in hardware virtualization, and use some of it key characteristics and concepts.
- Platform as a Service (PaaS) - consumer is provided with programming languages and tools, so as to be able to create and use various applications. PaaS is giving its consumers illusion of infinite resources, since it enables them to use expanding quantity of resources without requiring additional actions on consumer's part.
- Software as a Service (SaaS) – consumer is provided with provider's applications running on a cloud infrastructure. It represents the cloud computing from the end-user's point of view, used in everyday work, such as Gmail, Google Docs or Netflix.

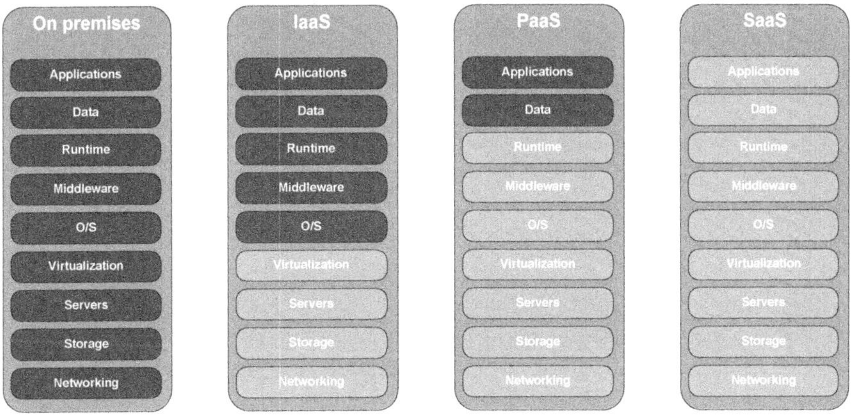

Fig. 2 Comparison between On-premises computing, IaaS, PaaS and SaaS.

Comparing differences between the three levels of Cloud computing and traditional, on-premises computing can be carried through deducting which resources or services are executed by customer and provider. We can see such comparison on Fig. 2 [11], where, resources and services in responsibility of the provider are shown in green boxes, while those in responsibility of the customer are shown in red.

Fig. 2 shows that SaaS solution may be used from anywhere and at any time, provided thin client (web browser) and internet connection. These features make SaaS software most attractive for small and medium enterprises, and require no additional resources and hardware on customer's part. Further research [9] showed that South-Eastern Balkan companies and government institutions represent most feasible customers for e-Procurement platform realized as SaaS.

3.2 Service Lifecycle

Looking at our profiles, we observed that each of our basic profiles belongs to exactly one state of produced documents' lifecycle. We also noticed that each of the advanced profiles belongs to two or more states of the lifecycle. This is shown on Fig. 3 [12]. For the purposes of electronic purchasing, we have defined four states for created documents, each of them uniquely named as:

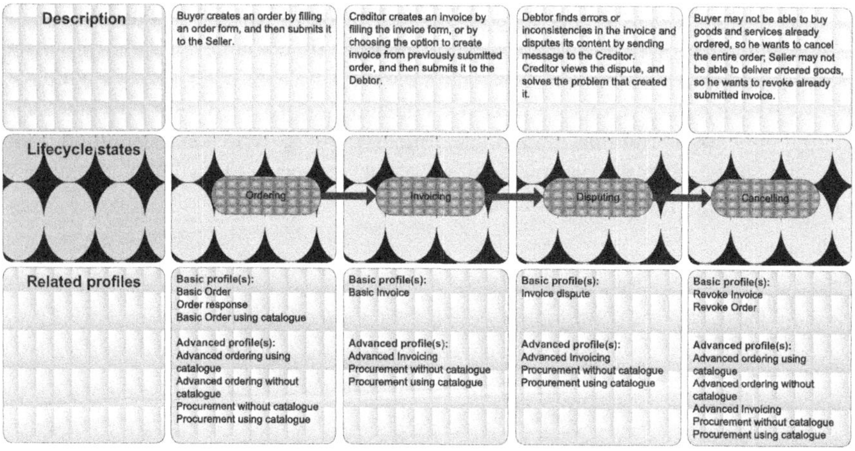

Fig. 3 Different states of e-Ordering and e-Invoicing solution lifecycle, and related profiles.

1. Ordering – In this state we place profiles used to create orders, and responses to orders, so the final outcome will be accepted order (contract between Buyer and Seller is established, and proceeds to next phase), or rejected order (there is no contract, and both parties have no further legal obligations). Basic profiles defined in this state are: Basic Order, Order response, and Basic Order using catalogue. Advanced profiles using one (or more) of basic profiles above are: Advanced ordering using catalogue, Advanced ordering without catalogue, Procurement using catalogue, and Procurement without catalogue.

2. Invoicing – In this state we have profiles concerning invoice creation, where Creditor creates an invoice (either by filling the invoice form, or creating it from previously submitted order) and sends it to the Debtor. Final outcome is an invoice received by the Debtor, claiming payment for delivered goods or services. The only basic profile defined in this state is Basic Invoice. Advanced profiles using it are: Advanced invoicing, Procurement using catalogue, and Procurement without catalogue.

3. Disputing – This state is concerned with disputing and solving the contents of an invoice. This means that if Debtor finds errors or inconsistencies in the invoice, he can dispute them by sending dispute message to the Creditor. Creditor can issue new, corrected invoice, or he can issue credit note to the Debtor. Basic profile defined in this state is Invoice dispute, and advanced profiles using it are: Advanced invoicing, Procurement using catalogue, and Procurement without catalogue.

4. Cancelling – This state is concerned with optional, more advanced problems, and are matter of agreement between concerned parties. If Buyer is not able or don't want to buy goods or services he previously ordered, he can revoke (cancel) the entire order. Similarly, if Seller can't deliver goods or services he billed, he can revoke the invoice. Outcome can be acceptance (with or without royalties) or rejection (contract is still valid, and both parties are obliged to drive it to the end). Basic profiles defined in this state are: Revoke Invoice and Revoke Order. All five advanced profiles use this state.

3.3 Modifying CEN/BII Model

In all these situations, the quantity of work required to make such modifications can be minimized if using completely segmented business profiles which do not include collaborations or messages used by another business profile. We named these profiles as Basic profiles. They include:

- Basic Order - Buyer makes and submits an order. Order is received and processed by the Seller.
- Order Response - Seller receives and processes an electronic Order. He accepts or rejects the order in question and sends the response to the Buyer.
- Basic Invoice - Supplier sends an electronic Invoice that can be received and processed by the Customer.
- Invoice Dispute - Customer processes an electronic Invoice sent by the Supplier, and notifies Supplier of inconsistencies in the Invoice. Authorized user on the Supplier side can take actions to correct the invoice in question, either by issuing credit statement, or a completely new invoice with corrected information.

Both Basic Order and Basic Invoice profiles, are already comprising part of CEN/BII specification, and we translate them into our model as such. Order Response and Invoice Dispute do not exist as separate profiles, instead they are combined into more complex profiles. Separating into distinct modules will make them easier to use, and give the model more advanced possibilities.

3.4 Introducing New Profiles to CEN/BII Model

In the CEN/BII model, there are a great number of profiles surpassing the scope of our solution, but also, not every profile is intended to be used according to our requirements – for example, profiles including creation and modification of online catalogues do not take in consideration using the same catalogues in the ordering process. Another issue is that CEN/BII profiles do not include the possibility of completely revoking already submitted invoices or orders. To address these issues, we created three new Basic profiles:

- Basic Order using Catalogue - Buyer uses catalogue to make an order and submits the order. Order is automatically rejected or accepted by Seller's system depending on two things: all ordered items are in the catalogue, and quantity per ordered item is smaller than quantity per catalogue item.
- Revoke Invoice - The Supplier creates a notification that he cannot deliver goods or services stated in the referenced invoice. This notification is send to the Customer and the Invoice in question is stored. If the referenced Invoice was created on the bases of previously created order, the Order in question is nullified the Supplier's system, and notification of its nullification is sent to the Customer to be processed.
- Revoke Order - Buyer creates a notification that he has no longer need or cannot purchase goods or services stated in the referenced order. Request is sent to the Seller and if the contract conditions allow it, and higher authorities on the Seller's side approve it, referenced Invoice is revoked, and the previously placed Order is nullified. A notification of Order nullification is sent to the Customer to be processed.

None of the newly introduced profiles exist as a part of the CEN/BII model, although there are platforms that support some of these functionalities [5], [10]. Main contributions to the CEN/BII model are that, through these three profiles, it becomes possible to:

- Create an order using supplier's catalogue
- Automate order response, without necessity for intervention on seller's part
- Establish more fluent workflow, including the possibility for automated invoice creation out of the order
- Cancel an order in a few guided steps
- Revoke faulty invoice, or cancel undeliverable order.

3.5 Advanced Profiles

After establishing our seven basic profiles, we discovered that all possible scenarios regarding purchasing process can be covered by composing the initial seven basic profiles into five new, advanced profiles.

Two of the advanced profiles involve only ordering process (Advanced ordering using catalogue, Advanced ordering without catalogue), one involves only invoicing (Advanced invoicing), and two involve both invoicing and ordering, thus enabling complete coverage of the process of purchasing goods or services (Procurement without catalogue, and Procurement using catalogue). It is noticeable that some profiles include common Basic profiles, but it was necessary to introduce them, since they also differ from each other. Following are brief descriptions of the five Advanced profiles:

- Advanced ordering using catalogue - The Customer uses catalogue to make an order and submits it, and the process follows the steps provided in Basic Order using Catalogue. This profile includes the possibility of revoking the order, as is described in Revoke Order profile. This profile consists of the following Basic profiles: Basic Order using Catalogue and Revoke Order.
- Advanced Ordering without Catalogue - The Customer makes an order and submits it. Steps of this document exchange are defined with Basic Order without Catalogue and Order Response. This profile also includes collaborations and transaction included in the Revoke Order profile. This profile uses following Basic profiles: Basic Order, Order Response, and Revoke Order.
- Advanced Invoicing - The Supplier sends an electronic Invoice that can be received and processed by the Customer. The Customer processes an electronic Invoice sent by the Supplier, and notifies Supplier of inconsistencies in the Invoice. Supplier resolves the dispute and sends corrected document(s) to be processed by the Customer. Advanced Invoicing uses following Basic profiles: Basic Invoice, Invoice Dispute, and Revoke Invoice.
- Procurement without Catalogue - The Customer creates and sends an electronic Order without using Supplier's electronic catalogue. The Supplier sends an electronic Order Response, stating whether the Order is accepted or rejected in full, and in case of positive response, it involves creation of an invoice. This profile also includes possibilities of disputing the invoice, revoking the invoice and revoking the order, as is described in the corresponding profiles. Procurement without Catalogue consists of the following Basic profiles: Basic Order, Order Response, Basic Invoice, Invoice Dispute, Revoke Invoice, and Revoke Order.
- Procurement using Catalogue - The Customer sends an electronic Order generated using Supplier's electronic catalogue that can be received and processed by the Supplier's system. The Supplier's system sends an electronic Order Response, stating whether the Order is accepted (invoice will be automatically created) or rejected in full (no further legal obligations between parties). This profile also facilitates optional use of Invoice Dispute, Revoke Invoice and Revoke Order profiles. This profile uses following Basic profiles: Basic Order using Catalogue, Basic Invoice, Invoice Dispute, Revoke Invoice, and Revoke Order.

4 Conclusion

Modification of the CEN/BII profiles for e-Ordering and e-Invoicing into seven Basic and five Advanced profiles will enable us to define and describe all possible scenarios involving ordering and invoicing processes. At the same time, we will be able to accommodate or create new Basic profiles. The fact that our Advanced profiles are completely created out of the Basic profiles will also be beneficiary, since it will require only minor changes or improvements to adjust to the new requirements.

The main benefit from this granulation of the business processes included in electronic ordering and invoicing, is that we can further expand our work on other processes included in e-Procurement, and create completely modular model. It will enable all future modelers and developers to intervene on a small part of the model or introduce new module, thus minimizing the efforts to modify or improve the solution.

We already mentioned that some of the existing platforms already include some of the features we introduced in our model, but since the current CEN/BII specification does not allow changes in the processes and requires thorough profile following, their potential cannot be entirely utilized. Developing and implementing a Basic profile requires far less effort than reworking entire process, and combining Basic profiles can facilitate the creation of long workflows and processes.

References

1. Kiroski, K., Gusev, M.: e-Invoicing and e-Ordering: Analysis and Comparison. In: Proceedings of The 7th International Conference for Informatics and Information Technology, CIIT 2010 (2010)
2. Kiroski, K., Gusev, M., Kostoska, M.: A new methodology to benchmark sophistication of e-Invoicing and e-Ordering. In: Proceedings of the 2nd ICT Innovations Conference, Ohrid, Macedonia (2010)
3. Pan-European Public Procurement OnLine project, `http://www.peppol.eu`
4. E-Business for All, `http://www.a-i-c.at/en/content1.php?gruppeID=14&bereichID=9&contentID=489&PHPSESSID=7b0cd340d5a48640eb80c6b04ec827d0`
5. BBG e-Shop, `http://e-shop.gv.at`
6. Capgemini Procurement Services, `http://www.ibxeurope.com/ibx-international.aspx`
7. CEN/ISSS Business Interoperability Interfaces for Public procurement in Europe (CENBII), `http://spec.cenbii.eu/`
8. CWA Annex C on Transport infrastructure, `http://spec.cenbii.eu/Tools/documents/BII3-C-TransportInfrastructure_d06.doc`
9. Sonntagbauer, P., Gusev, M., Tomic Rotim, S., Stefanovic, N., Kiroski, K., Kostoska M.: e-Government and e-Business in Western Balkans 2010. In: Proceedings of the 2nd ICT Innovations Conference, Ohrid, Macedonia (2010)
10. Introduction to OIOUBL (Offentlig Information Online Universal Business Language), `http://www.oioubl.info/classes/en/index.html`
11. Clayton, C.: Standard Cloud Taxonomies and Windows Azure, `http://blogs.msdn.com/b/cclayton/archive/2011/06/07/standard-cloud-taxonomies-and-windows-azure.aspx`
12. DMTF International: Interoperable Clouds, A White Paper from the Open Cloud Standards Incubator, DSP-IS0101 (November 2009)

Author Index

Made in the USA
Las Vegas, NV
08 November 2024

11362585R00234